NON-EUCLIDEAN GEOMETRY

First Edition 1942
Second Edition 1947
Third Edition 1957
Fourth Edition 1961
Fifth Edition 1965
Sixth Edition 1998

By the Same Author

REGULAR POLYTOPES
Dover, New York

THE REAL PROJECTIVE PLANE
Springer-Verlag, New York

INTRODUCTION TO GEOMETRY
Wiley, New York

PROJECTIVE GEOMETRY
Spring-Verlag, New York

TWELVE GEOMETRIC ESSAYS
Southern Illinois University Press

Library of Congress Catalog Card Number 98-85640

ISBN 0-88385-522-4

Printed in the United States of America

NON-EUCLIDEAN GEOMETRY

H. S. M. COXETER, C.C., F.R.S., F.R.S.C.
PROFESSOR EMERITUS OF MATHEMATICS
UNIVERSITY OF TORONTO

SIXTH EDITION

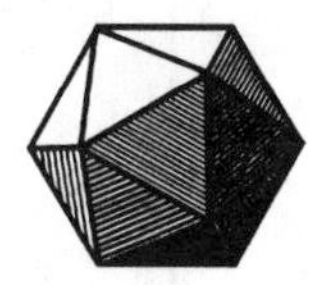

THE MATHEMATICAL ASSOCIATION OF AMERICA
Washington, D.C. 20036

A

B C

N M

O

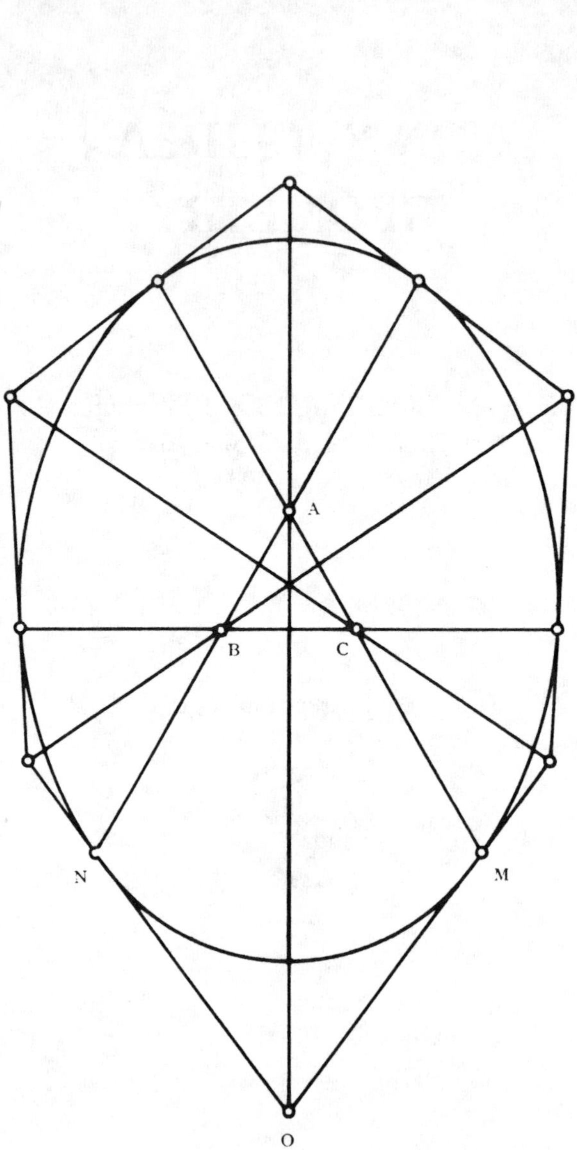

Brianchon's Theorem
and the concurrence of angle-bisectors,
(See pages 59 and 200)

SPECTRUM SERIES

Published by
THE MATHEMATICAL ASSOCIATION OF AMERICA

SPECTRUM SERIES

The Spectrum Series of the Mathematical Association of America was so named to reflect its purpose: to publish a broad range of books including biographies, accessible expositions of old or new mathematical ideas, reprints and revisions of excellent out-of-print books, popular works, and other monographs of high interest that will appeal to a broad range of readers, including students and teachers of mathematics, mathematical amateurs, and researchers.

All the Math That's Fit to Print, by Keith Devlin
Circles: A Mathematical View, by Dan Pedoe
Complex Numbers and Geometry, by Liang-shin Hahn
Cryptology, by Albrecht Beutelspacher
Five Hundred Mathematical Challenges, Edward J. Barbeau, Murray S. Klamkin, and William O.J. Moser
From Zero to Infinity, by Constance Reid
I Want to be a Mathematician, by Paul R. Halmos
Journey into Geometries, by Marta Sved
JULIA: A Life in Mathematics, by Constance Reid
The Last Problem, by E.T. Bell (revised and updated by Underwood Dudley)
The Lighter Side of Mathematics: Proceedings of the Eugène Strens Memorial Conference on Recreational Mathematics & its History, edited by Richard K. Guy and Robert E. Woodrow
Lure of the Integers, by Joe Roberts
Magic Tricks, Card Shuffling, and Dynamic Computer Memories: The Mathematics of the Perfect Shuffle, by S. Brent Morris
Mathematical Carnival, by Martin Gardner
Mathematical Circus, by Martin Gardner
Linear Algebra Problem Book, Paul R. Halmos
Mathematical Cranks, by Underwood Dudley
Mathematical Magic Show, by Martin Gardner
Mathematics: Queen and Servant of Science, by E.T. Bell
Memorabilia Mathematica, by Robert Edouard Moritz
New Mathematical Diversions, by Martin Gardner
Non-Euclidean Geometry, by H.S.M. Coxeter
Numerical Methods that Work, by Forman Acton
Numerology or What Pythagoras Wrought, by Underwood Dudley

Out of the Mouths of Mathematicians, by Rosemary Schmalz
Penrose Tiles to Trapdoor Ciphers ... and the Return of Dr. Matrix, by Martin Gardner
Polyominoes, by George Martin
The Search for E.T. Bell, also known as John Taine, by Constance Reid
Shaping Space, edited by Marjorie Senechal and George Fleck
Student Research Projects in Calculus, by Marcus Cohen, Edward D. Gaughan, Arthur Knoebel, Douglas S. Kurtz, and David Pengelley
The Trisectors, by Underwood Dudley
The Words of Mathematics, by Steven Schwartzman

MAA Service Center
P. O. Box 91112
Washington, DC 20090-1112
800-331-1MAA FAX 301-206-9789

PREFACE TO THE SIXTH EDITION

I am grateful to the Mathematical Association of America for rescuing my book from oblivion by agreeing to publish this new edition. I particularly appreciate Elaine Pedreira's careful editing and advice.

I welcome the opportunity to correct FIG. 7.2A on page 133 and to make other small improvements and especially to add the new §15.9. That section contains, in particular, an 'absolute' sequence of circles (FIG. 15.9A) which involves a surprising application of Fibonacci numbers (15.94). For further discussion of inversive geometry, the reader may like to look at "The inversive plane and hyperbolic space," *Abhandlungen aus dem Mathematischen Seminar der Universität Hamburg,* **28** (1965), pp. 217–242. The subject of §15.7 has been extended by E. B. Vinberg in his paper, "The volume of polyhedra on a sphere and in Lobachevsky space," *American Mathematical Society Translations* (2), **148** (1991), pp. 15–27. Vinberg draws attention to W. P. Thurston, "Three-dimensional manifolds, Kleinian groups and hyperbolic geometry," *Bulletin of the American Mathematical Society,* **6** (1986), pp. 357–382. When Hamlet exclaims (in Act II, Scene II) *"I could be bounded in a nutshell, and count myself a king of infinite space,"* he is providing a poetic anticipation of Poincaré's inversive model of the infinite hyperbolic plane, using a circular "nutshell" for the Absolute. In fact, the hyperbolic plane can be filled with infinitely many congruent equilateral triangles, seven at each vertex, to form the regular tessellation {3, 7}. One acquires an intuitive feeling for hyperbolic geometry by packing a disc with a multitude of curvilinear triangles, becoming smaller and smaller as they approach the peripheral circle, which represents the Absolute. This book's cover design shows the 63 largest of these triangular tiles, systematically coloured so that any two of them which are coloured alike are entirely disjoint, having no common vertex. In other words, every vertex belongs to seven or fewer triangles, all differently coloured.

H.S.M.C.

June, 1998.

PREFACE TO THE THIRD EDITION

Apart from a simplification of p. 206, most of the changes are additions. Accordingly, it has seemed best to put the extra material together as a new chapter (XV). This includes a description of the two families of "mid-lines" between two given lines, an elementary derivation of the basic formulae of spherical trigonometry and hyperbolic trigonometry, a computation of the Gaussian curvature of the elliptic and hyperbolic planes, and a proof of Schäfli's remarkable formula for the differential of the volume of a tetrahedron.

I gratefully acknowledge the help of L. J. Mordell and Frans Handest in the preparation of §15.6 (on quadratic forms) and §15.8 (on problems of construction), respectively.

H.S.M.C.

December, 1956.

PREFACE TO THE FIRST EDITION

The name *non-Euclidean* was used by Gauss to describe a system of geometry which differs from Euclid's in its properties of parallelism. Such a system was developed independently by Bolyai in Hungary and Lobatschewsky in Russia, about 120 years ago. Another system, differing more radically from Euclid's, was suggested later by Riemann in Germany and Schläfli in Switzerland. The subject was unified in 1871 by Klein, who gave the names *parabolic*, *hyperbolic*, and *elliptic* to the respective systems of Euclid, Bolyai-Lobatschewsky, and Riemann-Shläfli. Since then, a vast literature has accumulated, and it is with some diffidence that I venture to add a fresh exposition.

After a historical introductory chapter (which can be omitted without impairing the main development), I devote three chapters to a survey of real projective geometry. Although many text-books on that subject have appeared, most of those in English stress the connection with Euclidean geometry. Moreover, it is customary to define a conic and then derive the relation of pole and polar, whereas the application to non-Euclidean geometry makes it more desirable to define the polarity first and then look for a conic (which may or may not exist)! This treatment of projective geometry, due to von Staudt, has been found satisfactory in a course of lectures for undergraduates (Coxeter [6]).

In Chapters VIII and IX, the Euclidean and hyperbolic geometries are built up axiomatically as special cases of a more general "descriptive geometry." Following Veblen, I develop the properties of parallel lines (§8.9) *before* introducing congruence. For the introduction of *ideal* elements, such as points at infinity, I employ the method of Pasch and F. Schur. In this manner, hyperbolic geometry is eventually identified with the geometry of Klein's projective metric as applied to a real conic or quadric (Cayley's Absolute, §§8.1, 9.7). This elaborate process of identification is unnecessary in the case of *elliptic* geometry. For, the axioms of real projective geometry (§2.1) can be taken over as they stand. Any axioms of congruence that might be proposed would quickly lead to the absolute polarity, and so are conveniently replaced by the simple statement that one uniform polarity is singled out as a means for *defining* congruence.

Von Staudt's extension of real space to complex space is logically similar to Pasch's extension of descriptive space to projective space, but is far harder for students to grasp; so I prefer to deal with real space alone, expressing distance and angle in terms of real cross ratios. I hope this restriction to real space will remove some of the mystery that is apt to surround such concepts as Clifford parallels (§§7.2, 7.5). But Klein's complex treatment is given as an alternative (at the end of Chapters IV–VII).

In order to emphasize purely geometrical ideas, I introduce the various geometries synthetically. But coordinates are used for the derivation of trigonometrical formulae in Chapter XII.

Roughly speaking, the chapters increase in difficulty to the middle of the book. (Chapter VII may well be omitted on first reading, although it is my own favourite). Then they become progressively easier. For a rapid survey of the subject, just read the first and the last two.

For reading various parts of the manuscript in preparation, and making valuable suggestions for its improvement, I offer cordial thanks to my colleagues on the Editorial Board, especially Richard Brauer and G. de B. Robinson; also to N. S. Mendelsohn of the Department of Mathematics, to S. H. Gould of the Department of Classics, and to A. W. Tucker of Princeton University.

H.S.M. Coxeter

The University of Toronto,
May, 1942.

CONTENTS

I. THE HISTORICAL DEVELOPMENT OF NON-EUCLIDEAN GEOMETRY

II. REAL PROJECTIVE GEOMETRY: FOUNDATIONS

III. REAL PROJECTIVE GEOMETRY: POLARITIES, CONICS AND QUADRICS

CHAPTER I

THE HISTORICAL DEVELOPMENT OF NON-EUCLIDEAN GEOMETRY

1.1. Euclid. Geometry, as we see from its name, began as a practical science of measurement. As such, it was used in Egypt about 2000 B.C. Thence it was brought to Greece by Thales (640-546 B.C.), who began the process of abstraction by which positions and straight edges are idealized into points and lines. Much progress was made by Pythagoras and his disciples. Among others, Hippocrates attempted a logical presentation in the form of a chain of propositions based on a few definitions and assumptions. This was greatly improved by Euclid (about 300 B.C.), whose Elements became one of the most widely read books in the world. The geometry taught in high school today is essentially a part of the Elements, with a few unimportant changes.

According to the best editions, Euclid's basic assumptions consist of five "common notions" concerning magnitudes, and the following five Postulates:

I. *A straight line may be drawn from any one point to any other point.*

II. *A finite straight line may be produced to any length in a straight line.*

III. *A circle may be described with any centre at any distance from that centre.*

IV. *All right angles are equal.*

V. *If a straight line meet two other straight lines, so as to make the two interior angles on one side of it together less than two right angles, the other straight lines will meet if produced on that side on which the angles are less than two right angles.*

According to the modern view, these postulates are incomplete and somewhat misleading. (For the rigorous axioms that replace them, see §§ 8.3, 9.1, 9.5.) Still, they give some idea of the kind of assumptions that have to be made, and are of interest historically.

Postulate I is generally regarded as implying that any two points determine a unique line, Postulate II that a line is of infinite length. Euclid showed the great strength of his genius by introducing Postulate V, which is not self-evident like the others. (Moreover, his reluctance to introduce it provides a case for calling him the first non-Euclidean geometer!) Between his time and our own, hundreds of people, finding it complicated and artificial, have tried to deduce it as a proposition. But they only succeeded in replacing it by various equivalent assumptions, such as the following five:

1.11. *Two parallel lines are equidistant.* (Posidonius, first century B.C.)

1.12. *If a line intersects one of two parallels, it also intersects the other.* (Proclus, 410-485 A.D.)

1.13. *Given a triangle, we can construct a similar triangle of any size whatever.* (Wallis, 1616-1703.)

1.14. *The sum of the angles of a triangle is equal to two right angles.* (Legendre, 1752-1833.)

1.15. *Three non-collinear points always lie on a circle.* (Bolyai Farkas,* 1775-1856.)

According to Euclid's definition, two lines are parallel if they are coplanar without intersecting. (Following Gauss and Lobatschewsky, we shall modify this definition later.) The

*In Hungarian, the surname is put first. The "l" in "Bolyai" is mute.

existence of such pairs of lines follows from Euclid I, 27, which depends on Postulate II but not on Postulate V.

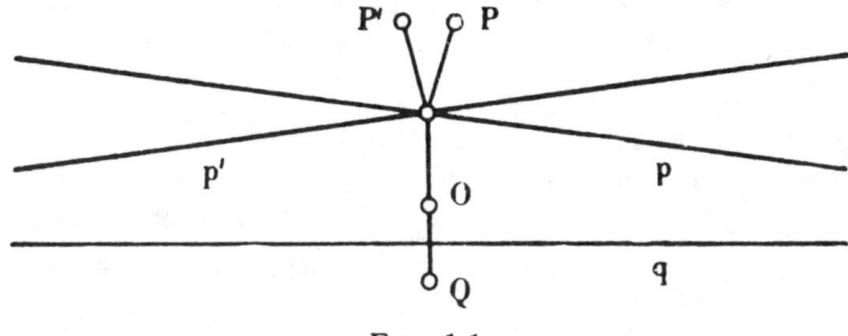

FIG. 1.1A

It is an interesting exercise to establish the equivalence of all the above statements, assuming Postulates I–IV (and the consequent propositions Euclid I, 1–28).* For instance, to deduce 1.12 from 1.15 we may proceed as follows.

Let $\mathbf{p}$ and $\mathbf{p}'$ be two intersecting lines, let $\mathbf{O}$ be any point on the perpendicular from $(\mathbf{p}, \mathbf{p}')$ to another line $\mathbf{q}$, and let $\mathbf{P}, \mathbf{P}', \mathbf{Q}$ be the reflected images of $\mathbf{O}$ in $\mathbf{p}, \mathbf{p}', \mathbf{q}$. (See Fig. 1.1A.) By Postulate IV, $\mathbf{p}$ and $\mathbf{p}'$ are not both perpendicular to $\mathbf{OQ}$; thus $\mathbf{P}$ and $\mathbf{P}'$ cannot both lie on $\mathbf{OQ}$. Suppose $\mathbf{P}$ does not lie on $\mathbf{OQ}$; then, by 1.15, $\mathbf{p}$ meets $\mathbf{q}$ at the centre of the circle $\mathbf{POQ}$. Hence $\mathbf{p}$ and $\mathbf{p}'$ cannot both be parallel to $\mathbf{q}$. Thus 1.15 implies that two intersecting lines cannot both be parallel to the same line. This statement (commonly known as Playfair's Axiom, though Playfair copied it from Ludlam) is clearly equivalent to 1.12.

The necessity of making some such assumption has been finally established only during the last hundred years. Nowadays, anyone who tries to prove Postulate V is classed with circle-squarers and angle-trisectors. For we know that there is a perfectly logical geometry in which the lines in question

*See Bonola [2], pp. 61, 119; Sommerville [1], pp. 288-293. All such references are to the bibliography on pp. 317-325.

may fail to meet, even when the interior angles are quite small. This remark is so easy to make today that we are apt to forget what a heresy it seemed to a generation brought up in the belief that Euclid's was the only true geometry. We now learn of many different geometries, but for historical reasons we reserve the name *non-Euclidean* for two special kinds: *hyperbolic* geometry, in which all the "self-evident" postulates I–IV are satisfied though Postulate V is denied, and *elliptic* geometry, in which the traditional interpretation of Postulate II is modified so as to allow the total length of a line to be infinite.

As a first glimpse of hyperbolic geometry, here are the statements that replace 1.11-1.15: Two lines cannot be equidistant; a line may intersect one of two parallels without intersecting the other; similar triangles are necessarily congruent; the sum of the angles of a triangle is less than two right angles; three points may be neither collinear nor concyclic. In elliptic geometry, on the other hand, any two coplanar lines intersect, so there are no parallels in Euclid's sense, and no equidistant lines in a plane. (We shall see, however, that equidistant lines are possible in space.)

Each of these geometries, Euclidean and non-Euclidean, is *consistent*, in the sense that the assumptions imply no contradiction. But which geometry is valid in physical space? It is important to realize that this question is meaningless until we have assigned physical equivalents for the geometrical concepts. Even the notion of a point, "position without magnitude," can only be realized by a process of approximation. Then, what is the physical counterpart for a straight line? The two most obvious answers are: a taut string, and a ray of light. According to recent developments in physics, these are not precisely the same! But the discrepancy is due to the presence of matter, and so a theoretical geometry of empty space remains significant. Consider, then, two rays of light, perpendicular to one plane. Certainly they remain equidistant according to all terrestrial experiments; but it is quite conceivable that they might ultimately diverge (as in hyperbolic geometry) or converge (as in elliptic).

1.2. Saccheri and Lambert. The most elaborate attempt to prove the "parallel postulate" was that of the Jesuit Saccheri (1667-1733), who based his work on an *isosceles birectangle,* i.e. a quadrangle **ABED** with **AD** = **BE** and right angles at **D** and **E**. It is obvious that the angles at **A** and **B** are equal. He considered the three hypotheses that they are obtuse, right, or acute, and showed that the assumption of any one of these hypotheses for a single isosceles birectangle implies the same for every isosceles birectangle. It was his intention to establish the hypothesis of the right angle by showing that either of the other hypotheses leads to a contradiction. He found that the hypothesis of the obtuse angle implies Postulate V, which in turn implies the hypothesis of the right angle. From the hypothesis of the acute angle he made many interesting deductions, always hoping for an eventual contradiction. We know now that his hope could never have been realized (without his making a mistake); but in the attempt he was unwittingly discovering many of the theorems of what was later to be known as hyperbolic geometry.

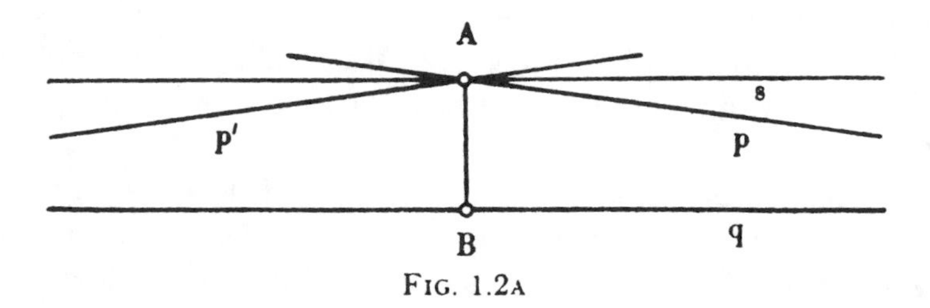

FIG. 1.2A

In particular, he considered a point **A** and a line **q** (not through **A**, and showed that, on the hypothesis of the acute angle, the flat pencil of lines through **A** contains two special lines **p** and **p′**, which divide the pencil into two parts, the first part consisting of the lines which intersect **q**, and the second of those which have a common perpendicular with **q**. (See Fig. 1.2A.) The line **p** (and likewise **p′**) is *asymptotic* to **q**, in the sense that the distance to **q** from a point proceeding along **p** continually diminishes, and eventually becomes smaller than any segment, taken as small as we please. The consequence which Saccheri imagined to be "contrary to the nature of a straight line" is that *the lines* **p** *and* **q** *have a common perpendicular at their common point at infinity*. (We shall see in §10.1 that this statement can in fact be justified.)

Fifty years later, Lambert (1728-1777) followed the same general program, using a *trirectangle* which can be regarded as one half of Saccheri's isosceles birectangle (divided along the join of the midpoints of **AB** and **DE**). He likewise rejected the hypothesis of the obtuse angle (for the fourth angle of his trirectangle), but he carried the consequences of the hypothesis of the acute angle still farther.* He defined the *defect* of a polygon as the difference between its angle-sum and that of a polygon of the same number of sides in the Euclidean plane; and, observing that the defect is additive for juxtaposed polygons, he concluded that, on the hypothesis of the acute angle, *the defect of a polygon is proportional to its area*. Comparing this with the well-known result concerning the angular excess of a spherical polygon, he suggested that the hypothesis of the acute angle would hold in the case of a sphere of imaginary radius.

In Euclidean geometry, on account of 1.13, *lengths* are measured in terms of an entirely arbitrary unit which has no structural significance. In measuring *angles*, on the other hand, we can make use of a natural unit, such as a right angle or a radian, which has particular geometrical properties. In this

*Stäckel and Engel [1], pp. 152-207.

sense we may say that Euclidean lengths are relative, whereas angles are absolute. Lambert made the notable discovery that, when Postulate V is denied, angles are still absolute, but lengths are absolute too. In fact, for every segment there is a corresponding angle, e.g. the angle of an equilateral triangle based on the given segment.

1.3. Gauss, Wachter, Schweikart, Taurinus. Gauss (1777-1855) was the first to take the modern point of view, that a geometry denying Postulate V should be developed for its intrinsic interest, without expecting any contradiction to arise. But, fearing ridicule, he kept these revolutionary ideas to himself until others had published them independently. From 1792 to 1813, he too tried to prove the parallel postulate; but after 1813 his letters show that he had overcome the customary prejudice, and developed an "anti-Euclidean" or "non-Euclidean" geometry, which is in fact the geometry of Saccheri's hypothesis of the acute angle. He discussed this with his pupil Wachter (1792-1817), who remarked in 1816 that the limiting form assumed by a sphere as its radius becomes infinite is a surface on which all the propositions of Euclid (including Postulate V) are valid; or, as we should say nowadays, that *the intrinsic geometry of a horosphere is Euclidean.*

Independently of Gauss, Schweikart (1780-1859) developed what he called "astral" geometry, in which the angle-sum of a triangle is less than two right angles and (consequently) diminishes as the area increases. In a Memorandum dated 1818 he observed that "the altitude of an isosceles right-angled triangle continually grows, as the sides increase, but it can never become greater than a certain length, which I call the Constant." Gauss complimented Schweikart on his results, and remarked that, if the Constant is called $k \log(1+\sqrt{2})$, the area of a triangle has the upper bound πk^2.

Thus encouraged, Schweikart persuaded his nephew

Taurinus (1794-1874) to devote himself to the subject; and it was in a letter written to this young man in 1824 that Gauss gave the fullest account of his own discoveries. Taurinus developed a "logarithmic-spherical" geometry by writing ik for the radius k in the formulae of spherical trigonometry. (Compare Lambert's suggestion about an imaginary sphere.) Thus, for a triangle with angles A, B, C and sides a, b, c, he found

1.31. $$\cos C = \sin A \sin B \cosh(c/k) - \cos A \cos B,$$

whence $$\cos C > -\cos(A+B), \text{ and}$$

1.32. $$A + B + C < \pi.$$

For the circumference and area of a circle of radius r, he obtained

$$2\pi k \sinh \frac{r}{k} \text{ and } 2\pi k^2 \left(\cosh \frac{r}{k} - 1\right),$$

and for the surface and volume of a sphere,

$$4\pi k^2 \sinh^2 \frac{r}{k} \text{ and } 2\pi k^3 \left(\sinh \frac{r}{k} \cosh \frac{r}{k} - \frac{r}{k}\right)$$

(Notice that, if k tends to ∞, these tend to the usual expressions in Euclidean geometry.)

1.4. Lobatschewsky. Formulae equivalent to these were derived rigorously (and quite independently) by the Russian mathematician Lobatschewsky (1793-1856), who shares with Gauss and the younger Bolyai the honour of having made the first really systematic study of what we now call hyperbolic geometry. His earliest paper was read in 1826, published in 1830, and a number of others followed.* Like Gauss, he defined parallelism in such a way that there are just two lines through a given point **A** parallel to a given line **q** (Fig. 1.2A), these being

*Lobatschewsky [1].

asymptomatic to **q**, as Saccheri had already shown. Drawing **AB** perpendicular to **q**, he defined the *angle of parallelism* $\Pi(\mathbf{AB})$ as the (acute) angle between **AB** and either of the parallels. In terms of a convenient form of the absolute unit of length (Taurinus's $k = 1$), he found that the angle of parallelism for the distance $\mathbf{AB} = c$ is

$$\Pi(c) = 2\arctan e^{-c} = \operatorname{arc}\cot(\sinh c) = \operatorname{arc}\cos(\tanh c),$$

which decreases from $\frac{1}{2}\pi$ to 0 as c increases from 0 to ∞. The same result, in the form $\sin \Pi(c) \cosh c = 1$, could have been obtained by Taurinus also, if he had put $A = \Pi(c)$, $B = \frac{1}{2}\pi$, $C = 0$, $k = 1$ in 1.31.

Lobatschewsky derived his trigonometrical formulae from a study of the horocycle (circle of infinite radius) and horosphere (sphere of infinite radius), in the course of which he rediscovered Wachter's theorem that the geometry of horocycles on a horosphere is identical with the geometry of straight lines in the Euclidean plane. Having also rediscovered Lambert's formula $\pi - A - B - C$ for the area of a triangle **ABC**, he proceeded to calculate the volume of a tetrahedron,* expressing the result in terms of his famous transcendental function

$$L(x) = \int_0^x \log \sec y \, dy.$$

Observing that the trigonometrical formulae of Euclidean geometry are valid in the infinitesimal neighbourhood of a point in the new geometry, Lobatschewsky considered the possibility that his geometry might replace Euclid's in the exploration of astronomical space. The crucial experiment would consist in finding a positive lower bound for the parallax of stars. For, if c is a diameter of the Earth's orbit, measured in terms of the (unknown) absolute unit, the parallax of any star should exceed $\frac{1}{2}\pi - \Pi(c)$. But this remains an open question, since such a lower bound, if it exists, is smaller than the

*Cf. Schläfli [1], p. 97; Richmond [1]; Coxeter [1].

allowance for experimental error. The failure of the experiment merely tells us that, if space is in fact hyperbolic, the absolute unit must be many millions of times as long as the diameter of the Earth's orbit.

In this connection we must bear in mind that, although a geometry may seem more interesting if we can compare it with the real world, its validity as a logical structure is not affected, but depends only on its internal consistency. In order to show that his "imaginary" geometry or "pangeometry" is as consistent as Euclidean geometry, Lobatschewsky pointed out that it is all based on his formulae for a triangle, which lead to the familiar formulae for a spherical triangle when the sides a, b, c are replaced by ia, ib, ic. Any inconsistency in the new geometry could be "translated" into an inconsistency in spherical geometry (which is part of Euclidean geometry). Thus, after two thousand years of doubt, the independence of Euclid's Postulate V was finally established.

1.5. Bolyai. Many of the same results were discovered about the same time by Bolyai János (1802-1860), who wrote to his father, Bolyai Farkas, in 1823: "I have resolved to publish a work on the theory of parallels, as soon as I shall have put the material in order. . . . The goal is not yet reached, but I have made such wonderful discoveries that I have been almost overwhelmed by them. . . . *I have created a new universe from nothing.*"

Bolyai Farkas expressed the wish to insert his son's discoveries in his own book, as an Appendix.* In making this offer, he remarked, more appropriately than he realized, that "many things have an epoch, in which they are found at the same time in several places, just as the violets appear on every side in spring."

The younger Bolyai's speciality was the "absolute science

*Bolyai [1], [2].

of space" (or *absolute geometry*), consisting of those propositions which are independent of Postulate V, so that they hold in both Euclidean and hyperbolic geometry. For instance, he expressed the "sine rule" for a triangle **ABC** in the form

$$\bigcirc a : \bigcirc b : \bigcirc c :: \sin A : \sin B : \sin C,$$

where $\bigcirc a$ denotes the circumference of a circle of radius a. He observed that such formulae hold also in *spherical* geometry.

1.6. Riemann. The full recognition that spherical geometry is itself a kind of non-Euclidean geometry, without parallels, is due to Riemann (1826-1866). He realized that Saccheri's hypothesis of the obtuse angle becomes valid as soon as Postulates I, II and V are modified to read:

I. *Any two points determine at least one line.*
II. *A line is unbounded.*
V. *Any two lines in a plane will meet.*

For a line to be unbounded and yet of finite length, it merely has to be re-entrant, like a circle. The great circles on a sphere provide a model for the finite lines on a finite plane, and, when so interpreted, satisfy the modified postulates. But if a line and a plane can each be finite and yet unbounded, why not also an n-dimensional manifold, and in particular the three-dimensional space of the real world? In Riemann's words of 1854: "The unboundedness of space possesses a greater empirical certainty than any other external experience. But its infinite extent by no means follows from this; on the other hand, if we assume independence of bodies from position, and therefore ascribe to space constant curvature, it must necessarily be finite provided this curvature has ever so small a positive value."*

According to the General Theory of Relativity, astronomical space has positive curvature locally (wherever there is matter), but we cannot tell whether the curvature of

*Riemann [1], p. 36.

"empty" space is exactly zero or has a very small positive or negative value. In other words, we still cannot decide whether the real world is approximately Euclidean or approximately non-Euclidean.

Riemann employed the "infinitesimal approach" to geometry, wherein the differential of distance is expressed as the square root of a quadratic form in the differentials of the coordinates. In the special case of constant curvature, his formula is

1.61. $$ds = \frac{\sqrt{(\Sigma dx^2)}}{1+\frac{1}{4}K\Sigma x^2}$$

A year or two before Riemann read his epoch-making *Habilitationsschrift*, quoted above, Schläfli (1814-1895) developed the analytical geometry of n-dimensional Euclidean space,* and considered in particular the *hyper-sphere* $\Sigma x^2 = k^2$, which provides a model for Riemann's $(n-1)$-dimensional spherical space.

In the differential geometry of a surface in ordinary space, the product of the maximum and minimum "normal curvatures" is usually denoted by K. (Thus $K = k^{-2}$ for a sphere of radius k.) Gauss made the notable discovery that this *specific curvature* can be expressed in terms of quantities measured on the surface itself, without using properties of the underlying Euclidean space (e.g. normals). Thus it could still be defined if the underlying space did not exist. Riemann's "constant curvature" is the n-dimensional analogue of this K. Although the geometry of astronomical space, according to his hypothesis, may be identical with that of a hyper-sphere in four-dimensional Euclidean space, it does not follow that there is in any physical sense a Euclidean four-space in which the spherical three-space is imbedded.† Thus spherical space is like

*Schläfli [1].

†Sommerville [2], p. 199 (§7).

the substance of a balloon with an extra dimension; but the simile breaks down if we seek a meaning for the air inside or outside the balloon.

1.7. Klein. Riemann developed the differential geometry of spherical space. On the other hand, Cayley (1821-1895) considered space "in the large," defining distance in terms of homogeneous coordinates. But it was Klein (1849-1925) who first saw clearly how to rid spherical geometry of its one blemish: the fact that two coplanar lines (being two great circles of a sphere) have not just one but *two common points.* Since every point determines a unique antipodal point, and every figure is thus duplicated at the antipodes, he realized that nothing would be lost, but much gained, by abstractly identifying each pair of antipodal points, i.e. by changing the meaning of the word "point" so as to call such a pair *one point.*

The word "line" will then be used for a great circle with every pair of diametrically opposite points identified (or a great semicircle with its two ends identified). So also, the word "plane" will be used for a great sphere with every pair of antipodal points identified, and analogous definitions can be made in any number of dimensions. With this meaning for the words, *any* two points determine a unique line; for, antipodal points are no longer two but one. Thus the traditional form of Postulate I is restored. As for Postulate II, a line is still unbounded, though finite, its length being half that of the great circle. Right angles retain their ordinary meaning, but a circle appears as a pair of antipodal circles.

It was to this modification of spherical geometry that Klein gave the name *elliptic* geometry. It is in many ways simpler than either spherical or Euclidean geometry, and can be developed quite independently. The geometry of pairs of antipodal points is merely a *model* for it, a convenient representation in terms of more familiar concepts. Another model is obtained by considering the diameters which join such pairs of points. In this manner the points and lines in elliptic space of n dimensions are represented by the lines and planes through a fixed point O in Euclidean space of $n+1$ dimensions. In particular, elliptic geometry of two dimensions is represented as the geometry of a *bundle* in ordinary space. To interpret the elliptic concepts in terms of Euclidean concepts, we translate them according to the following "dictionary":

The elliptic plane	Euclidean space, in the neighbourhood of a fixed point O
Point	Line through O
Line	Plane through O
Segment	Angle
Angle	Dihedral angle
Perpendicular lines	Perpendicular planes
Triangle	Trihedron
Circle	Right circular cone
Rotation about a point	Rotation about a line through O
Reflection in a line	Reflection in a plane through O
etc.	etc.

A third model (for elliptic plane geometry) can be derived from this second model by considering the section of the bundle by an arbitrary plane, not passing through O. This has the advantage of representing points by points, and lines by lines.* But distances and angles are inevitably distorted, since the distance between two points has to be re-defined as the angle subtended at O. Moreover, certain points of the elliptic geometry are left out, since certain lines of the bundle are parallel to the chosen plane. In order to accommodate these extra points, it is natural to augment the Euclidean plane by postulating *points at infinity*, one for every direction, in the manner advocated by Kepler (1571-1630) and Desargues (1593-1662). When this is done, we have the *projective* plane,

*Klein [1], p. 604; [3], p. 148.

in which every two lines intersect (either at an ordinary point or at infinity). Thus, if metrical ideas are left out of consideration, elliptic geometry is the same as real projective geometry.

Conversely, real projective geometry (which we shall develop in Chapters II, III, IV) contains certain correspondences which enable us to define the elliptic metric in the whole space (see Chapters V, VI, VII), and to define either the Euclidean or the hyperbolic metric in a suitable part of space (Chapter IX).

The study of elliptic geometry is almost forced upon us as soon as we have added points and lines at infinity to Euclidean space. For, such points and lines form a plane—the *plane at infinity*—whose intrinsic geometry is elliptic. (See §9.5.)

To sum up, the metrical geometries with which we are concerned are Euclidean, hyperbolic, spherical, and elliptic. Our preoccupation with these four, as against all other continuous geometries, is justified by the fact that only in these cases is space *completely isotropic*, in the sense that all the lines through each point are alike. It is an interesting result in differential geometry that, if space is continuous and isotropic, it is also homogeneous or, as Riemann would say, of constant curvature.

CHAPTER II

REAL PROJECTIVE GEOMETRY: FOUNDATIONS

2.1. Definitions and axioms. In any geometry, logically developed, each definition of an entity or relation involves other entities and relations; therefore certain particular entities and relations must remain undefined. Similarly, the proof of each proposition uses other propositions; therefore certain particular propositions must remain unproved; these are the *axioms*. We take for granted the machinery of logical deduction, and the primitive concept of a *class* (or "set of all").

Unless the contrary is stated, the word *correspondence* will be used in the sense of *one-to-one* correspondence. Thus a set of entities is said to correspond to another set if every entity in each set is associated with a unique entity in the other set. In geometry the entities are usually points or lines, and the set of entities is called a *figure*. Thus we speak of a correspondence between two figures. It is often convenient to regard the correspondence as an operation which changes the first figure into the second. (Familiar instances are rotation, reflection, inversion, and reciprocation.) The general technique for discussing correspondences belongs properly to the theory of groups; but the following outline will suffice for our purposes.

We shall find it convenient to denote a correspondence by a capital Greek letter, such as Θ, writing $F\Theta = F'$ to mean that Θ relates the figure F to F' (or that the figure corresponding to F is F'). If a second correspondence Φ relates the figure F' to F_1', we write $F'\Phi = F_1'$, or $F\Theta\Phi = F_1'$, and say that the *product* $\Theta\Phi$ relates F to F_1'. The trivial correspondence that relates every entity to itself is called the *identity*, and is denoted by 1 (since its product with Θ is Θ itself). If $\Theta\Phi = 1$, we call Φ the

inverse of Θ, writing $\Phi = \Theta^{-1}$. (Thus the relation $F\Theta = F'$ is equivalent to $F = F'\Theta^{-1}$.) Some authors write $\Theta(F)$ instead of $F\Theta$, so as to exhibit it as a ***function*** of F. (Note that, in one of the accepted notations, we write $x = \sin^{-1}x'$ when $\sin x = x'$.)

If a correspondence Θ relates F to F′, while another correspondence Φ relates the pair of figures (F, F′) to (F_1, F_1'), we say that Φ ***transforms*** Θ into the correspondence between F_1 and F_1'. Since

$$F_1' = F'\Phi = F\Theta\Phi = F_1\Phi^{-1}\Theta\Phi,$$

this transformed correspondence is $\Phi^{-1}\Theta\Phi$. It may happen that Θ itself relates F_1 to F_1', so that Φ transforms Θ into itself. We then say that Θ is ***invariant*** under transformation by Φ. Since the relation $\Phi^{-1}\Theta\Phi = \Theta$ may be written $\Theta\Phi = \Phi\Theta$, an equivalent statement is that Θ and Φ are ***permutable.*** (As a familiar example of correspondences which are ***not*** permutable, consider the reflections in two planes not perpendicular to one another.)

According to Klein, the character of any geometry is determined by the type of correspondence under which its relations are invariant; e.g. Euclidean geometry is invariant under "similarity transformations."* The title of this book refers strictly to just two geometries, elliptic and hyperbolic; but certain others are so closely interwoven with these as to compel our attention. The concept of ***similarity***, which plays such a vital role in Euclidean geometry, has no analogue in either of the non-Euclidean geometries. On the other hand, the concept of ***parallelism*** (for lines in one plane) belongs to both Euclidean and hyperbolic geometry, but is lacking in elliptic. Bolyai János (§1.5) gave the name ***absolute*** geometry to the large body of propositions common to Euclidean and hyperbolic geometry. Some of these propositions will be used in Chapter IX, before

*Veblen and Young [2], I, pp. 64-68; II, pp. 78, 119.

we split absolute geometry into its two parts by affirming or denying the uniqueness of parallelism (or Postulate V).

The contrast between absolute and elliptic geometry is clearly seen in the theory of *order*.* In either geometry we can describe a "four point" order, saying that four collinear points fall into two pairs which *separate* each other. In absolute geometry this can be derived from the stronger "three point" order, in which we say that one of three collinear points lies *between* the other two. But in elliptic geometry all lines are closed, and so the notion of order does not specialize one of three points: order is no longer "serial" but "cyclic."

Throughout the ages, from the ancient Egyptians and Euclid to Poncelet and Steiner, geometry has been based on the concept of measurement, which is defined in terms of the relation of *congruence*. It was von Staudt (1798-1867) who first saw the possibility of constructing a logical geometry without this concept. Since his time there has been an increasing tendency to focus attention on the much simpler relation of *incidence*,† which is expressed by such phrases as "The point **A** lies on the line **p**" or "The line **p** passes through the point **A**."

Euclidean geometry with congruence left out is called *affine* geometry. As so many figures in Euclidean geometry are defined in terms of congruence (e.g. equilateral triangle, circle, conic section), it might seem that in affine geometry there would be little left to talk about. It is true that the content of affine geometry is less rich than that of Euclidean, but it is still possible to define conics, for example, and to distinguish the three types: ellipse, parabola, hyperbola. Postulate V, in its non-metrical form 1.12, allows us to define an attenuated kind of congruence, by which we can compare certain segments (namely those which are parallel to one another) and measure area.‡ But the notion of "perpendicularity" is entirely lack-

*Vailati [1]. See also Veblen and Young [2], II, p. 44, and Russell [1], IV. †Pieri [1]; Baker [1], I, p. 4. ‡Heffter and Koehler [1], I, p. 219.

ing. By suitably *defining* perpendicularity, we can restore the whole of Euclidean geometry. By modifying the definition, we can derive instead *Minkowskian* geometry,* the four-dimensional case of which is used in the Special Theory of Relativity.

Similarly, elliptic geometry with congruence left out is *real projective* geometry. This was developed (*qua* Euclidean geometry augmented by points at infinity) long before elliptic geometry itself, and is still widely studied for its own interest. It excels affine geometry in the symmetry of its propositions of incidence, which occur in pairs in accordance with the "principle of duality." Moreover, it *includes* all the other geometries that have been mentioned. For, by suitably defining perpendicularity we can restore the metrical properties of elliptic geometry, and by modifying the definition we can derive instead hyperbolic geometry. Again, by specializing a plane (in the three-dimensional case) or a line (in the two-dimensional), we can derive affine geometry, and thence either Euclidean or Minkowskian.

In the following "genealogy," each geometry (save the first) is derived from its parent by some kind of specialization:

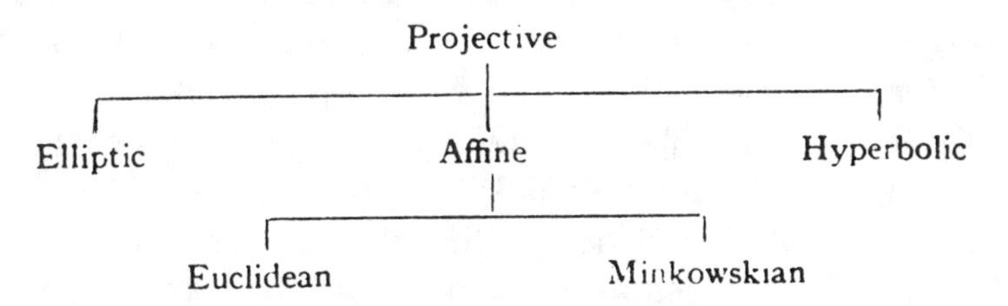

In view of the above remarks, we shall set aside all metrical considerations till Chapter v, and survey the foundations of real projective geometry, using the axioms of Pieri, Vailati, and Dedekind. In this case there are two undefined entities,

*Minkowski [1]; Robb [1].

point and *line*; two undefined relations, *incidence* and *separation.*

AXIOMS OF INCIDENCE

2.111. *There are at least two points.*

2.112. *Any two points are incident with just one line.*

The line thus determined by two points **A** and **B** is said to *join* the points, and is denoted by **AB**.

2.113. *The line* **AB** *is incident with at least one point besides* **A** *and* **B**.

Points incident with a line are said to lie *on* the line, or to be *collinear*. The class of points on a line is called a *range*.

2.114. *There is at least one point not incident with the line* **AB**.

Lines incident with a point are said to pass *through* the point, or to be *concurrent*. Two such lines are said to meet or *intersect*. By joining the points of a range to a point **C**, not belonging to this range, we obtain a *flat pencil* of lines, with centre **C**. A *plane* is the class of points on the lines of a flat pencil, together with the class of lines joining pairs of these points.

2.115. *If* **A**, **B**, **C** *are three non-collinear points, and* **D** *is a point on* **BC** *distinct from* **B** *and* **C**, *while* **E** *is a point on* **CA** *distinct from* **C** *and* **A**, *then there is a point* **F** *on* **AB** *such that* **D**, **E**, **F** *are collinear.* (See Fig. 8.8A on page 173.)

It follows that a plane contains all the points on each of its lines, and can be defined equally well by any pencil contained in it. In terms of three non-collinear points, or a non-incident point and line, we denote a plane by **ABC** or **Ap**. Points or lines in one plane are said to be *coplanar*.

2.116. *There is at least one point not in the plane* **ABC**.

2.117. *Any two planes intersect in a line.*

So far, we may appear to have been underlining the obvious. Every student of elementary geometry is familiar with the terms just used. But the real importance of the foregoing remarks lies less in what we have said than in what we have not said. We have ***not*** mentioned the possibility of the distance **AB** being equal to the distance **CD**; we shall never make such a remark as long as we are dealing with projective geometry only.

From the above axioms it is possible to deduce all the propositions of incidence for points, lines, and planes. If lines **p** and **q** intersect, we denote their common point by (**p**, **q**) and their plane by **pq**. Two lines which do not intersect are said to be *skew*. Planes incident with a line are said to pass through the line, or to be *coaxial*. The class of planes through a line **p** is called an *axial pencil*, with axis **p**. The class of lines and planes through a point **O** is called a *bundle* (or sheaf, or star), with centre **O**.

By considering "degrees of freedom" we are led to speak of a point as having *no dimension*, a line *one dimension*, and a plane *two dimensions*. The lines of a flat pencil, and likewise the planes of an axial pencil, correspond (by incidence) to the points of a range, which is the *section* of the pencil by a line. For this reason, ranges and pencils together are described as *one-dimensional primitive forms*. Similarly, planes and bundles are *two-dimensional* forms, the points and lines of a plane being sections of the lines and planes of a bundle; and the whole space is *three-dimensional.*

One of the most important correspondences in projective geometry is *perspectivity*. This is the correspondence established between two coplanar lines, or two planes, by regarding them as different sections of the same flat pencil or bundle, respectively. In the case of lines, we say that the flat pencil *projects* the one range into the other, and the two ranges are said to be *in perspective*. (See Fig. 2.1A, where **O** is the centre of the pencil.) Corresponding points are indicated by formulae such as

$$\mathbf{ABC} \ldots \doublebarwedge \mathbf{A'B'C'} \ldots, \text{ or } \mathbf{ABC} \ldots \overset{\mathbf{O}}{\doublebarwedge} \mathbf{A'B'C'} \ldots .$$

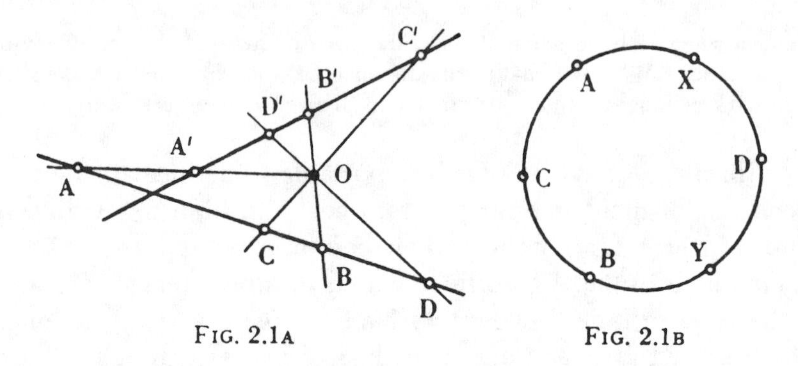

FIG. 2.1A FIG. 2.1B

Our second undefined relation refers to two pairs of points on a line. Following Vailati, we use the symbol **AB || CD** to mean that **A** and **B** *separate* **C** and **D**. (The significance of this relation is most clearly seen by representing the line as a circle. See Fig. 2.1B.)

AXIOMS OF SEPARATION

2.121. *If* **A, B, C** *are three collinear points, there is at least one point* **D** *such that* **AB || CD**.

2.122. *If* **AB || CD**, *then* **A, B, C, D** *are collinear and distinct.*

2.123. *If* **AB || CD**, *then* **AB || DC**.

2.124. *If* **A, B, C, D** *are four collinear points, then either* **AB || CD** *or* **AC || BD** *or* **AD || BC**.

2.125. *If* **AB || CD** *and* **AD || BX**, *then* **AB || CX**.

2.126. *If* **AB || CD** *and* **ABCD** $\overline{\wedge}$ **A′B′C′D′**, *then* **A′B′ || C′D′**.

From the last of these axioms we can deduce* that the relation **AB || CD** implies **CD || AB**. Putting **C** for **X** in 2.125, and using 2.122, we see that the relation **AB || CD** excludes **AC || BC**. The existence-axiom 2.121 has been inserted as the most natural way to secure an infinity of points.

---

*Robinson [1], p. 119 (second footnote). Following Veblen, Robinson uses the symbol $\overline{\wedge}$ for the combination of several perspectives. Following von Staudt, who invented that symbol, I prefer to define it differently, though later the two definitions will be seen to be equivalent (in **real** geometry).

Accordingly, the above axioms differ slightly from those given by Vailati (and quoted by Veblen and Russell).

If **A**, **B**, **C** are three collinear points, we define the *segment* **AB/C** as the class of points **X** for which **AB** ∥ **CX**. Thus the segment **AB/C** does not contain **C**. The segment with its end points **A** and **B** is called an *interval*, and is denoted by $\overline{\mathbf{AB}}/\mathbf{C}$. If **X** and **Y** belong to $\overline{\mathbf{AB}}/\mathbf{C}$, the interval $\overline{\mathbf{XY}}/\mathbf{C}$ is said to be *interior* to $\overline{\mathbf{AB}}/\mathbf{C}$, and a point **D** lies *between* **X** and **Y** in $\overline{\mathbf{AB}}/\mathbf{C}$ if it belongs to **XY/C**, i.e. if **XY** ∥ **CD**. (Either **X** or **Y** may coincide with either **A** or **B**. In other cases we have either **AX** ∥ **BY** or **AY** ∥ **BX**.) Thus "three point" order becomes valid when we restrict consideration to an interval, or to a segment.

AXIOM OF CONTINUITY

2.13. *For every partition of all the points of a segment into two non-vacuous sets, such that no point of either lies between two points of the other, there is a point of one set which lies between every other point of that set and every point of the other set.*

This final axiom will be used in §2.7.

2.2. Models.

When we say that a system of axioms is consistent, we mean that no two theorems, logically deduced from them, can be contradictory (like the statements "All right angles are equal" and "Some lines are self-perpendicular"). Clearly, there is no direct test for consistency, since we cannot follow up the infinite number of possible chains of deduction to see whether any two of them lead to a contradiction. For an indirect test we use a *model*, which is a set of objects satisfying the same axioms as the undefined entities of the original system. Any contradiction implied by the original system would be represented by a contradiction in the model, and this cannot occur so long as the objects unquestionably exist. These objects may be (defined or undefined) entities in another abstract system whose consistency is taken for granted, or they may be physical objects whose reality is accepted for reasons outside the domain of mathematics. In the former case the assumption of consistency can only be justified by means of a model of the model; so it may well be argued that every question of consistency is ultimately based on properties of the physical world as interpreted by our senses. (For those who dislike this materialistic conclusion, a possible loophole is offered by the recent attempts to prove the consistency of arithmetic in a direct fashion.*)

We give here three models for real projective geometry. The first is in terms of affine geometry, whose consistency is usually established by means of Cartesian coordinates (the "model of the model"). The second refers directly to the number system. The third is in terms of absolute geometry, which, being based on the "self-evident" postulates I—IV, is amenable to direct comparison with physical space.

To construct the first model, we define axial pencils and bundles in affine (or, if preferred, Euclidean) space, in such a way as to include pencils and bundles of parallels, and then set up a "dictionary," as follows:

Real projective space	Affine space
Point	Bundle
Line	Axial pencil
A point lies on a line	The planes of a bundle include those of a pencil
Two points determine a line	The common planes of two bundles form a pencil

Setting up this model is effectively equivalent to the classical derivation of projective space from affine (or Euclidean) space by adding the "ideal" points and lines of a postulated

*Hilbert and Bernays [1].

"plane at infinity." For, such points arise as centres of bundles of parallel lines, and such lines arise as axes of pencils of parallel planes.

The direct appeal to arithmetic is made by using homogeneous coordinates. The appropriate dictionary this time is as follows:

Real projective space	The real number system
Point **A**	The class of ordered tetrads of real numbers, proportional to a given tetrad (a_0, a_1, a_2, a_3)
Line **p**	The class of ordered hexads of real numbers, proportional to a given hexad $\{P_{23}, P_{31}, P_{12}, P_{01}, P_{02}, P_{03}\}$ satisfying the equation $P_{23}P_{01}+P_{31}P_{02}+P_{12}P_{03}=0$ (For convenience we define P_{32} etc., so that $P_{\nu\mu}=-P_{\mu\nu}$ and therefore $P_{\nu\nu}=0$)
Point **A** lies on line **p**	$a_0P_{0\nu}+a_1P_{1\nu}+a_2P_{2\nu}+a_3P_{3\nu}=0$ for two (and therefore all four) values of ν
Line **AB**	$\{a_0b_1-a_1b_0,\ a_0b_2-a_2b_0,\ a_0b_3-a_3b_0,$ $a_2b_3-a_3b_2,\ a_3b_1-a_1b_3,\ a_1b_2-a_2b_1\}$
AB ‖ **CD**	The a's, and likewise the b's, c's, and d's, satisfy two independent linear homogeneous equations; and the respective ratios a_μ/a_ν, c_μ/c_ν, b_μ/b_ν, d_μ/d_ν, or some cyclic permutation thereof, are in strictly ascending order of magnitude for at least one choice of μ and ν

For further details of this model, see §4.6. The verification of all the axioms is an interesting exercise.

If we are content to consider the two-dimensional projective geometry of a single plane, a third model consists of the lines and planes through one point in ordinary space (as in §1.7, but without the metrical concepts):

The real projective plane	Euclidean or non-Euclidean space, in the neighbourhood of a fixed point **O**
Point **A**	Line **a** through **O**
Line **AB**	Plane **ab**
Range	Flat pencil
Flat pencil	Axial pencil
Triangle	Trihedron
Conic	Quadric cone

This has the advantage of symmetry, which the first model lacks. Moreover, the geometry of a bundle can be developed without using "Postulate V"; in fact a bundle is essentially the same thing in absolute geometry as in projective. But in order to adapt this model to three dimensions, we would have to consider the "hyper-bundle" of lines and planes through a point in four dimensions—which is quite satisfactory for anyone who has become familiar with the properties of absolute (or Euclidean) hyper-space.

2.3. The principle of duality. Three non-collinear points **A**, **B**, **C** are called the *vertices* of a *triangle* **ABC**; its *sides* are the three lines **BC**, **CA**, **AB**. Analogously, four non-coplanar points **A**, **B**, **C**, **D** are the vertices of a *tetrahedron* **ABCD**; its *edges* and *faces* are the six lines **AD**, **BD**, **CD**, **BC**, **CA**, **AB**, and the four planes **BCD**, **CDA**, **DAB**, **ABC**.

The principle of duality in the plane affirms that every definition remains significant, and every theorem remains true, when we interchange "point" and "line," and make a few consequent alterations in wording. This means that the geometry of lines forms a model for the geometry of points. The following definition provides an example:

Four coplanar points, **A**, **B**, **C**, **D**, of which no three are collinear, are the vertices of a *complete quadrangle** **ABCD**, with the six lines **AD**, **BD**, **CD**, **BC**, **CA**, **AB** for sides. The points of intersection of "opposite" sides, namely (**AD**, **BC**), (**BD**, **CA**), (**CD**, **AB**), are called diagonal points, and are the vertices of the diagonal triangle.

Four coplanar lines **a**, **b**, **c**, **d**, of which no three are concurrent, are the sides of a *complete quadrilateral** **abcd**, with the six points (**a**, **d**), (**b**, **d**), (**c**, **d**), (**b**, **c**), (**c**, **a**), (**a**, **b**) for vertices. The joins of "opposite" vertices, namely (**a**,**d**)(**b**,**c**),(**b**, **d**)(**c**, **a**),(**c**, **d**)(**a**, **b**), are called diagonal lines, and are the sides of the diagonal triangle.

The principle of duality in space allows the analogous interchange of "point" and "plane." Thus 2.117 is the space-dual of 2.112. Here is another example:

Five points **A**, **B**, **C**, **D**, **E**, of which no four are coplanar, are the vertices of a *complete pentagon* **ABCDE**, with the ten lines **AB**, . . . , **DE** for edges, and the ten planes **ABC**, . . . , **CDE** for faces. Each edge lies in three faces.

Five planes $\alpha, \beta, \gamma, \delta, \epsilon$, of which no four are concurrent, are the faces of a *complete pentahedron* $\alpha\beta\gamma\delta\epsilon$, with the ten lines (α, β), . . . , (δ, ϵ) for edges, and the ten points (α, β, γ), . . . , $(\gamma, \delta, \epsilon)$ for vertices. Each edge contains three vertices.

To justify the principle of duality, we observe that the axioms imply their own duals. For instance, 2.115 enables us to prove the plane-dual of 2.112, namely†

2.31. *Any two coplanar lines intersect.*

(This is the result that most clearly distinguishes projective geometry from affine geometry. It rules out the possibility of parallels.)

Having proved a theorem, we can state the space-dual theorem without more ado; for a proof could in fact be written down mechanically by dualizing every step in the proof of the

*When there is no danger of confusion, we shall omit the word "complete."

†Veblen and Young [2], I, p. 19.

original theorem. The same remark applies to the *plane*-dual of any theorem which can be proved without using points outside the plane. Consider, however, Desargues' Theorem and its converse (which are easily proved with the help of 2.116):

2.32. *If the vertices of two coplanar triangles correspond in such a way that the joins of corresponding vertices are concurrent, then the intersections of corresponding sides are collinear.*

2.33. *If the sides of two coplanar triangles correspond in such a way that the intersections of corresponding sides are collinear, then the joins of corresponding vertices are concurrent.*

Either of these dual theorems can be deduced from the other, without leaving the plane;* but it is not legitimate to invoke the principle of duality in the plane for this purpose, since the initial proof is essentially three-dimensional.

To obtain a sufficient set of axioms for projective geometry in two dimensions, we can replace 2.115—2.117 by 2.31 and 2.32 (omitting the word *coplanar*). The principle of duality will then hold without reservation.

2.4. Harmonic sets. A large part of our investigation (e.g. Chapter v) will be concerned with the geometry of points on a single line, where there is no scope for incidences. This deficiency is compensated by the possibility of defining the *harmonic conjugate* of a given point with respect to two given points. (We think of this as a one-dimensional concept, even though it requires incidences in two dimensions for its construction and in three dimensions for the proof of its uniqueness.) We shall use the abbreviation H(**AB**, **CD**) for the state-

*See, for instance, Baker [1] I, p. 181, or Robson [1], p. 211. Cf. Veblen and Young [2], I, p. 41.

ment that **D** is the harmonic conjugate of **C** with respect to **A** and **B**, which means* that there is a quadrangle **IJKL** such that one pair of opposite sides intersect at **A**, and a second pair at **B**, while the third pair meet **AB** at **C** and **D**. This relation is clearly symmetrical between **A** and **B**, and between **C** and **D**. Given three collinear points **A**, **B**, **C**, we can obtain **D** by taking two points **I**, **J**, collinear with **C**, and constructing the intersections **K** = (**AJ**, **BI**), **L** = (**AI**, **BJ**), **D** = (**AB**, **KL**). It is a simple consequence of 2.32 and 2.33 that the position of **D** is independent of the choice of **I** and **J**. But can we be sure that **D** is distinct from **C**? (This is important for certain applications.) The following proof is due to Enriques.

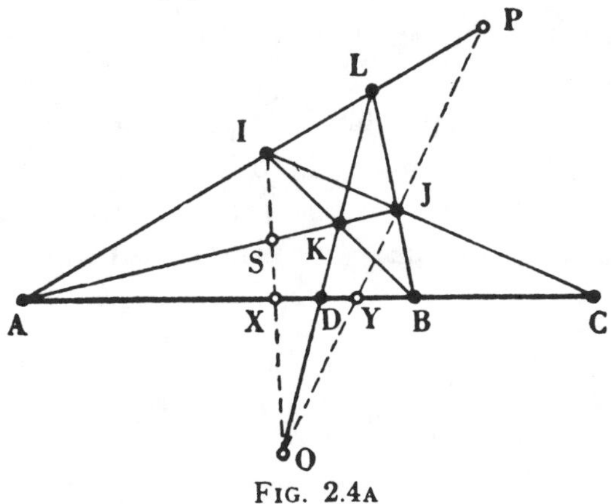

FIG. 2.4A

By 2.121 and 2.123, we can take a point **S** such that **AK** || **SJ**, as in Fig. 2.4A. Let **IS** meet **AB** at **X**, and **KL** at **O**; let **JO** meet **AB** at **Y**, and **AI** at **P**. Then

$$\mathbf{AKSJ} \overset{\mathbf{I}}{\underset{\wedge}{=}} \mathbf{ABXC},^{\dagger} \quad \mathbf{AKSJ} \overset{\mathbf{O}}{\underset{\wedge}{=}} \mathbf{ALIP} \overset{\mathbf{J}}{\underset{\wedge}{=}} \mathbf{ABCY}, \quad \mathbf{AKSJ} \overset{\mathbf{O}}{\underset{\wedge}{=}} \mathbf{ADXY}.$$

By 2.126, we therefore have **AB** || **XC**, **AB** || **CY**, **AD** || **XY**.

*De la Hire [1], lib. i, prop. xx; Enriques [1], p. 51.

†If **X** happens to coincide with **D**, the argument ends here; for then the given relation **AK**||**SJ** implies **AB**||**DC**.

Thus both **X** and **Y** are in the segment **AB/C**, and **D** lies between them. In other words,

2.41. H(**AB**, **CD**) *implies* **AB** || **CD**.

By applying the plane-dual of the above construction for the fourth harmonic point of three collinear points, we obtain a unique "fourth harmonic line" for any three lines of a flat pencil. The figure involved is almost the same as before; in fact **ID** is the harmonic conjugate of **IC** with respect to **IA** and **IB**. Thus a harmonic set of points is joined to any external point by a harmonic set of lines. Dually, any section of a harmonic set of lines is a harmonic set of points. Hence

2.42. *If* H(**AB**, **CD**) *and* **ABCD** $\barwedge$ **A'B'C'D'**, *then* H(**A'B'**, **C'D'**).

Two-dimensional geometry admits three alternative analogues for harmonic conjugacy. One of these, which plays an important part in the introduction of coordinates (§4.3), is the so-called *trilinear polarity*. (The other two are the harmonic homology of §3.1, and the true polarity of §3.2.) The trilinear pole of a line **g**, with respect to a triangle **ABC**, is constructed as follows.

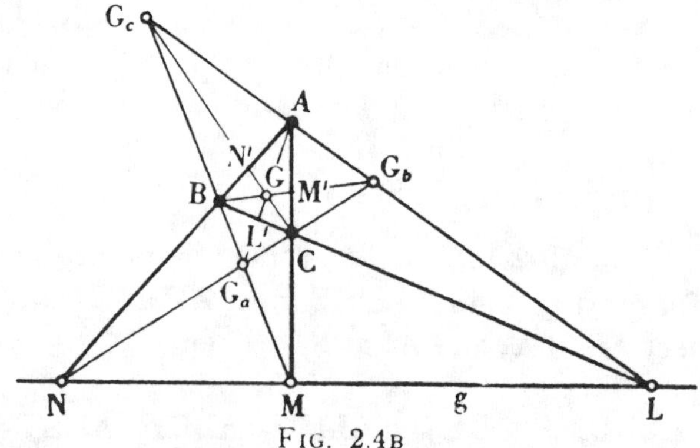

FIG. 2.4B

Let the sides of **ABC** meet **g** in **L**, **M**, **N**, and let $\mathbf{G_aG_bG_c}$ be the triangle formed by the lines **AL**, **BM**, **CN**. By 2.33, the three lines $\mathbf{AG_a}$, $\mathbf{BG_b}$, $\mathbf{CG_c}$ are concurrent; their common point **G** is the trilinear pole of **g**. These lines meet the sides of the triangle **ABC** in points **L′**, **M′**, **N′**, which are the harmonic conjugates of **L, M, N** with respect to the point-pairs **BC**, **CA**, **AB**. Conversely,* we may define **L′**, **M′**, **N′** as the harmonic conjugates of **L**, **M**, **N**, and obtain **G** as the point of concurrence of **AL′**, **BM′**, **CN′**. So also $\mathbf{G}_a$ is the trilinear pole of the line **M′N′**, which therefore contains **L**; and similarly $\mathbf{G}_b$ and $\mathbf{G}_c$ are the trilinear poles of **N′L′M** and **L′M′N**.

In affine geometry, we recognize the trilinear pole of the line at infinity as the centroid of the triangle.

2.5. Sense. The intuitive idea of the two opposite directions along a line, or round a circle, is so familiar that we are apt to overlook the niceties of its theoretical basis. Some authors regard sense as an undefined relation, and define separation in terms of it. But the comparison of sense requires six points, whereas separation involves only four; therefore we prefer to deduce sense from separation. As a first step we observe that Axioms 2.12 imply the following theorem:†

2.51. *If* **AB** || **CD**, *the two points* **A** *and* **B** *divide the rest of their line into just two segments*, **AB/C** *and* **AB/D**.

Two such segments, and likewise the corresponding intervals, are said to be *supplementary*.

It is an immediate consequence of 2.124 that three collinear points **A**, **B**, **C** divide the rest of their line into three segments **BC/A**, **CA/B**, **AB/C**; and it follows by induction that the notation $\mathbf{A}_0, \mathbf{A}_1, \ldots, \mathbf{A}_{n-1}$ can be assigned to n collinear points in such a way that they divide the rest of their line into n segments $\mathbf{A}_r\mathbf{A}_{r+1}/\mathbf{A}_{r-1}$ (with suffixes reduced modulo n). This division of the line into segments is maintained if we change each symbol $\mathbf{A}_r$ into either $\mathbf{A}_{s+r}$ or $\mathbf{A}_{s-r}$, for a fixed residue s (mod n). By means of one of these changes, any particular three

*Poncelet [1], II, p. 34.

†Robinson [1], p. 120.

of the n points may be named $\mathbf{A}_0, \mathbf{A}_b, \mathbf{A}_c$, where $0 < b < c < n$. This notation facilitates the definition of one-dimensional *sense*.

Let **ABC** and **DEF** be two triads of distinct points on one line. (Any of **D**, **E**, **F** may happen to coincide with any of **A**, **B**, **C**.) Let the distinct points of this set be named $\mathbf{A}_0, \mathbf{A}_1, \ldots,$ $\mathbf{A}_{n-1}$ ($n = 3$, 4, 5, or 6) in such a way that $\mathbf{A} = \mathbf{A}_0$, $\mathbf{B} = \mathbf{A}_b$, $\mathbf{C} = \mathbf{A}_c$, with $b < c$. Suppose that then $\mathbf{D} = \mathbf{A}_d$, $\mathbf{E} = \mathbf{A}_e$, $\mathbf{F} = \mathbf{A}_f$. If $d < e < f$ or $e < f < d$ or $f < d < e$, we say that the two triads have the same sense, and write

$$\mathrm{S}(\mathbf{DEF}) = \mathrm{S}(\mathbf{ABC}).$$

If, on the other hand, $f < e < d$ or $d < f < e$ or $e < d < f$, we say that the two triads have opposite senses, and write

$$\mathrm{S}(\mathbf{DEF}) \neq \mathrm{S}(\mathbf{ABC}).$$

(The arithmetical ideas employed here do not involve any fresh assumptions, but merely avoid separate consideration of the 228 possible ways of distributing **D**, **E**, **F** among **A**, **B**, **C**.)

We easily verify that the relation of having the same sense is reflective, symmetric, and transitive, and that

$$\mathrm{S}(\mathbf{ABC}) = \mathrm{S}(\mathbf{BCA}) = \mathrm{S}(\mathbf{CAB}) \neq \mathrm{S}(\mathbf{ACB}).$$

All triads which have the same sense as **ABC** are said to belong to the *sense-class* S(**ABC**). It follows that

2.52. *There are two sense-classes in the line:*

$$\mathrm{S}(\mathbf{ABC}) \textit{ and } \mathrm{S}(\mathbf{ACB}).$$

In other words, the line is *orientable.*

The direct connection between sense and separation is given by the following theorem:

2.53. *The relation* $\mathbf{AB} \parallel \mathbf{CD}$ *is equivalent to* $\mathrm{S}(\mathbf{ABC}) \neq \mathrm{S}(\mathbf{ABD})$.

PROOF. If $\mathbf{AB} \parallel \mathbf{CD}$, the line is divided into four segments $\mathbf{AD}/\mathbf{C}$, $\mathbf{DB}/\mathbf{A}$, $\mathbf{BC}/\mathbf{D}$, $\mathbf{CA}/\mathbf{B}$, which enable us to write (as above, with $n = 4$):

$$\mathbf{A} = \mathbf{A}_0, \quad \mathbf{D} = \mathbf{A}_1, \quad \mathbf{B} = \mathbf{A}_2, \quad \mathbf{C} = \mathbf{A}_3$$

Hence $S(\mathbf{ABC}) = S(\mathbf{ADB}) \neq S(\mathbf{ABD})$. Conversely, given $S(\mathbf{ABC}) \neq S(\mathbf{ABD})$, we can reverse the argument and deduce $\mathbf{AB} \parallel \mathbf{CD}$.

It is now easy to justify the intuitive consequence of using circular diagrams such as Fig. 2.1B, where clockwise and counter-clockwise senses can be indicated by an arrow pointing one way or the other.

In virtue of 2.126, all of the above theory of sense in one dimension can be applied to the lines or planes of a pencil: three lines of a flat pencil determine two sense-classes $S(\mathbf{abc})$ and $S(\mathbf{acb})$, and three planes of an axial pencil determine two sense-classes $S(\alpha\beta\gamma)$ and $S(\alpha\gamma\beta)$. Moreover, the notion of sense can be extended from one to two dimensions, where a sense-class is defined by the vertices of a quadrangle. (This is roughly equivalent to the statement that a sense of rotation is defined by the centre of a circle and three points on its circumference.) But the conclusion is different: *all quadrangles in the plane have the same sense, so there is only one sense-class; the plane is ***non-orientable.***

The projective plane, with its single sense-class, is not very easy to visualize. In ordinary space a non-orientable surface is "one-sided," and must cross itself if unbounded. But the impossibility of distinguishing two senses of rotation is easily seen in the geometry of a bundle (which is the "third model" of §2.2). For, any rotation about a line of the bundle (i.e. about a point of the projective plane) is clockwise when we look along the line in one direction, and counter-clockwise when we look along it in the opposite direction.

In projective geometry of three dimensions, we might define a sense-class by means of the vertices of a complete pentagon, but it is easier to use Veblen's notion of a *doubly*

*Veblen and Young [2], II, pp. 67, 422; Klein [3], pp. 12-17.

oriented line $(\mathbf{ABC}, \alpha\beta\gamma)$. This is a line associated with one sense-class $S(\mathbf{ABC})$ among the points on it and one sense-class $S(\alpha\beta\gamma)$ among the planes through it. Thus the line **AB** provides four doubly oriented lines: $(\mathbf{ABC}, \alpha\beta\gamma)$, $(\mathbf{ACB}, \alpha\gamma\beta)$, $(\mathbf{ABC}, \alpha\gamma\beta)$, $(\mathbf{ACB}, \alpha\beta\gamma)$. Two doubly oriented lines are said to be *doubly perspective* if they can be named $(\mathbf{ABC}, \alpha\beta\gamma)$ and $(\mathbf{A'B'C'}, \alpha'\beta'\gamma')$ in such a way that $\mathbf{A}, \mathbf{B}, \mathbf{C}, \mathbf{A}', \mathbf{B}', \mathbf{C}'$ lie on $\alpha', \beta', \gamma', \alpha, \beta, \gamma$, respectively. Two doubly oriented lines are said to be *similarly oriented* if they are related by a sequence of such "double perspectivities." It can be proved† that $(\mathbf{ABC}, \alpha\beta\gamma)$ is similarly oriented with $(\mathbf{ACB}, \alpha\gamma\beta)$, but not with $(\mathbf{ABC}, \alpha\gamma\beta)$, and that

2.54. *There are just two classes of doubly oriented lines, such that any two doubly oriented lines are similarly oriented if and only if they belong to the same class.*

Intuitively, this is the distinction between right-handed and left-handed *screws*. The general result is that a projective space is orientable or non-orientable according as its number of dimensions is odd or even.

2.6. Triangular and tetrahedral regions. The plane-dual and space-dual of 2.51 may be stated as follows:

Two coplanar lines (or two planes) divide the rest of their plane (or of space) into two classes of points, such that two points in different classes are separated by the points in which their join meets the given lines (or planes), whereas two points in the same class are not so separated.

Such classes of points are called *regions*.* A third line (or plane) will in general subdivide each region. Hence

2.61. *The sides of a triangle* **ABC** *divide the rest of the plane* **ABC** *into four regions.*

---

†Veblen and Young [2], II, p. 449. Cf. Russell [1], p. 232.

*Veblen and Young [2], II, pp. 51-54, 385-400.

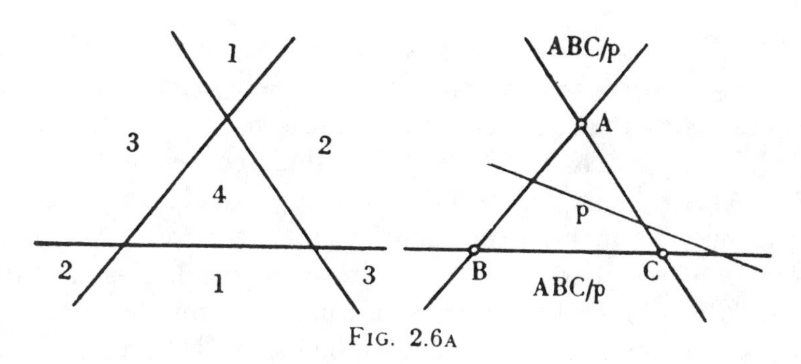

FIG. 2.6A

To distinguish one of these, we may proceed as follows. A line **p**, not passing through any of **A, B, C**, is divided by **BC, CA, AB** into three segments, lying respectively in three of the four regions. The remaining region, to which **p** is *exterior*, is then denoted by **ABC/p**, as in Fig. 2.6A.

So also, three non-coaxial planes divide the rest of space into four regions, each of which will be subdivided by a fourth plane (not concurrent with the others). Hence

2.62. *The faces of a tetrahedron* **ABCD** *divide the rest of space into eight regions.*

We distinguish any one of these as **ABCD**/ϖ, where ϖ is an exterior plane.

The above results become quite obvious when we apply them to affine space with ideal elements, taking one side of the triangle (or one face of the tetrahedron) to be the line (or plane) at infinity. then 2.61 describes the four "quadrants," and 2.62 the eight "octants."

2.7. Ordered correspondences. A correspondence between two ranges is said to be *ordered* if it preserves the relation of separation. The points of the second range which correspond to **A**, **B**... in the first will be denoted by **A**′, **B**′,.... Thus the correspondence is ordered if **A**′**B**′ ∥ **C**′**D**′ whenever

AB || **CD**. If the two ranges are on the same line ("superposed"), the correspondence is said to be *direct* or *opposite* according as it preserves or reverses sense, i.e. according as S(**ABC**) = or ≠ S(**A′B′C′**). A point which corresponds to itself (**A′** = **A**) is called a *double point*. As a particular direct correspondence, we include the identity, in which every point is a double point. Direct and opposite correspondences are easily seen to combine like positive and negative numbers; e.g. the product of two opposite correspondences is direct. These notions extend in an obvious manner to correspondences between any kind of one-dimensional primitive forms; e.g., in a correspondence between two flat pencils, a line which corresponds to itself is called a double line.

Our first application of the Axiom of Continuity is in proving the following lemma* (which we shall need several times, notably in 2.84, where we enumerate the double points of a projectivity):

2.71. *If an ordered correspondence relates an interval* $\overline{\mathbf{AB}}/\mathbf{C}$ *to an interior interval* $\overline{\mathbf{A'B'}}/\mathbf{C}$, *then the latter interval contains a double point* **M**, *such that there is no double point between* **A** *and* **M** *(in* $\overline{\mathbf{AB}}/\mathbf{C}$*).*

PROOF. It is convenient to say, of points **X** and **Y** in $\overline{\mathbf{AB}}/\mathbf{C}$, that **X** *precedes* **Y** (and **Y** *follows* **X**) if S(**XYC**) = S(**ABC**). If **A′** coincides with **A**, then **A** is itself the desired point **M**. If **B** coincides with **B′**, while every other point of $\overline{\mathbf{AB}}/\mathbf{C}$ precedes its corresponding point, then **B** is the desired point **M**. Setting aside these two extreme cases, we divide the segment **AB**/**C** into two sets of points:

(i) Points **P** such that every point **H** which precedes **P** precedes its corresponding point **H′**,

(ii) Points **Q** which follow at least one point **K** which does not precede its corresponding point **K′**.

*Enriques [1], pp. 71–75. See also Coxeter [6], pp. 168–170 (§10.6).

We are assuming the existence of such a point **K**; the second set includes every point between this and **B**. To see that the first set likewise contains some points, we observe that every point between **A** and **B**, and in particular every point between **A** and **A′**, is related to a point between **A′** and **B′**. Hence, if the correspondence is direct, so that **A′** precedes **B′** (as in Fig. 2.7A), the first set certainly contains **A′**. If, on the other hand, the correspondence is opposite, so that **A′** follows **B′** (as in Fig. 2.7C), consider any point **P** between **A** and **B**. If **P** precedes (or coincides with) its corresponding point **P′**, then every point which precedes **P** is related to a point which follows **P′**; hence the first set contains this **P**. But if **P** follows **P′**, then **P′** precedes its corresponding point **P″**, and by the same argument the first set contains this **P′**.

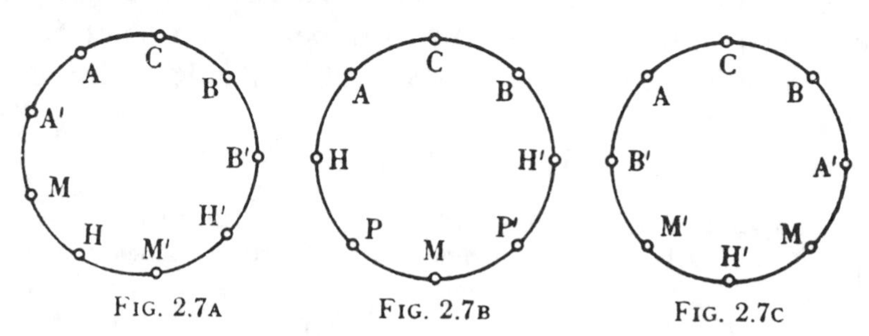

FIG. 2.7A FIG. 2.7B FIG. 2.7C

Clearly, every point of the first set precedes every point of the second. Hence, by 2.13, there is a point **M** such that every point which precedes **M** belongs to the first set, while every point which follows **M** belongs to the second. (Actually, **M** itself belongs to the first.) By (i), every point **H** which precedes **M** precedes its corresponding point **H′**; hence there is no double point preceding **M**. To see that **M** itself is a double point, we again consider the two types of correspondence separately.

If the correspondence is direct, suppose if possible that **M** is not a double point. If **M** follows **M′**, then **M′** follows **M″**, contradicting (i). But if **M** precedes **M′**, as in Fig. 2.7A, then every point **H** between **M** and **M′** precedes its corresponding point **H′** (which follows **M′**); this makes **M′** belong to the first set, which is absurd.

If, on the other hand, the correspondence is opposite, every point **Q** of the second set follows its corresponding point **Q′**, since **Q′** precedes the point **K′** of (ii). Let **H** be a fixed point between **A** and **M**, and **P** a variable point between **H** and **M**, as in Fig. 2.7B. Then, the correspondence being opposite, **H′** follows **P′**, which follows **P**. Thus **H′** follows every point which precedes **M**, and so either follows or coincides with **M**. Similarly, for any point **K** between **B** and **M**, the corresponding point **K′** either precedes or coincides with **M**. If **M′** precedes **M**, as in Fig. 2.7C, then every point **H′** between **M** and **M′**, being also between **A′** and **M′**, corresponds to a point **H** between **A** and **M**, which is absurd. Similarly, it is absurd to suppose that **M′** follows **M**.

Hence, finally, **M** is a double point.

Although this rigorous proof (due to Enriques) is quite subtle, the result is intuitively obvious when we think of a correspondence between *moving* points. On a circular race-track, Tom runs from **A** to **B** while Dick runs (over part of the same ground) from **A′** to **B′**; **M** is the place where Tom first passes Dick.

In the case when the correspondence is *opposite*, we easily see that **M** is the *only* double point in $\overline{\mathbf{AB}}/\mathbf{C}$. By applying the same theorem to the inverse correspondence, we deduce that there is likewise just one double point in the supplementary interval. Conversely, every opposite correspondence relates a pair of intervals, one containing the other; for, if **A′** is related to **A″** (which may or may not coincide with **A**), then the two intervals $\overline{\mathbf{AA'}}$ are related to the two intervals $\overline{\mathbf{A'A''}}$, one of which is interior. Hence

2.72. *Every opposite correspondence has exactly two double points.*

We shall see, in §4.1, that the points of a line can be represented by the real numbers and ∞. The relation $x' = x^{-3}$ is an example of an opposite correspondence; its two double points are ± 1. On the other hand, the direct correspondence $x' = x^3$ has four double points: $0, \infty, \pm 1$. In fact, a direct correspondence may have any number of double points, from none to infinitely many.

By 2.126, a perspectivity is an ordered correspondence; so also is the result of any sequence of perspectivities. Thus, if **A**, **B**, **I**, **K** in Fig. 2.4A are fixed, the correspondence between **C** and **D** is ordered, since it is given by the sequence of perspectivities

$$\mathbf{ABC} \overset{\mathbf{I}}{\underset{\wedge}{=}} \mathbf{AKJ} \overset{\mathbf{B}}{\underset{\wedge}{=}} \mathbf{AIL} \overset{\mathbf{K}}{\underset{\wedge}{=}} \mathbf{ABD}.$$

By 2.41 and 2.53, this correspondence reverses sense. In other words,

2.73. *The correspondence between harmonic conjugates with respect to two fixed points is opposite.*

If H(**AB**, **CD**) and H(**AB**, $\mathbf{C}_1\mathbf{D}_1$), so that $\mathbf{C}_1\mathbf{D}_1$ is another pair in the same correspondence, we have $S(\mathbf{CDC}_1) \neq S(\mathbf{DCD}_1)$, i.e.

$$S(\mathbf{CDC}_1) = S(\mathbf{CDD}_1).$$

By 2.53, this means that $\mathbf{C}_1$ and $\mathbf{D}_1$ do not separate **C** and **D**. Hence

2.74. *If two pairs of points on a line are each harmonic conjugates with respect to a third pair, they do not separate each other.*

By applying 2.71 to the product of two correspondences of the kind just considered, it is easy to prove* the converse theorem (which plays an important part in the theory of projectivities):

*Enriques [1], p. 77; Holgate [1], p. 36.

2.75. *If two pairs of points on a line do not separate each other, there is ai least one pair of points which are harmonic conjugates with respect to each of the given pairs.*

2.8. One-dimensional projectivities. Following von Staudt, we define a *projectivity* between two ranges (or two pencils, or a range and a pencil) as a correspondence which preserves the harmonic relation,* so that $H(\mathbf{A'B'}, \mathbf{C'D'})$ whenever $H(\mathbf{AB}, \mathbf{CD})$; and we indicate corresponding elements by formulae such as

$$\mathbf{A\,B\,C}\ldots \barwedge \mathbf{A'B'C'}\ldots.$$

This relation is clearly reflexive, symmetric, and transitive. (The analogous two- and three-dimensional correspondences will constitute the main topic of Chapter III.)

By 2.42, every perspectivity is a projectivity; so also is any product of perspectivities. We shall see (in 2.86) that, conversely, every projectivity can be constructed as a product of perspectivities; but for the present we shall be content to treat such a product as a special case of a projectivity. In this manner it is easily verified that, for any four collinear points,

2.81. $$\mathbf{A\,B\,C\,D} \barwedge \mathbf{B\,A\,D\,C}.$$

This means that there is at least one projectivity which interchanges **A** with **B**, and **C** with **D**. Hence

$$\mathbf{A\,B\,C\,D} \barwedge \mathbf{B\,A\,D\,C} \barwedge \mathbf{D\,C\,B\,A} \barwedge \mathbf{C\,D\,A\,B}.$$

In particular,

2.82. $H(\mathbf{AB}, \mathbf{CD})$ *implies* $H(\mathbf{CD}, \mathbf{AB})$.

2.83. *Every projectivity is an ordered correspondence.*

PROOF. Given $\mathbf{A\,B\,C\,D} \barwedge \mathbf{A'B'C'D'}$, where $\mathbf{AB} \parallel \mathbf{CD}$, we have to show that $\mathbf{A'B'} \parallel \mathbf{C'D'}$. Suppose, if possible, that this

*Von Staudt [1], pp. 49, 59; Enriques [1], pp. 78, 81, 84, 89, 101, 108.

were not the case. Then, by 2.75, the second range would contain points **M′** and **N′**, such that H(**A′B′**, **M′N′**) and H(**C′D′**, **M′N′**). These would correspond to points **M** and **N** in the first range, such that H(**AB**, **MN**) and H(**CD**, **MN**). By 2.74 and 2.82, this contradicts our hypothesis that **AB**||**CD**.

In particular, a projectivity between the points of one line is either direct or opposite. However, in contrast to the general ordered correspondence, this special kind has the following property:

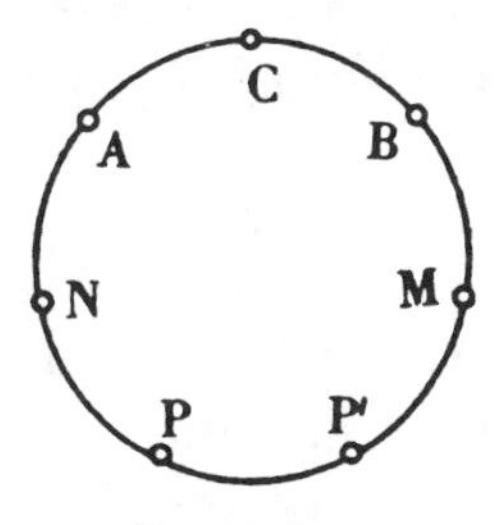

FIG. 2.8A

2.84. *A projectivity which has more than two double points is the identity.*

PROOF. Suppose, if possible, that we can have a non-trivial projectivity with three double points **A**, **B**, **C**, so that for some point **P**,

$$\mathbf{ABCP} \barwedge \mathbf{ABCP'}$$

with **P′** distinct from **P**. (Fig. 2.8A.) Since the double points were named arbitrarily, we may suppose that **AB** || **CP**. Then, by 2.83, **AB** || **CP′**; and there is no loss of generality in supposing that, in the interval $\overline{\mathbf{AB}}$/**C**, **P′** lies between **P** and **B**. Since the projectivity relates the interval $\overline{\mathbf{PB}}$/**C** to the interior interval $\overline{\mathbf{P'B}}$/**C**, 2.71 reveals the presence of a double point **M**, such that there is no double point between **P** and **M**. Applying the same theorem to the inverse projectivity, which relates

$\overline{\mathbf{P'A}}/\mathbf{C}$ to $\overline{\mathbf{PA}}/\mathbf{C}$, we find another double point **N**, such that there is no double point between **P′** and **N**. Since the segments **PM**/**C** and **P′N**/**C** overlap, there is no double point in **MN**/**C**. But the harmonic conjugate of **C** with respect to **M** and **N** is clearly a double point, and lies in **MN**/**C**. Thus we have proved the theorem by ***reductio ad absurdum***.

The following has been called the Fundamental Theorem of Projective Geometry:

2.85. *A projectivity between two ranges is uniquely determined when we are given three points of one and the corresponding three points of the other.*

PROOF. A product of perspectivities by which $\mathbf{ABC} \,\overline{\wedge}\, \mathbf{A'B'C'}$ can be chosen in many ways, such as the following. If the two ranges are on distinct lines, as in Fig. 2.8B, take any point $\mathbf{C_0}$ on **AB′**, and use centres $\mathbf{O_1} = (\mathbf{BB'}, \mathbf{CC_0})$, $\mathbf{O_2} = (\mathbf{AA'}, \mathbf{C_0C'})$. Then

$$\mathbf{ABC} \overset{\mathbf{O_1}}{\underset{\wedge}{=}} \mathbf{AB'C_0} \overset{\mathbf{O_2}}{\underset{\wedge}{=}} \mathbf{A'B'C'}.$$

If **AB** and **A′B′** are skew lines, this construction may be described more simply by saying that the related ranges are traced out by a pencil of planes (with axis $\mathbf{O_1O_2}$). If the two ranges are on one line, a range related to one of them can be obtained by applying an arbitrary perspectivity, and then we proceed as before (thus using three perspectivities in all). Denote this product of perspectivities (in either case) by Φ, and let Θ be *any* projectivity having the same effect on **A**, **B**, **C**. Then, since

$$\mathbf{A}\Theta\Phi^{-1} = \mathbf{A'}\Phi^{-1} = \mathbf{A},$$

and similarly for **B** and **C**, the "quotient" projectivity $\Theta\Phi^{-1}$ has three double points. By 2.84, $\Theta\Phi^{-1} = 1$, and $\Theta = \Phi$. Thus the projectivity is unique.

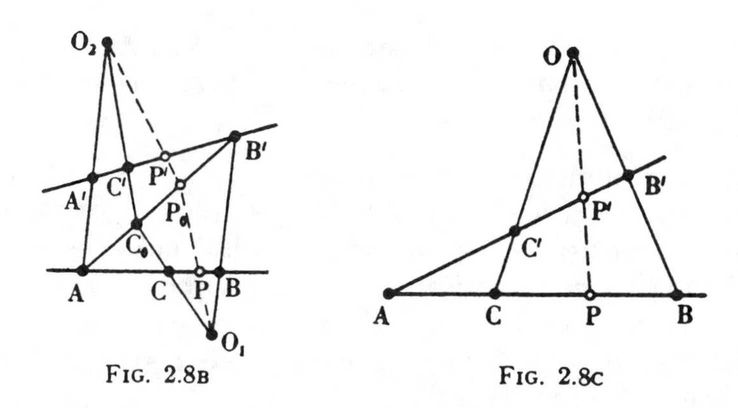

FIG. 2.8B FIG. 2.8C

As an immediate corollary, we have

2.86. *Every projectivity can be constructed as the product of two or three perspectivities.*

In one important case, a single perspectivity suffices:

2.87. *If a projectivity between the points of two distinct lines has a double point, it is a perspectivity.*

PROOF. By 2.85, there is only one projectivity by which $\mathbf{A\,B\,C} \barwedge \mathbf{A\,B'C'}$. (See Fig. 2.8C.) Using the centre $\mathbf{O} = (\mathbf{BB'}, \mathbf{CC'})$, we have $\mathbf{A\,B\,C} \doublebarwedge \mathbf{A\,B'C'}$.

A projectivity in one line is said to be *elliptic*, *parabolic*, or *hyperbolic*, according as the number of double points is 0, 1, or 2 (these being the numbers of points at infinity on the three types of conic in Euclidean geometry). As a special case of 2.72, we have

2.88. *Every opposite projectivity is hyperbolic.*

Therefore every elliptic or parabolic projectivity is direct. On the other hand,

2.89. *A hyperbolic projectivity is opposite or direct according as the double points do or do not separate a pair of corresponding points.*

PROOF. Let **M**, **N** be the double points, and **AA′** any pair of corresponding points. We merely have to compare S(**MNA**) with S(**MNA′**), using 2.53.

2.9. Involutions. We now consider a concept which will be seen to have a fundamental bearing on the subject of non-Euclidean geometry. By its aid we shall define coordinates in Chapter IV, and metrical notions in Chapter V. An *involution** is a projectivity of period two ($\Theta^2 = 1$), i.e. a non-trivial projectivity which is its own inverse ($\Theta^{-1} = \Theta \neq 1$). From 2.81 and 2.85 we easily deduce

2.91. *Any projectivity which has one doubly-corresponding pair* ($\mathbf{A\,A'} \barwedge \mathbf{A'A}$) *is an involution.*

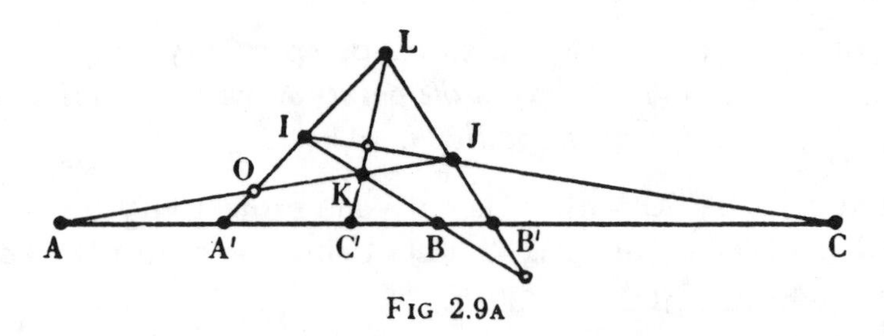

FIG 2.9A

By a further application of 2.81, we obtain the relation

$$\mathbf{A\,A'B\,C} \barwedge \mathbf{A\,A'C'B'}$$

as a necessary and sufficient condition for the pairs **AA′**, **BB′**, **CC′** to belong to an involution. Three such pairs of points are said to form a "quadrangular set," for the following reason. Let **O** denote the diagonal point (**IL**, **JK**) of a quadrangle **IJKL**, whose sides meet a line in points **A**, **B**, **C**, **A′**, **B′**, **C′**, as in Fig. 2.9A. Then

$$\mathbf{A\,A'B\,C} \overset{\mathbf{I}}{\doublebarwedge} \mathbf{A\,O\,K\,J} \overset{\mathbf{L}}{\doublebarwedge} \mathbf{A\,A'C'B'}.$$

*Von Staudt [1], pp. 118-122; Enriques [1], pp. 124-129, 138.

Hence

2.92. *The pairs of opposite sides of a quadrangle meet any coplanar line (not through a vertex) in three pairs of an involution.*

An involution may be elliptic or hyperbolic; but it cannot be parabolic, as the following theorem shows:

2.93. *If an involution has one double point it has another, and any two corresponding points are harmonic conjugates with respect to the two double points.*

PROOF. Let **AA′** be any pair of an involution in which **M** is a double point. Then the harmonic conjugate of **M** with respect to **A** and **A′**, being also the harmonic conjugate of **M** with respect to **A′** and **A**, is a second double point.

The following two corollaries are easily deduced:

2.94. *If a hyperbolic projectivity has a pair of corresponding points which are harmonic conjugates with respect to the double points, it is an involution.*

2.95. *The relation* $\mathbf{MNAA'} \barwedge \mathbf{MNA'A}$ *is equivalent to* H(**MN**, **AA′**).

By 2.88, every opposite involution is hyperbolic. By 2.73, every hyperbolic involution is opposite. Hence

2.96. *An involution is elliptic or hyperbolic according as it is direct or opposite.*

If the involution determined by pairs **AA′** and **BB′** is elliptic, and so direct, we have S(**AA′B**) = S(**A′AB′**) ≠ S(**AA′B′**), and **AA′** || **BB′**. Similarly, if the involution is hyperbolic we have S(**AA′B**) = S(**AA′B′**). Hence

2.97. *An involution is elliptic or hyperbolic according as two of its pairs do or do not separate each other.*

If Θ and Φ are two involutions in one line, we have $\Theta^{-1} = \Theta$, $\Phi^{-1} = \Phi$, and $(\Theta\Phi)^{-1} = \Phi^{-1}\Theta^{-1} = \Phi\Theta$. Hence

2.98. *Two involutions are permutable if and only if their product is an involution.*

We are now ready to prove a theorem which will enable us to define reflections and translations in Chapter V:

2.99. *Given two points* **A** *and* **B**, *and an elliptic involution* Ω *in the line* **AB**, *there are just two projectivities which are permutable with* Ω *and relate* **A** *to* **B**. *One of these is a hyperbolic involution, and the other is elliptic (but is not an involution unless* **A** *and* **B** *are a pair in* Ω, *in which case it coincides with* Ω).

PROOF. If **A** and **B** are not a pair in Ω, we define $\mathbf{A}' = \mathbf{A}\Omega$, $\mathbf{B}' = \mathbf{B}\Omega$. If there is a hyperbolic or parabolic projectivity Φ, such that $\Omega\Phi = \Phi\Omega$ and $\mathbf{A}\Phi = \mathbf{B}$, suppose that it has a double point **M**, and define $\mathbf{M}' = \mathbf{M}\Omega$. Then

$$\mathbf{M}'\Phi = \mathbf{M}\Omega\Phi = \mathbf{M}\Phi\Omega = \mathbf{M}\Omega = \mathbf{M}'$$

is another double point, and Φ is hyperbolic (not parabolic). Since

$$\mathbf{M}\Phi\Omega = \mathbf{M}' \text{ and } \mathbf{M}'\Phi\Omega = \mathbf{M},$$

the product $\Phi\Omega$ is an involution (by 2.91). Hence

$$\Phi^2 = \Phi\Omega \,.\, \Omega\Phi = (\Phi\Omega)^2 = 1,$$

and Φ is itself an involution.

Since $\mathbf{A}'\Phi = \mathbf{B}'$, the only possible involution which is permutable with Ω, and relates **A** to **B**, is that determined by the pairs **AB**, **A′B′**. Conversely, the involution Φ so defined is in fact permutable with Ω; for, since

$$\mathbf{A}\Phi\Omega = \mathbf{B}\Omega = \mathbf{B}' \text{ and } \mathbf{B}'\Phi\Omega = \mathbf{A}'\Omega = \mathbf{A},$$

$\Phi\Omega$ is an involution. Moreover, Φ is in fact hyperbolic; for, by 2.97, $\mathbf{AA}' \parallel \mathbf{BB}'$, and therefore **A** and **B** do not separate **A′** and **B′**.

If there is a non-involutory (and therefore elliptic) projectivity Ψ, such that $\Omega\Psi = \Psi\Omega$ and $\mathbf{A}\Psi = \mathbf{B}$, we define $\mathbf{C} = \mathbf{A}\Psi^{-1}$, $\mathbf{C}' = \mathbf{C}\Omega$. Then $\mathbf{C}'\Psi = \mathbf{A}'$, and $\mathbf{A}'\Psi = \mathbf{B}'$. Thus

$$\mathbf{A\,A'C\,C'} \barwedge \mathbf{B\,B'A\,A'} \barwedge \mathbf{A'A\,B'B},$$

AA′, **CB′**, **BC′** are three pairs of an involution, and

$$\mathbf{A\,A'C\,B} \barwedge \mathbf{A'A\,B'C'} \barwedge \mathbf{A\,A'B\,C}.$$

(The last projectivity used here is Ω.) It follows, by 2.95, that H(**AA′**, **BC**). Thus Ψ is determined as the projectivity by which **A A′C** $\barwedge$ **B B′A**, where **C** is the harmonic conjugate of **B** with respect to **A** and **A′**. Conversely, the projectivity Ψ so defined is in fact permutable with Ω; for, $\Omega\Psi$ and $\Psi\Omega$ have the same effect on each of three points, namely

$$(\mathbf{AA'C})\Omega\Psi = (\mathbf{A'AC'})\Psi = (\mathbf{B'BA'}) = (\mathbf{BB'A})\Omega = (\mathbf{AA'C})\Psi\Omega.$$

If, on the other hand, **A** and **B** are a pair in Ω (so that **A′** coincides with **B**, and the previous construction breaks down), let Θ be a projectivity such that $\Omega\Theta = \Theta\Omega$ and $\mathbf{A}\Theta = \mathbf{B}$. Then, since

$$\mathbf{B}\Theta = \mathbf{A}\Omega\Theta = \mathbf{A}\Theta\Omega = \mathbf{B}\Omega = \mathbf{A},$$

Θ must be an involution. Since **A** and **B** are interchanged by both Ω and Θ, they are unchanged by the product $\Omega\Theta$. Hence either $\Omega\Theta = 1$, in which case $\Theta = \Omega$, or $\Omega\Theta$ is the hyperbolic involution with double points **A** and **B**, in which case Θ is itself hyperbolic, by 2.96.

CHAPTER III

REAL PROJECTIVE GEOMETRY: POLARITIES, CONICS AND QUADRICS

3.1. Two-dimensional projectivities. The history of conics begins about 430 B.C., when Hippocrates of Chios expressed the "duplication of the cube" as a problem which his followers could solve by means of intersecting curves. Some seventy years later, Menaechmus showed that these curves can be defined as sections of a right circular cone by a plane perpendicular to a generator. Their metrical properties (such as the theorem regarding the ratio of the distances to focus and directrix) were described in great detail by Aristaeus, Euclid, and Apollonius.* Apollonius introduced the names ellipse, parabola, and hyperbola, and discovered the harmonic property of pole and polar. But the earliest genuinely non-metrical property is the theorem of Pascal (1623-1662), who obtained it at the age of sixteen. (See 3.35.) A hundred years later, Maclaurin used similar ideas in one of his constructions for the conic through five given points.† The first systematic account of projective properties is due to Steiner (1796-1863). But his definition in terms of related pencils (3.34) lacks symmetry, as it specializes two points on the conic (the centres of the pencils); moreover, several steps have to be taken before the *self-duality* of a conic becomes apparent. Von Staudt (1798-1867) made the important discovery that the relation which a conic establishes between poles and polars is really more fundamental than the conic itself, and can be set up independently (§3.2). This "polarity" can then be used to *define* the conic, in a manner that is perfectly symmetrical and

*See Zeuthen [1]. †Maclaurin [1], p. 350.

immediately self-dual: a conic is simply the locus of points which lie on their polars, or the envelope of lines which pass through their poles. Von Staudt's treatment of quadrics is analogous, in three dimensions.

We shall find projectivities easier to define in two or three dimensions than in one (as there is no need to mention the preservation of harmonic sets). But we have to distinguish two kinds: collineations and correlations.

A *collineation* between two planes (which may coincide) is a correspondence which relates collinear points to collinear points, and consequently concurrent lines to concurrent lines; in other words, it is a point-to-point and line-to-line correspondence preserving incidence. Since this transforms a quadrangle into a quadrangle, it automatically preserves the harmonic relation; and the correspondence "induced" between two corresponding ranges is a one-dimensional projectivity.

3.11. *Any collineation which leaves a quadrangle or quadrilateral invariant is the identity.*

PROOF. If the four sides of a quadrilateral are invariant, so are its six vertices. Since then three points on each side are invariant, 2.84 shows that every point on each side is invariant. Hence every line in the plane (joining points on two distinct sides) is invariant, and so also every point. The dual argument establishes the same result for a quadrangle.

In other words, any collineation which leaves four independent points (or lines) invariant is the identity.

3.12. *A collineation is uniquely determined when a pair of corresponding quadrangles or quadrilaterals is assigned.*

PROOF. Such a collineation is easily constructed by means

of a sequence of perspectivities* (three or four, according as the two planes are distinct or coincident). Its uniqueness follows from 3.11.

It can be proved that the only *involutory* collineations in a plane are *harmonic homologies*, having any given point **O** as centre and any line **o**, not through **O**, as axis. Such a correspondence relates each point **A** in the plane to its harmonic conjugate with respect to the two points **O** and (**o**, **OA**), and consequently relates each line **a** to its harmonic conjugate with respect to the two lines **o** and **O**(**o**, **a**).

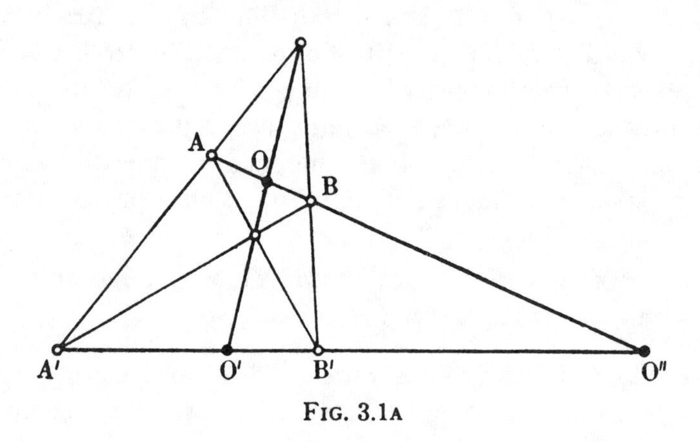

FIG. 3.1A

3.13. *The product of three harmonic homologies, whose centres and axes are the vertices and sides of a triangle, is the identity.*

PROOF. Let Φ, Φ′, Φ″ be the homologies, with centres **O**, **O′**, **O″** and axes **O′O″**, **O″O**, **OO′**. Take four points, **A**, **B**, **A′**, **B′**, such that H(**OO″**, **AB**) and H(**O′O″**, **A′B′**), as in Fig. 3.1A. Then Φ interchanges **A** and **B**, leaving the other points invariant; similarly Φ′ interchanges **A′** and **B′**; but Φ″ inter-

*Robinson [1], pp. 133-137. Cf. von Staudt [1], pp. 60-66, 125.

changes both pairs. Hence $\Phi\Phi'\Phi''$ leaves all four points invariant and (by 3.11) is the identity.

We come now to the second kind of projectivity. A *correlation* between two planes (which may coincide) is a correspondence which relates collinear points to concurrent lines, and consequently concurrent lines to collinear points; in other words, it is a point-to-line and line-to-point correspondence which preserves incidence in accordance with the principle of plane-duality. Thus, if it relates a point **A** to a line **a'**, and a line **b** to a point **B'**, then **B'** lies on **a'** if and only if **b** passes through **A**. Since the correlation transforms a harmonic set of four points into a harmonic set of four lines, it induces a projectivity between the points of **b** and the lines through **B'**. Clearly, the product of two correlations is not a correlation but a collineation; in fact, collineations and correlations combine like positive and negative numbers.

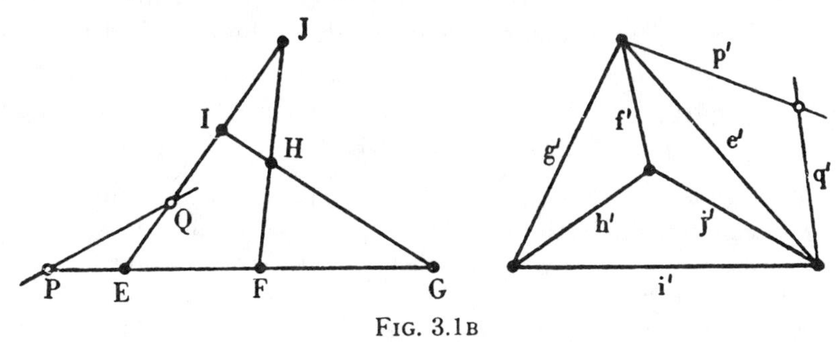

FIG. 3.1B

3.14. *A correlation is uniquely determined when a quadrilateral and the corresponding quadrangle are assigned.*

PROOF. By associating the vertices **E**, **F**, **G**, **H**, **I**, **J** of the quadrilateral with the sides **e'**, **f'**, **g'**, **h'**, **i'**, **j'** of the quadrangle, as in Fig. 3.1B, we establish one-dimensional projectivities $\mathbf{EFG} \barwedge \mathbf{e'f'g'}$ and $\mathbf{EIJ} \barwedge \mathbf{e'i'j'}$, which enable us to associate any points **P** on **EF**, and **Q** on **EI**, with definite lines **p'** through

$(\mathbf{e}',\mathbf{f}')$, and $\mathbf{q}'$ through $(\mathbf{e}',\mathbf{i}')$. In this way we obtain a definite point $(\mathbf{p}',\mathbf{q}')$ corresponding to any line $\mathbf{PQ}$. Let $(\mathbf{p}_1,\mathbf{q}_1)$, $(\mathbf{p}_2,\mathbf{q}_2)$, $(\mathbf{p}_3,\mathbf{q}_3)$ be the points corresponding to three concurrent lines $\mathbf{P}_1\mathbf{Q}_1$, $\mathbf{P}_2\mathbf{Q}_2$, $\mathbf{P}_3\mathbf{Q}_3$. Then

$$\mathbf{e}'\mathbf{p}_1\mathbf{p}_2\mathbf{p}_3 \overline{\wedge}\, \mathbf{E}\ \mathbf{P}_1\mathbf{P}_2\mathbf{P}_3 \overline{\overline{\wedge}}\, \mathbf{E}\ \mathbf{Q}_1\mathbf{Q}_2\mathbf{Q}_3 \overline{\wedge}\, \mathbf{e}'\mathbf{q}_1\mathbf{q}_2\mathbf{q}_3.$$

Hence, by the plane-dual of 2.87, these three points are collinear. This shows that the correspondence we have set up is in fact a correlation. It is unique since, if Γ and Γ' were two such, we would have $\Gamma'\Gamma^{-1}=1$, by 3.11. (We could have proved 3.12 similarly, without using a sequence of perspectivities.)

3.2. Polarities in the plane. A *polarity* is a correlation of period two, i.e. a correlation which is its own inverse, so that if it relates any point $\mathbf{A}$ to a line $\mathbf{a}$, it also relates $\mathbf{a}$ to the same point $\mathbf{A}$. We call $\mathbf{A}$ the *pole* of $\mathbf{a}$, and $\mathbf{a}$ the *polar* of $\mathbf{A}$. If $\mathbf{B}$ lies on $\mathbf{a}$, its polar, $\mathbf{b}$, passes through $\mathbf{A}$. We call $\mathbf{A}$ and $\mathbf{B}$ *conjugate points*, $\mathbf{a}$ and $\mathbf{b}$ *conjugate lines*. The polarity induces an involution of conjugate points on any line which is not self-conjugate, and an involution of conjugate lines through any point which is not self-conjugate.*

We shall see in §3.8 that, in certain three-dimensional polarities, *every* point is self-conjugate. However, there are no such "null" polarities in two dimensions; in fact, it is impossible for a line to contain more than two self-conjugate points.†

A polarity is a correlation, a correlation is a correspondence, and we consistently use this last word in the strict sense of "one-to-one correspondence." Thus, in the definition of a correlation between two planes, we mean that every point in the first has a definite corresponding line in the second, and that every line in the second corresponds to a definite point in the first. The "trilinear polarity" of §2.4 is not a correspondence in this sense (for, the trilinear pole of a side of the triangle is partially indeterminate). Moreover, it does not relate concurrent lines to collinear points. Thus it is not a true polarity, and to that extent its name is unfortunate.

---

*Von Staudt [1], pp. 131-136.

†Enriques [1], pp. 184-185.

Several theorems, originally stated for conics, are really properties of polarities; e.g.*

3.21. Hesse's Theorem. *If two pairs of opposite sides of a quadrangle are pairs of conjugate lines in a given polarity, so is the third pair.*

PROOF. Let the sides of the quadrangle **IJKL** meet the polar of **L** in the points **A**, **B**, **C**, **A′**, **B′**, **C′**, as in Fig. 2.9A. If the sides **IL** and **JL** are respectively conjugate to **JK** and **IK**, their poles are **A** and **B**. Hence the involution of conjugate points on the polar of **L** contains the pairs **AA′**, **BB′**, which suffice to determine it. By 2.92, it also contains the pair **CC′**. Hence **KL** is the polar of **C**, and is conjugate to **IJ**.

By considering the triangle formed by the poles of **JK**, **KI**, **IJ**, we immediately deduce

3.22. Chasles's Theorem. *If the vertices of one triangle are the poles of the sides of another, the joins of corresponding vertices are concurrent* (as in 2.32).

The two-dimensional analogue of the one-dimensional theorem 2.91 is as follows:

3.23. *Any correlation which transforms each vertex of one triangle into the opposite side is a polarity.*

PROOF. Let the vertices and opposite sides be **A**, **B**, **C** and **a**, **b**, **c**. A correlation which transforms **B** into **b**, and **C** into **c**, also transforms **BC** = **a** into (**b**, **c**) = **A**, and so *interchanges* the vertices with the opposite sides. Let **P** be any point not on a side, so that the corresponding line **p** does not pass through a vertex. By 3.14, **ABCP** and **abcp** suffice to determine the correlation. Construct the six points

*Hesse [1], p. 301; Chasles [2], p. 98.

$$\mathbf{P_a} = (\mathbf{AP}, \mathbf{a}),\ \mathbf{P_b} = (\mathbf{BP}, \mathbf{b}),\ \mathbf{P_c} = (\mathbf{CP}, \mathbf{c}),$$
$$\mathbf{A_p} = (\mathbf{a}, \mathbf{p}),\quad \mathbf{B_p} = (\mathbf{b}, \mathbf{p}),\quad \mathbf{C_p} = (\mathbf{c}, \mathbf{p}).$$

For each point **D** on **a**, there is a corresponding line **d** through **A**, and the correlation induces in **a** a projectivity between points **D** and (**d**, **a**). This projectivity, in which $\mathbf{B C P_a} \barwedge \mathbf{C B A_p}$, is an involution, and so transforms $\mathbf{A_p}$ into $\mathbf{P_a}$. Thus, when **D** is $\mathbf{A_p}$, **d** is $\mathbf{AP_a}$ or **AP**; similarly the line corresponding to $\mathbf{B_p}$ is **BP.** Hence the correlation transforms $\mathbf{p} = \mathbf{A_p B_p}$ into $\mathbf{P} = (\mathbf{AP}, \mathbf{BP})$, and is a polarity.

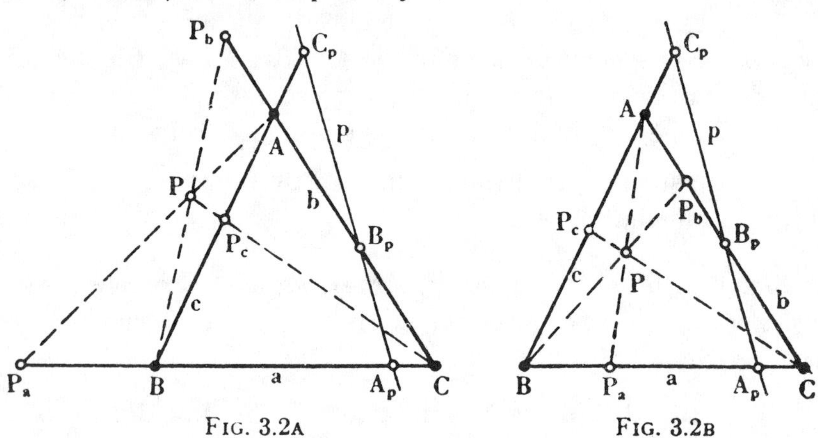

FIG. 3.2A FIG. 3.2B

Given **ABC**, **P**, and **p**, the polarity is determined. Such a triangle **ABC** is said to be *self-polar*. Any two vertices or sides are conjugate. A given polarity has infinitely many self-polar triangles; for, **A** and **B** may be any two points which are conjugate, but neither self-conjugate, and then **C** is determined as (**a**, **b**).

A polarity is said to be *hyperbolic* or *elliptic* according as it does or does not admit a self-conjugate point (i.e. a point which lies on its own polar). The following theorem shows that both types exist:

3.24. *If* **p** *is a line not through any vertex of a self-polar triangle*

ABC, *and* **P** *is its pole, the polarity is elliptic or hyperbolic according as* **P** *lies in the region* **ABC/p** *or in one of the other three regions determined by the triangle.*

PROOF. If **P** lies in **ABC/p**, as in Fig. 3.2A, we have

$$\mathbf{BC} \parallel \mathbf{P_aA_p},\quad \mathbf{CA} \parallel \mathbf{P_bB_p},\quad \mathbf{AB} \parallel \mathbf{P_cC_p},$$

and the involutions of conjugate points on the sides of the triangle are all elliptic (by 2.97). Hence, if we take a new position for **p**, the same separations must hold, and again **P** lies in **ABC/p**. Since it is then impossible for **P** to lie on **p**, the polarity is elliptic. But if **P** lies in one of the other regions, as in Fig. 3.2B, two of the above separations cease to hold, and two of the three involutions are hyperbolic. The double points of these involutions being self-conjugate, the polarity is hyperbolic.

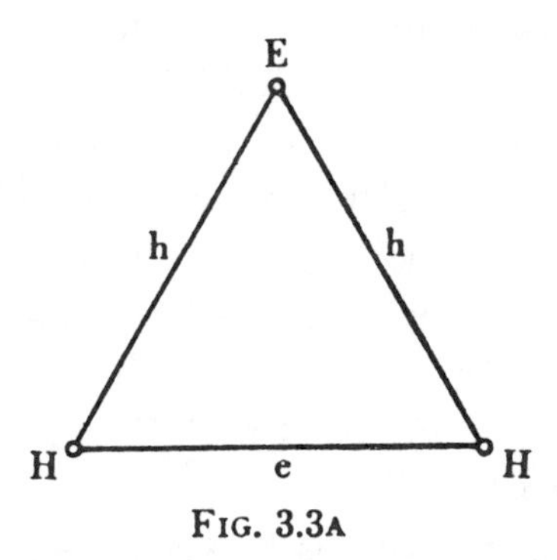

FIG. 3.3A

3.3. Conics. With reference to a given polarity, a non-self-conjugate line (or point) is said to be elliptic or hyperbolic according to the nature of the involution of conjugate points on it (or lines through it). Thus, if the polarity is elliptic, *all* lines and points are elliptic; but if it is hyperbolic, every self-polar triangle has one elliptic side and vertex, two hyperbolic sides and vertices, as in Fig. 3.3A. Hence, in the latter case, every point on an elliptic line is hyperbolic, and likewise every line through an elliptic point. But every hyperbolic line contains two self-conjugate points, which separate* its elliptic points from its hyperbolic points; and every hyperbolic point lies on two self-conjugate lines, which separate the elliptic and hyperbolic lines through it. On the other hand, the only self-conjugate point on a self-conjugate line is its pole, and the only self-conjugate line through a self-conjugate point is its polar.

Following von Staudt, we define a *conic* as the class of self-conjugate points and lines in a hyperbolic polarity. The points are said to lie *on* the conic, and the lines are called *tangents*. The pole of a tangent is its *point of contact*. Hyperbolic lines (each containing two points on the conic) are called *secants*. Hyperbolic points (each lying on two tangents) and elliptic lines (containing no points on the conic) are said to be *exterior*. Elliptic points (through which no tangent passes) are said to be *interior*.

(These notions are particularly relevant to the subject of this book, since the interior of a conic provides the most important model for hyperbolic geometry.)

The polar of an exterior point **P** joins the points of contact of the two tangents through **P**. For, if these tangents are **m** and **n**, with points of contact **M** and **N**, then **MN** is the polar of (**m**, **n**) = **P**. (See Fig. 3.3B.)

The polar of an interior or exterior point **C** contains the harmonic conjugate of **C** with respect to the two points in which any secant through **C** meets the conic. For, these two points are the double points of the involution of conjugate points on the secant. It follows that the conic (as a whole) is invariant under any harmonic homology whose centre is the pole of its axis, and that the diagonal triangle of an inscribed

*Strictly, this separation needs further discussion. See Enriques [1], p. 262.

quadrangle (or of a circumscribed quadrilateral) is self-polar.*

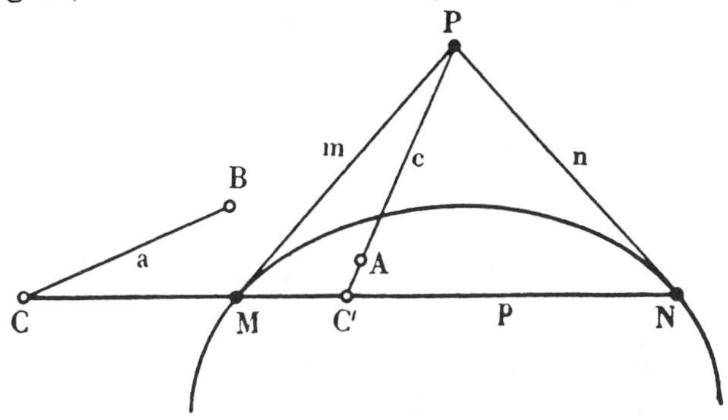

FIG. 3.3B

3.31. *A definite conic is determined when two tangents, their points of contact, and a pair of conjugate points are given.*

PROOF. Let **MP** and **NP** be given as the tangents at **M** and **N**, let **A** and **B** be the given conjugate points, and let **C** be the harmonic conjugate of **C′** = (**AP**, **MN**) with respect to **M** and **N**, as in Fig. 3.3B. Consider the definite correlation which transforms **M**, **N**, **P**, **A** into **MP**, **NP**, **MN**, **BC**, and consequently transforms **MN** into **P**, **PA** into **C**, and **C′** into **PC**. This induces in **MN** a projectivity which relates **C′** to **C** and has double points **M** and **N**. By 2.94, the projectivity is an involution; so the correlation transforms **C** into **PC′**. We can now apply 3.23 to the triangle **PCC′**.

3.32. Seydewitz's Theorem. *If a triangle is inscribed in a conic, any line conjugate to one side meets the other two sides in conjugate points.*

PROOF. Let the sides **LM** and **LN** of the inscribed triangle **LMN** meet the tangents at **N** and **M** in **N′** and **M′** respectively, as in Fig. 3.3C. For every point **A** on **LM**, there is a conjugate point **A′** = (**a**, **LN**) on **LN**. Since the range of points **A** is projective with the pencil of polars **a**, this correspondence between **A** and **A′** is a projectivity, and is determined by **L M N′** $\barwedge$

*Von Staudt [1], pp. 137-143.

L M′N. By 2.87, the projectivity is a perspectivity with centre **P** = (**MM′**, **NN′**). Hence the pairs of conjugate points on **LM** and **LN** (respectively) are cut out by the pencil of lines through **P**, the pole of **MN**.

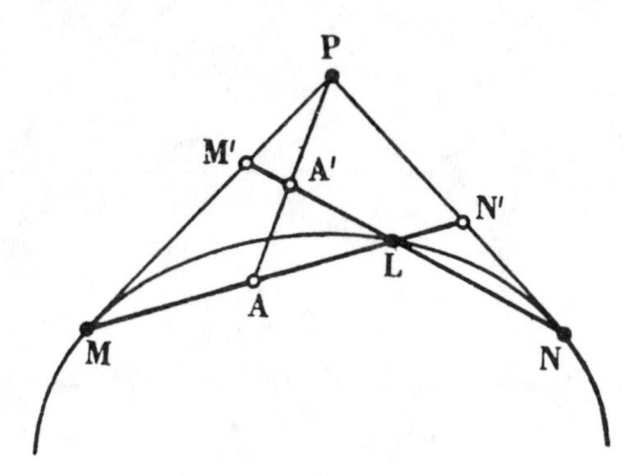

FIG. 3.3C

3.33. Steiner's Theorem.* *By joining all the points on a conic to any two fixed points on the conic, we obtain two projectively related pencils.*

PROOF. Let **M** and **N** be the two fixed points, and **L** a variable point on the conic. On any fixed line conjugate to **MN**, the lines **ML** and **NL** of the two pencils determine a pair of conjugate points, **A** and **A′**, as in Fig. 3.3C. Since such pairs of points belong to an involution, the pencils are projectively related. In particular, the lines **MN** and **MP** of the first pencil are related to the lines **NP** and **NM** of the second.

Conversely,

3.34. *The points of intersection of corresponding lines of two projectively related pencils in one plane, with distinct centres* **M** *and* **N**, *are the points on a conic through* **M** *and* **N**, *provided* **MN** *is not a double line in the projectivity.*

*Steiner [1], p. 139.

PROOF. Suppose the lines **MN, MP, ML** of the first pencil are related to the lines **NP, NM, NL** of the second. Then the desired conic is determined by the polarity in which **L** is self-conjugate while the polars of **M, N, P** are **MP, NP, MN**. (Use 3.31, with the conjugate points coincident.)

Most text-books on Projective Geometry give 3.34 as the *definition* of a conic. Steiner's Theorem reconciles the two alternative treatments, and allows us to use the customary proof* for the dual theorems of Pascal and Brianchon:

3.35. Pascal's Theorem. *If* **L, M′, N, L′, M, N′** *are any six points on a conic, the three points* **(M′N, MN′), (N′L, NL′), (L′M, LM′)** *are collinear.*

3.36. Brianchon's Theorem. *If* **l, m′, n, l′, m, n′** *are any six tangents to a conic, the three lines* **(m′, n) (m, n′), (n′, l) (n, l′), (l′, m) (l, m′)** *are concurrent.*

3.4. Projectivities on a conic. Since the two pencils of 3.33 are in ordered correspondence, any such pencil defines a definite order for all the points on the conic. Thus the Axioms of Separation hold for points on a conic, as well as for points on a line. Accordingly, we extend the meaning of the word *range* to include such a class of points, and say that two ranges on the conic are projectively related if the pencils joining them to any fixed point on the conic are projectively related.† Any such projectivity, relating pairs of points on the conic, carries with it a projectivity relating pairs of tangents (the polars of the points). A projectivity on a conic, as on a line, may be direct or opposite, elliptic or parabolic or hyperbolic, and it is an involution if it admits a doubly-corresponding pair.

*Cremona [1], p. 121, or Robinson [1], p. 39.

†Similarly, we can define projectivities between two distinct conics. See Bellavitis [1], p. 270; von Staudt [1], pp. 149, 158.

It is a simple consequence of 3.35 that every projectivity on a conic has an *axis*, which contains the intersection of "cross-joins" of any two pairs of corresponding points, as in Fig. 3.4A. The double points, if any exist, lie on the axis; therefore the projectivity is hyperbolic, parabolic, or elliptic, according as its axis is a secant, a tangent, or an exterior line. A hyperbolic projectivity may be opposite or direct, as in 2.89.

By applying the polarity which defines the conic, we deduce that the projectivity of tangents has a *centre*, which lies on the join of "cross-intersections" of any two pairs of corresponding tangents.

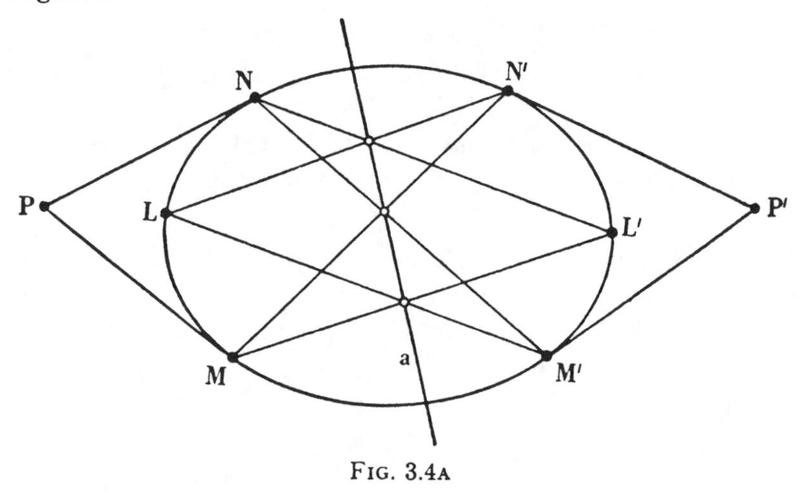

FIG. 3.4A

3.41. *Any projectivity on a conic determines a collineation of the whole plane.*

PROOF. If the projectivity is given by **L M N** $\barwedge$ **L′M′N′**, let **P** and **P′** be the poles of **MN** and **M′N′**. By 3.12, the quadrangles **L M N P** and **L′M′N′P′** are related by a definite collineation, which transforms the given conic, as determined by points **L**, **M**, **N** and tangents **PM**, **PN**, into the same conic as determined by **L′**, **M′**, **N′**, **P′M′**, **P′N′**. In other words, this collineation preserves the conic and induces the projectivity.

In particular, a harmonic homology whose centre is the pole of its axis induces on the conic an involution, elliptic or hyperbolic according as the centre is interior or exterior.

3.5. The fixed points of a collineation. In one dimension, an elliptic projectivity leaves no point invariant. It is remarkable, then, that a two-dimensional collineation (in one plane) always has an invariant point somewhere. (It may have more than one; e.g. a homology has infinitely many.) To see this, we need the following lemma.

3.51. *If two coplanar conics have a common point, at which their tangents are distinct, they have at least one other common point.*

To save space, we omit the formal proof.* The result is intuitively obvious if we think of a variable point on one conic, moving so as to cross the other. Since the first conic includes both interior and exterior points of the second, considerations of continuity show that there must be at least two points of intersection.

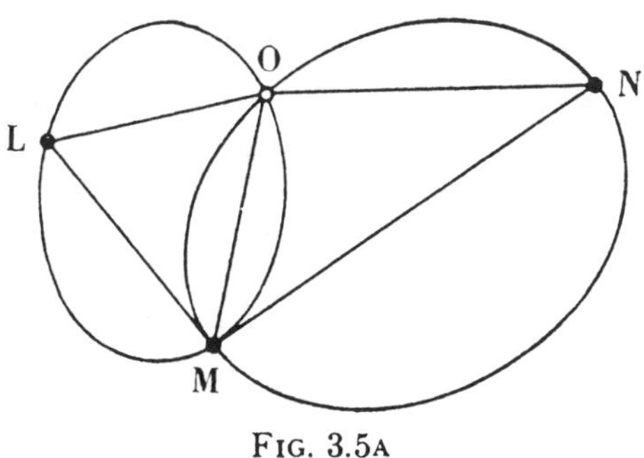

FIG. 3.5A

3.52. *Every collineation of a plane into itself has at least one invariant point.*

*Enriques [1], pp. 294-298.

PROOF. If the collineation has two invariant lines, their point of intersection is an invariant (i.e. self-corresponding) point. If not, let **L** be a point not lying on any invariant line, so that, if **L** is transformed into **M**, and **M** into **N**, the line **LM** is distinct from **MN**, as in Fig. 3.5A. The collineation induces a projectivity between the lines through **L** and **M**, and between the lines through **M** and **N**. Since **LM** and **MN** are not invariant, the loci of points of intersection of corresponding lines are conics: one through **L** and **M** touching **MN**, and one through **M** and **N** touching **ML**. These two conics, having distinct tangents at their common point **M**, have at least one other common point, say **O**. The collineation transforms **LO** into **MO**, and **MO** into **NO**; therefore **O** is an invariant point.

The two conics may have four common points. Then the collineation has three invariant points.

3.6. Cones and reguli. By applying space-duality to §§ 3.2 and 3.3, we obtain the analogous theory of polarities in a *bundle*, i.e., of involutory correspondences between lines and planes through one fixed point. Such a polarity may be elliptic or hyperbolic, and in the latter case its self-conjugate lines and planes are the *generators* and *tangent planes* of a *quadric cone.* Every property of conics leads to a corresponding property of cones; e.g. the space-dual of Brianchon's Theorem 3.36 is the following: If **l**, **m′**, **n**, **l′**, **m**, **n′** are any six generators of a cone, the three lines (**m′n**, **mn′**), (**n′l**, **nl′**), (**l′m**, **lm′**) are coplanar.

A more interesting system of lines may be defined as follows.* A *regulus* is the class of lines which meet each of three skew lines **a**, **b**, **c**. The lines are again called *generators.*

Every plane through **c** meets **a** and **b** in two points whose join, **AB**, is a transversal of the three skew lines. The generator **AB**, meeting **c** at **C**, may also be described as the intersection

*Enriques [1], p. 333.

of planes **Ca**, **Cb**. Thus the regulus is a self-dual figure, each generator being both the join of corresponding points of projectively related ranges on **a** and **b**, and also the intersection of corresponding planes of projectively related axial pencils through **a** and **b**.

The transversal to two generators **AB** and **A′B′**, from any point **D″** on a third generator **A″B″**, determines points **D** and **D′**, such that $\mathbf{ABCD} \barwedge \mathbf{A'B'C'D'}$. By 2.85, the projectivity thus established between **AB** and **A′B′** is independent of the choice of **A″B″**. Hence the given regulus determines another *associated* regulus, such that every generator of either regulus meets every generator of the other. (The generators of the associated regulus include the original lines **a**, **b**, **c**.)

The above remarks suffice to prove the following theorem:

3.61. *If four skew lines have two transversals on which they determine projectively related ranges, they belong to a regulus, and so have an infinity of transversals.*

The remaining possibilities for four skew lines are as follows:

3.62. *If four skew lines do not belong to a regulus, the number of their transversals may be* 0, 1, *or* 2.

PROOF. The regulus determined by three of the four skew lines is generated by corresponding planes of two axial pencils, which meet the fourth line in projectively related ranges. There will be a transversal for each double point in this projectivity (i.e., for each point in which the fourth line meets the regulus).

3.7. Three-dimensional projectivities. In space, a *collineation* can again be defined as a correspondence which relates collinear points to collinear points; it consequently relates flat pencils to flat pencils, and planes to planes. It clearly induces a collineation between any two corresponding planes, and a projectivity between any two corresponding lines or pencils.

3.71. *Any collineation which leaves a complete pentagon or pentahedron invariant is the identity.*

PROOF. If the five faces of a complete pentahedron are invariant, so are its ten edges and ten vertices. Since then three points on each edge are invariant, 2.84 shows that every point on each edge is invariant. Hence every plane (joining points on three distinct edges) is invariant, and so also every line and point. The dual argument establishes the same result for a complete pentagon.

Any given point **O** and plane ω, not incident, are the centre and axial plane of a *harmonic homology*, which relates each point **A** to its harmonic conjugate with respect to the points **O** and (ω, **OA**).

3.72. *The product of four harmonic homologies, whose centres and axial planes are the vertices and faces of a tetrahedron, is the identity.*

PROOF. Let Φ, Φ', Φ'', Φ''' be the homologies, with centres **O**, **O**$'$, **O**$''$, **O**$'''$. Take six points **A**, **B**, **A**$'$, **B**$'$, **A**$''$, **B**$''$, such that H(**OO**$'$, **AB**), H(**O**$'$**O**$''$, **A**$'$**B**$'$), H(**O**$''$**O**$'''$, **A**$''$**B**$''$). Then **A** and **B** are interchanged by Φ and again by Φ', **A**$'$ and **B**$'$ by Φ' and again by Φ'', **A**$''$ and **B**$''$ by Φ'' and again by Φ'''. Hence $\Phi\Phi'\Phi''\Phi''' = 1$, preserving all six points, of which five form a complete pentagon (**ABA**$'$**A**$''$**B**$''$ or **ABB**$'$**A**$''$**B**$''$).

In three dimensions, as in one dimension, we can distinguish two kinds of collineation: *direct* (preserving sense) and *opposite* (reversing sense). For, by the definition of a "double perspectivity" on p. 34, any collineation preserves similarity of orientation (for a pair of doubly oriented lines). In other words, to see whether a given collineation is direct or opposite, we merely have to test its effect on one doubly oriented line. Consider, for instance, the harmonic homology Φ and the doubly oriented line (**OAB**, $\alpha\beta\gamma$), where **B** is the harmonic homologue of **A**. This is transformed into (**OBA**, $\alpha\beta\gamma$). Hence

3.73. *A harmonic homology is an opposite collineation.*

A *correlation* in space is a correspondence which relates collinear points to coaxial planes, and so preserves incidence in accordance with the principle of space-duality. It clearly induces a projectivity between the points of a line and the planes through the corresponding line. Theorem 3.14 is readily extended as follows:*

3.74. *A correlation is uniquely determined when we are given a complete pentahedron and the corresponding complete pentagon.*

3.8. Polarities in space. A polarity is a correlation of period two, so that if it relates any point **A** to a plane α, it also relates α to **A**. We call **A** the pole of α, and α the polar plane of **A**. If **B** lies on α, its polar plane, β, passes through **A**. We call **A** and **B** conjugate points, α and β conjugate planes. Since **AB** lies in the polar plane of any point on (α, β), two such lines are symmetrically related; we call them *polar lines* (of each other). Two lines are said to be *conjugate* if either meets the polar line of the other. If there is a plane α which is not self-conjugate, the polar planes and polar lines of the points and lines in α will meet α in lines and points according to a plane polarity, which we say is induced by the polarity in space. Similarly, an involution is induced in any line which is neither self-conjugate nor self-polar. If **BCD** is a self-polar triangle for the polarity induced in the polar plane of a non-self-conjugate point **A**, the tetrahedron **ABCD** is said to be self-polar. Any two vertices or faces are conjugate; any two opposite edges are polar lines.

*Von Staudt [1], pp. 60-69.

Arguments similar to those used in proving 3.23 and 3.24 suffice for the following analogous theorems:*

3.81. *Any correlation which transforms each vertex of one tetrahedron into the opposite face is a polarity.*

3.82. *If ϖ is a plane not through any vertex of a self-polar tetrahedron* **ABCD**, *and* **P** *is its pole, the polarity does not or does admit a self-conjugate point, according as* **P** *lies in the region* **ABCD**/ϖ *or in one of the other seven regions determined by the tetrahedron.*

In the former case the polarity is said to be *uniform* (or elliptic), since the involution of conjugate points on *any* line is elliptic; there are no self-conjugate points, lines, or planes, and no self-polar lines. In the latter case there are still two alternative possibilities, since "the other seven regions" consist of four of one kind and three of another. In fact, the four faces of the tetrahedron meet the plane ϖ in a quadrilateral, thereby dividing the rest of the plane into seven regions (which are

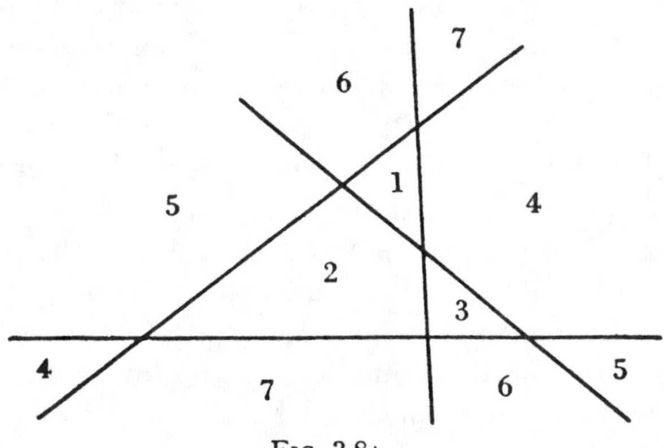

FIG. 3.8A

sections of the seven regions in space); of these plane regions, four are three-sided and three four-sided. (See the odd and

*Von Staudt [1], pp. 190-196.

even numbers in Fig. 3.8A.) According as **P** lies in one of the regions whose section is three- or four-sided, *three* or *four* of the six edges of the tetrahedron will contain *hyperbolic* involutions of conjugate points, the involutions in the remaining edges being elliptic. Thus, putting the number of elliptic involutions first, we may distinguish polarities of types

$$(6, 0), \ (3, 3), \ (2, 4),$$

though it is not yet established that the last two are mutually exclusive, since we have considered only one self-polar tetrahedron.

The analogous symbols for elliptic and hyperbolic polarities in two dimensions are (3, 0) and (1, 2). Since the polarities induced in the faces of the tetrahedron must each be of one of these two types, the distribution of elliptic and hyperbolic involutions is as indicated in Fig. 3.8B, where the edges containing such involutions are drawn in full and broken lines, respectively. Thus, in the (3, 3) case, opposite edges contain unlike involutions, whereas in both the other cases opposite edges contain like involutions.

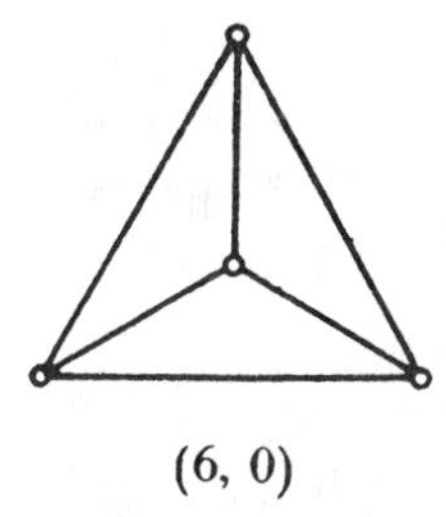
(6, 0)

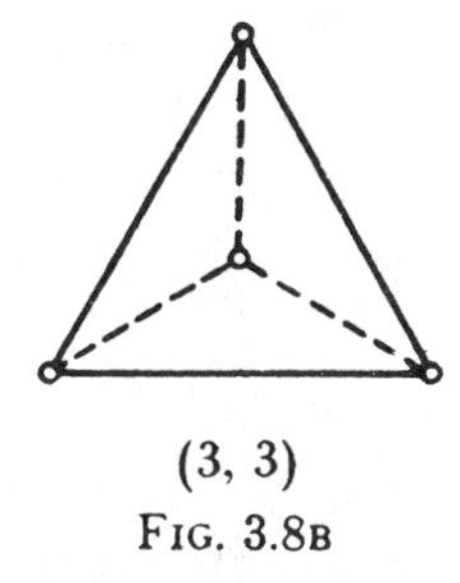
(3, 3)

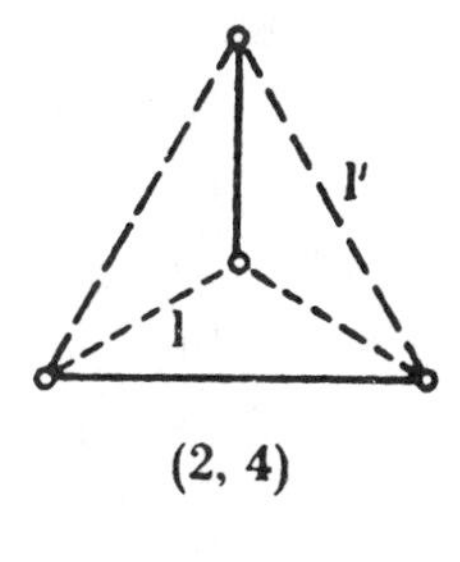

(2, 4)

FIG. 3.8B

In the (2, 4) case, let **l** and **l**′ be a pair of opposite edges containing hyperbolic involutions, with double points **A**, **B** and **A**′, **B**′. Since **l** and **l**′ are polar lines, the polar planes of the self-conjugate points **A** and **A**′ are **Al**′ and **A**′**l**, and the polar line of **AA**′ is (**Al**′, **A**′**l**) = **AA**′. Thus a (2, 4) polarity admits *self-polar lines.*

Conversely, if a polarity admits a self-polar line and a non-self-conjugate point (or plane), it is of type (2, 4) with respect to every self-polar tetrahedron. For, since every non-self-conjugate plane contains a self-conjugate point (where it meets the self-polar line), the polarity induced in such a plane is necessarily hyperbolic; thus every face of any self-polar tetrahedron is of type (1, 2), and the tetrahedron itself is of type (2, 4). It follows that a (2, 4) polarity cannot be also a (3, 3) polarity. Hence

3.83. *If a polarity admits a self-conjugate point and a non-self-conjugate point, it is of type* (2, 4) *or* (3, 3) *according as it does or does not admit a self-polar line.*

We define a *quadric* as the class of self-conjugate points and planes in such a polarity: in the (2, 4) case, a *ruled* (or ring-shaped) quadric; and in the (3, 3) case, an *oval* (or non-ruled) quadric. The self-conjugate points are said to lie on the quadric, their polar planes are called tangent planes, the self-conjugate lines are called tangent lines, and, in the (2, 4) case, the self-polar lines are called generators.

A tangent line **t**, whose polar line is **t**′, meets the quadric only at its point of contact (**t**, **t**′), whose polar plane is the tangent plane **tt**′. In each tangent plane, the flat pencil of tangent lines contains an involution of such pairs. In the case of the ruled quadric, this involution is hyperbolic, its double lines being two generators. Lines which are neither tangents nor generators fall into two categories, according as the involutions of conjugate points on them are elliptic or hyperbolic: *exterior* lines, which do not meet the quadric at all; and *secant* lines, which meet it twice.

The properties of an oval quadric are closely analogous to those of a conic. Such a quadric divides the points of space, and so also the planes, into three categories, as follows:

(i) Interior points, through which no tangents pass;
(ii) Points on the quadric, each the centre of a flat pencil of tangent lines;
(iii) Exterior points, the tangent lines through each forming a cone;

(i) Exterior planes, which do not meet the quadric;
(ii) Tangent planes, each containing a flat pencil of tangent lines;
(iii) Secant planes, each meeting the quadric in a conic.

3.84. *The generators of a ruled quadric form two associated reguli.*

PROOF. Let **a** and **b** be two skew generators of the quadric. The polar plane of each point **A** on **a** meets **b** in a point **B** whose polar plane is **Ab**. Since the range of points **A** is projectively related to the range of points **B**, their joins form a regulus of self-polar lines (i.e. generators of the quadric). Similarly, relating the points on any two of these generators, say **m** and **n**, we infer that also the generators of the associated regulus belong to the quadric. These account for all the generators of the quadric, since the two generators in any tangent plane γ, whose pole (or point of contact) is **C**, may be described as (**Ca**, γ) and (**Cm**, γ).

Returning to the general classification of polarities, we still have to consider the possibility that *every* point (and every plane) is self-conjugate, so that there is no self-polar tetrahedron, and the only appropriate type-symbol is (0, 0). Then we have what is known as a *null polarity* (or null system), and the class of self-polar lines is called a *linear complex*. Clearly, every plane contains a flat pencil of such lines, with its centre at the pole of the plane; and conversely, a polarity which admits a flat pencil of self-polar lines can only be a null polarity.

The following theorem establishes the existence of such a polarity:*

3.85. *The correlation which transforms the vertices of a complete pentagon* **ABCDE** *into the respective planes* **EAB, ABC, BCD, CDE, DEA** *is a null polarity.*

PROOF. The given correlation transforms the plane **EAB** into the point (**DEA, EAB, ABC**) = **A**, and so is the same as its inverse; i.e., it is a polarity. Each of the lines **AB, BC, CD, DE, EA** is self-polar; e.g. **DE** = (**CDE, DEA**). The polar plane of **P** = (**ABC, DE**) is **BDE**; therefore the line **BP** = (**ABC, BDE**) is self-polar. The flat pencil of lines through **B** in **ABC** is transformed into itself according to a projectivity with three double lines: **BA, BC, BP**. Hence, by the plane-dual of 2.84, every line of the pencil is self-polar.

The following summary gives, in the customary notation, the number of points and lines which play a special role in the four types of polarity.

Polarity	Uniform	With respect to an oval quadric	With respect to a ruled quadric	Null
Type-symbol	(6, 0)	(3, 3)	(2, 4)	(0, 0)
Self-conjugate points (or planes)	0	∞^2	∞^2	∞^3
Self-conjugate lines	0	∞^3	∞^3	0
Self-polar lines	0	0	$2\infty^1$	∞^3

*Von Staudt [1], p. 194.

CHAPTER IV

HOMOGENEOUS COORDINATES

4.1. The von Staudt-Hessenberg calculus of points. Projective geometry might well be described as "What we can do with an ungraduated straight edge or ruler, without compasses." It is hoped that Chapters II and III have shown what a wealth of elegant theorems can be obtained without any appeal to measurement. It is one of von Staudt's greatest discoveries that even such apparently metrical notions as coordinates and cross ratios can be introduced non-metrically. The following is a brief outline of his method, as revised by Hessenberg,* and a summary of the standard results in analytical projective geometry.

We saw, in §2.9, that an involution cannot be parabolic. It is sometimes convenient, however, to extend the meaning of the word "involution" so as to admit the relation which associates every point on a line with one particular point on the same line. We call this a *degenerate* involution. It enables us to omit the proviso (in parentheses) in 2.92.

Let $(\mathbf{AA'})(\mathbf{BB'})$ denote the involution determined by the pairs $\mathbf{AA'}$ and $\mathbf{BB'}$. Then if $\mathbf{A} \neq \mathbf{B}$, $(\mathbf{AM})(\mathbf{BM})$ is the degenerate involution with double point $\mathbf{M}$.

Let $\mathbf{P}_0$, $\mathbf{P}_1$, $\mathbf{P}_\infty$ be any three fixed collinear points, and $\mathbf{X}$, $\mathbf{Y}$, $\mathbf{Z}$ variable points on the same line. We define the *sum* $\mathbf{X}+\mathbf{Y}$ as the point corresponding to $\mathbf{P}_0$ in $(\mathbf{XY})(\mathbf{P}_\infty\mathbf{P}_\infty)$, and the *difference* $\mathbf{Z}-\mathbf{Y}$ as the point corresponding to $\mathbf{Y}$ in $(\mathbf{ZP}_0)(\mathbf{P}_\infty\mathbf{P}_\infty)$. Thus $\mathbf{X}+\mathbf{Y}=\mathbf{Y}+\mathbf{X}$, $(\mathbf{Z}-\mathbf{Y})+\mathbf{Y}=\mathbf{Z}$, $\mathbf{X}\pm\mathbf{P}_0=\mathbf{X}$, and if $\mathbf{X}\neq\mathbf{P}_\infty$, $\mathbf{X}\pm\mathbf{P}_\infty=\mathbf{P}_\infty$. Also $\mathbf{P}_0-\mathbf{X}$ is the harmonic

*Von Staudt [2], pp. 15-18, 166-176, 256-283; Hessenberg [1]; Robinson [1], pp. 90-104.

conjugate of $\mathbf{X}$ with respect to $\mathbf{P}_0$ and $\mathbf{P}_\infty$. By means of incidences in a plane,* it is easily proved that

4.11. $$\mathbf{X}+(\mathbf{Y}+\mathbf{Z})=(\mathbf{X}+\mathbf{Y})+\mathbf{Z}.$$

Accordingly, if we define $\mathbf{P}_2=\mathbf{P}_1+\mathbf{P}_1$, $\mathbf{P}_x=\mathbf{P}_{x-1}+\mathbf{P}_1$, and $\mathbf{P}_{-x}=\mathbf{P}_0-\mathbf{P}_x$, we shall have $\mathbf{P}_x\pm\mathbf{P}_y=\mathbf{P}_{x\pm y}$ for all integers x and y.

Next, we define the *product* $\mathbf{X}\cdot\mathbf{Y}$ as the point corresponding to $\mathbf{P}_1$ in $(\mathbf{X}\mathbf{Y})(\mathbf{P}_0\mathbf{P}_\infty)$, and the *quotient* $\mathbf{Z}:\mathbf{Y}$ as the point corresponding to $\mathbf{Y}$ in $(\mathbf{Z}\mathbf{P}_1)(\mathbf{P}_0\mathbf{P}_\infty)$. Thus

$$\mathbf{X}\cdot\mathbf{Y}=\mathbf{Y}\cdot\mathbf{X},\quad (\mathbf{Z}:\mathbf{Y})\cdot\mathbf{Y}=\mathbf{Z},\quad \mathbf{X}\cdot\mathbf{P}_1=\mathbf{X}=\mathbf{X}:\mathbf{P}_1,$$
$$\mathbf{X}\cdot\mathbf{P}_0=\mathbf{P}_0=\mathbf{X}:\mathbf{P}_\infty \text{ (if } \mathbf{X}\neq\mathbf{P}_\infty\text{), and}$$
$$\mathbf{X}:\mathbf{P}_0=\mathbf{P}_\infty=\mathbf{X}\cdot\mathbf{P}_\infty \text{ (if } \mathbf{X}\neq\mathbf{P}_0\text{).}$$

The associative law for multiplication, namely $\mathbf{X}\cdot(\mathbf{Y}\cdot\mathbf{Z})=(\mathbf{X}\cdot\mathbf{Y})\cdot\mathbf{Z}$, can be proved in the same manner as the associative law for addition, 4.11. But the distributive law can be proved "in one dimension," as follows.

By the definition of $\mathbf{X}\cdot\mathbf{Y}$, we have

4.12. $$\mathbf{P}_\infty\mathbf{P}_0\mathbf{P}_1\mathbf{Y} \barwedge \mathbf{P}_0\mathbf{P}_\infty(\mathbf{X}\cdot\mathbf{Y})\mathbf{X},$$

whence, by 2.81,

4.13. $$\mathbf{P}_\infty\mathbf{P}_0\mathbf{P}_1\mathbf{Y} \barwedge \mathbf{P}_\infty\mathbf{P}_0\mathbf{X}(\mathbf{X}\cdot\mathbf{Y}).$$

This last projectivity, which is the same for all positions of $\mathbf{Y}$, transforms the involution $(\mathbf{Y}\mathbf{Z})(\mathbf{P}_\infty\mathbf{P}_\infty)$ into $(\mathbf{X}\cdot\mathbf{Y}\,\mathbf{X}\cdot\mathbf{Z})(\mathbf{P}_\infty\mathbf{P}_\infty)$. But $\mathbf{P}_0$ is paired with $\mathbf{Y}+\mathbf{Z}$ in the former involution, and with $\mathbf{X}\cdot\mathbf{Y}+\mathbf{X}\cdot\mathbf{Z}$ in the latter. Hence

$$\mathbf{X}\cdot(\mathbf{Y}+\mathbf{Z})=\mathbf{X}\cdot\mathbf{Y}+\mathbf{X}\cdot\mathbf{Z}.$$

It follows that, if we define $\mathbf{P}_x=\mathbf{P}_n:\mathbf{P}_d$ for all rational numbers $x=n/d$, we shall have

$$\mathbf{P}_x+\mathbf{P}_y=\mathbf{P}_{x+y},\qquad \mathbf{P}_x-\mathbf{P}_y=\mathbf{P}_{x-y},$$
$$\mathbf{P}_x\cdot\mathbf{P}_y=\mathbf{P}_{xy},\qquad \mathbf{P}_x:\mathbf{P}_y=\mathbf{P}_{x/y}.$$

---

*Veblen and Young [2], I, pp. 143, 146.

Such points $\mathbf{P}_x$, along with $\mathbf{P}_\infty$, are said to form a ***net of rationality*** in the line.

Let x be any rational number other than 0. Since $\mathbf{P}_{x+y} = \mathbf{P}_x + \mathbf{P}_y$, $\mathbf{P}_{2x}$ is the point corresponding to $\mathbf{P}_0$ in $(\mathbf{P}_x\mathbf{P}_x)(\mathbf{P}_\infty\mathbf{P}_\infty)$; hence $\mathrm{H}(\mathbf{P}_\infty\mathbf{P}_x, \mathbf{P}_0\mathbf{P}_{2x})$, and $\mathbf{P}_\infty\mathbf{P}_x \,||\, \mathbf{P}_0\mathbf{P}_{2x}$. We can now prove by induction that

$$\mathbf{P}_\infty \mathbf{P}_x \,||\, \mathbf{P}_0 \mathbf{P}_{mx}$$

for all integers $m > 1$. For, by the definition of $\mathbf{X}+\mathbf{Y}$,

$$\mathbf{P}_\infty\mathbf{P}_0\, \mathbf{X}\, \mathbf{Y} \;\overline{\wedge}\; \mathbf{P}_\infty\, (\mathbf{X}+\mathbf{Y})\, \mathbf{Y}\, \mathbf{X}.$$

In particular, $\mathbf{P}_\infty \mathbf{P}_0 \mathbf{P}_x \mathbf{P}_{(m-1)x} \;\overline{\wedge}\; \mathbf{P}_\infty \mathbf{P}_{mx} \mathbf{P}_{(m-1)x} \mathbf{P}_x$. Hence the separation $\mathbf{P}_\infty\mathbf{P}_x || \mathbf{P}_0 \mathbf{P}_{(m-1)x}$ implies $\mathbf{P}_\infty\mathbf{P}_{(m-1)x} || \mathbf{P}_{mx}\mathbf{P}_x$, which (by 2.123 and 2.125) implies $\mathbf{P}_\infty\mathbf{P}_x \,||\, \mathbf{P}_0\mathbf{P}_{mx}$.

Thus the points $\mathbf{P}_x, \mathbf{P}_{2x}, \mathbf{P}_{3x}, \ldots$ occur in order in one of the two segments $\mathbf{P}_0\,\mathbf{P}_\infty$. Replacing x by $-x$, we see that the same holds for $\mathbf{P}_{-x}, \mathbf{P}_{-2x}, \mathbf{P}_{-3x}, \ldots$. Moreover, these lie in the ***other*** segment $\mathbf{P}_0\mathbf{P}_\infty$, since $\mathrm{H}(\mathbf{P}_0\mathbf{P}_\infty, \mathbf{P}_x\mathbf{P}_{-x})$. Hence all the points

$$\ldots, \mathbf{P}_{-3x}, \mathbf{P}_{-2x}, \mathbf{P}_{-x}, \mathbf{P}_0, \mathbf{P}_x, \mathbf{P}_{2x}, \ldots, \mathbf{P}_\infty$$

occur in order. In other words, if m and n are any integers, while d is a positive integer, we have $\mathrm{S}(\mathbf{P}_{mx}\mathbf{P}_{nx}\mathbf{P}_\infty) = \mathrm{S}(\mathbf{P}_0\mathbf{P}_{dx}\mathbf{P}_\infty)$ if and only if $m<n$. Writing $1/d$ for x, and observing that any two rational numbers can be expressed as m/d and n/d (in terms of a "common denominator"), we deduce that

$$\mathrm{S}(\mathbf{P}_a\, \mathbf{P}_b\, \mathbf{P}_\infty) = \mathrm{S}(\mathbf{P}_0\, \mathbf{P}_1\, \mathbf{P}_\infty)$$

if and only if $a<b$. Hence*

4.14. *The order of the points* $\mathbf{P}_x$ *of a net of rationality agrees with the order of the rational numbers* x.

The step from a net of rationality to the whole line is made by means of Axiom 2.13.* Let y be any real number, which we first suppose to be positive. We divide all the points of the

*Veblen and Young [1], p. 368.

"positive" segment $\mathbf{P}_0\,\mathbf{P}_\infty/\mathbf{P}_{-1}$ into two sets: (i) all points which lie between $\mathbf{P}_0$ and every "rational" point $\mathbf{P}_x$ with $x>y$, and (ii) the rest of the segment. The point determined by this dichotomy is denoted by $\mathbf{P}_y$. For a negative y, we treat the "negative" segment $\mathbf{P}_0\,\mathbf{P}_\infty/\mathbf{P}_1$ similarly. Thus

4.15. *There is a definite point for every real number, the order of the points agreeing with the order of the numbers.*

Conversely, every point of the line can be so numbered. For, any point divides the numbered points, and thence the real numbers, into two sets, to which we can apply the arithmetical axiom of Dedekind. The real number x is called the *abscissa* of the point P_x.

4.2. One-dimensional projectivities. Since $\mathbf{P}_{a-x}=\mathbf{P}_a-\mathbf{P}_x$, the above definition for the difference of two points may be expressed in the form

$$\mathbf{P}_\infty\mathbf{P}_a\mathbf{P}_0\mathbf{P}_x \barwedge \mathbf{P}_\infty\mathbf{P}_0\mathbf{P}_a\mathbf{P}_{a-x}.$$

Replacing x by $-x$, and applying the involution $(\mathbf{P}_0\mathbf{P}_0)(\mathbf{P}_\infty\mathbf{P}_\infty)$, we deduce that

$$\mathbf{P}_\infty\mathbf{P}_{-a}\mathbf{P}_0\mathbf{P}_x \barwedge \mathbf{P}_\infty\mathbf{P}_0\mathbf{P}_a\mathbf{P}_{a+x}.$$

Considering $\mathbf{P}_x$ as a variable point, we thus see that the transformation

4.21. $$x'=a+x$$

defines a projectivity between points $\mathbf{P}_x$ and $\mathbf{P}_{x'}$.

By 4.12 and 4.13, we have

$$\mathbf{P}_\infty\mathbf{P}_0\mathbf{P}_1\mathbf{P}_x \barwedge \mathbf{P}_0\mathbf{P}_\infty\mathbf{P}_1\mathbf{P}_{1/x}$$

and $$\mathbf{P}_\infty\mathbf{P}_0\mathbf{P}_1\mathbf{P}_x \barwedge \mathbf{P}_\infty\mathbf{P}_0\mathbf{P}_a\mathbf{P}_{ax}.$$

Thus the transformations

4.22. $$x'=1/x,$$

4.23. $$x'=ax \qquad (a\neq 0)$$

define two further projectivities.

It follows that the combined transformation

4.24. $$x' = \beta + \cfrac{\gamma}{\delta + \cfrac{1}{\epsilon + x}} = \frac{(\beta\delta+\gamma)x + (\beta+\beta\delta\epsilon+\gamma\epsilon)}{\delta x + (1+\delta\epsilon)} \qquad (\gamma \neq 0)$$

likewise defines a projectivity. Moreover, the values ∞, 0, 1 for x give for x' the values $\beta+\gamma/\delta$, $\beta+\gamma/(\delta+1/\epsilon)$, $\beta+\gamma/\{\delta+1/(\epsilon+1)\}$, which may be arbitrarily assigned, provided they are distinct. Hence, by 2.85,

The general one-dimensional projectivity is given by the linear fractional transformation

4.25. $$x' = (px+q)/(rx+s) \qquad (ps-qr \neq 0)$$

or by the bilinear relation

4.26. $$axx' + bx + cx' + d = 0 \qquad (ad-bc \neq 0).$$

By observing their effect on the points $\mathbf{P}_0$, $\mathbf{P}_1$, $\mathbf{P}_\infty$, we see that the projectivities 4.21 and 4.22 are direct and opposite, respectively, while 4.23 is direct or opposite according as a is positive or negative. Hence 4.24, 4.25, and 4.26 are direct or opposite according to the sign of

$$\gamma = ps - qr = ad - bc.$$

By considering the discriminant of the quadratic equation

$$ax^2 + (b+c)x + d = 0,$$

we see that the projectivity 4.26 is elliptic, parabolic, or hyperbolic according as $ad-bc$ is greater than, equal to, or less than $\frac{1}{4}(b-c)^2$. The projectivity is an involution if it is symmetrical with respect to x and x', i.e. if $b=c$. (Cf. 2.88 and 2.96.)

Any involution can be expressed in the form $xx'+d=0$, by assigning the symbols $\mathbf{P}_0$ and $\mathbf{P}_\infty$ to one of its pairs. For any non-vanishing number c, we can change the notation by assigning the symbol $\mathbf{P}_x$ to the point previously called $\mathbf{P}_{cx}$. The given involution then becomes $c^2xx'+d=0$. Finally, taking $c=\sqrt{|d|}$,

Any elliptic or hyperbolic involution can be expressed in the "canonical form"

4.27. $$xx' \pm 1 = 0.$$

The hyperbolic involution $xx'-1=0$ has double points $\mathbf{P}_{\pm 1}$. More generally, for any unequal finite numbers s and t, $(\mathbf{P}_s\mathbf{P}_s)(\mathbf{P}_t\mathbf{P}_t)$ is

4.28. $$xx' - \tfrac{1}{2}(s+t)(x+x') + st = 0;$$

for, this relation becomes $(x-s)(x-t)=0$ when we put $x'=x$. On the other hand, $(\mathbf{P}_s\mathbf{P}_s)(\mathbf{P}_\infty\mathbf{P}_\infty)$ is

4.29. $$x + x' = 2s.$$

4.3. Coordinates in one and two dimensions. The sum, difference, product, and quotient of points (§4.1) depends on our choice of $\mathbf{P}_0$, $\mathbf{P}_1$, $\mathbf{P}_\infty$. The transformation 4.25 enables us to get rid of this particular choice by using the symbol $\mathbf{P}_x$ to denote the point previously called $\mathbf{P}_{x'}$.

The interpretation of 4.25 when $x=\infty$ requires a little care. Such difficulties are avoided if we replace the abscissae x by pairs of *homogeneous coordinates* x_0, x_1, such that $x_1/x_0 = x$. Then if $x_0 \neq 0$, (x_0, x_1) denotes the point $\mathbf{P}_{x_1/x_0}$, but $(0, 1)$ denotes $\mathbf{P}_\infty$. Thus every ordered pair of real numbers, not both zero, defines a point on the line, but each point is equally well defined by every pair $(\rho x_0, \rho x_1)$, where $\rho \neq 0$; hence "homogeneous." In particular, the harmonic conjugate of (x_0, x_1) with respect to $(1, 0)$ and $(0, 1)$ is $(-x_0, x_1)$ or $(x_0, -x_1)$.

By the fundamental theorem 2.85, any three collinear points are projectively related to any other three collinear points. But the case of four collinear points is different, as some tetrads are harmonic while others are not. Accordingly, we seek a number which will distinguish a given tetrad from all unrelated tetrads.

For any four collinear points $\mathbf{P}_x$, $\mathbf{P}_y$, $\mathbf{P}_s$, $\mathbf{P}_t$, not necessarily distinct, we define the *cross ratio*

4.31. $$\{P_xP_y,\ P_zP_t\} = \frac{(x-z)(y-t)}{(x-t)(y-z)} = \frac{(x_1z_0-x_0z_1)(y_1t_0-y_0t_1)}{(x_1t_0-x_0t_1)(y_1z_0-y_0z_1)}.$$

The former expression (in terms of abscissae) is convenient to use when the point P_∞ is not involved. The cross ratio is clearly unchanged by each of the transformations 4.21, 4.22, 4.23, and is therefore a projective invariant. In other words, the relation $ABCD \barwedge A'B'C'D'$ is equivalent to the equation $\{AB, CD\} = \{A'B', C'D'\}$. But there is no projective invariant depending on fewer than four points. (The *simple* ratio **AC/BC**, which occurs in *affine* geometry, has no place in the present theory, though it can actually be derived from $\{AB, CD\}$ by calling **D** the "point at infinity.")

Clearly

$$\{AB, CD\} = \{BA, DC\} = \{CD, AB\} = \{DC, BA\},$$

in agreement with 2.81. In particular,

$$\{AB, AD\} = \{BA, DA\} = 0,$$
$$\{AA, CD\} = \{CD, AA\} = 1,$$
$$\{AB, CA\} = \{BA, AC\} = \infty.$$

Also

4.32. $$\{AB, DC\} = \{AB, CD\}^{-1}, \quad \{AC, BD\} = 1 - \{AB, CD\}.$$

Another useful formula is

4.33. $$\{P_xP_y,\ P_0P_\infty\} = x/y,$$

which shows that the relation H(**AB, CD**) is equivalent to $\{AB, CD\} = -1$. By 4.33 again, we have

$$\{P_yP_z,\ P_0P_\infty\}\ \{P_zP_x,\ P_0P_\infty\}\ \{P_xP_y,\ P_0P_\infty\} = 1.$$

Thus, for any five points **A, B, C, I, X**, on a line or conic,

4.34. $$\{BC, IX\}\ \{CA, IX\}\ \{AB, IX\} = 1.$$

By 3.12, any four coplanar points, no three collinear, are related (by a collineation) to any other such set of four points. On the other hand, each of *five* coplanar points is joined to the remaining four by lines which have a definite cross ratio. We proceed to use three of these five cross ratios to determine a point **X** with reference to a quadrangle **ABCI** in the simplest possible way. Applying 4.34 to the conic **ABCIX**, we infer the existence of real numbers x_0, x_1, x_2, such that

$$\{\mathbf{BC}, \mathbf{IX}\} = \frac{x_1}{x_2}, \quad \{\mathbf{CA}, \mathbf{IX}\} = \frac{x_2}{x_0}, \quad \{\mathbf{AB}, \mathbf{IX}\} = \frac{x_0}{x_1}.$$

These, or any numbers proportional to them (not all zero), are called the *coordinates* of **X** with respect to the *triangle of reference* **ABC** and the *unit point* **I**, and the point **X** is denoted by (x_0, x_1, x_2). To make this definition significant, we must stipulate that the unit point (1, 1, 1) shall not lie on any side of the triangle.

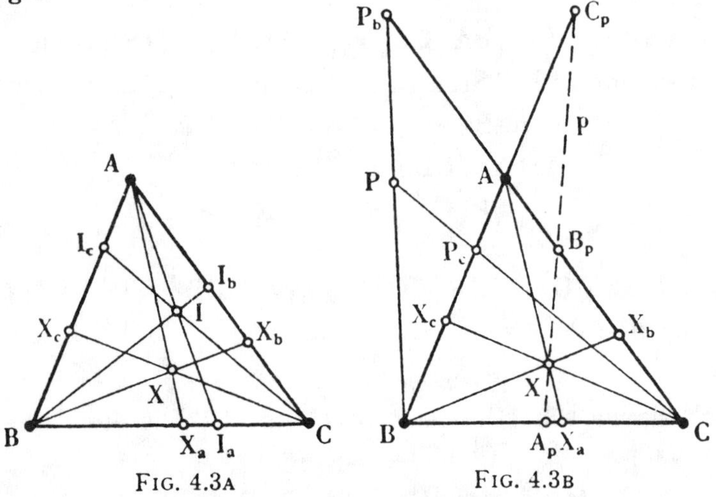

FIG. 4.3A FIG. 4.3B

Each of the above cross ratios may be interpreted as applying to the four lines obtained by joining the four points mentioned to the remaining one of **A**, **B**, **C**, **I**, **X**, as in Fig. 4.3A. Thus, if **X** lies on **BC**, we have $\{\mathbf{CA}, \mathbf{IX}\} = \infty$ and $\{\mathbf{AB}, \mathbf{IX}\} = 0$, so the line **BC** has the *equation* $x_0 = 0$. Similarly **CA** and **AB** have equations $x_1 = 0$ and $x_2 = 0$; therefore **A** is $(1, 0, 0)$.

Similarly **B** and **C** are (0, 1, 0) and (0, 0, 1). It is important to notice that the point $(x_0, x_1, 0)$ on **AB** has one-dimensional coordinates x_0, x_1, referred to $\mathbf{P}_0 = \mathbf{A}$, $\mathbf{P}_\infty = \mathbf{B}$, $\mathbf{P}_1 = \mathbf{I}_c$. The line joining **C** to $(y_0, y_1, 0)$ evidently has the equation $x_1/x_0 = y_1/y_0$ or

$$y_1 x_0 - y_0 x_1 = 0.$$

To see that a line not through any vertex of the triangle has likewise a linear equation, consider such a line **p** (as in Fig. 3.2B) and let $\mathbf{P} = (p_0, p_1, p_2)$ be its trilinear pole. In other words, defining $\mathbf{P}_b$ and $\mathbf{P}_c$ so that $H(\mathbf{P}_b\mathbf{B}_p, \mathbf{CA})$ and $H(\mathbf{P}_c\mathbf{C}_p, \mathbf{AB})$, let **P** be the point where $\mathbf{BP}_b$ meets $\mathbf{CP}_c$. (See Fig. 4.3B.) Then, for any point **X** on **p**, we have

$$\mathbf{C\,B_p\,A\,X_b} \overset{\mathbf{B}}{\underset{\wedge}{=}} \mathbf{A_p\,B_p\,C_p X} \overset{\mathbf{C}}{\underset{\wedge}{=}} \mathbf{B\,A\,C_p\,X_c}.$$

Since the points $\mathbf{X}_b$, $\mathbf{X}_c$, $\mathbf{B}_p$, $\mathbf{C}_p$ are respectively

$$(x_0, 0, x_2),\ (x_0, x_1, 0),\ (-p_0, 0, p_2),\ (p_0, -p_1, 0),$$

we easily calculate the cross ratios

$$\{\mathbf{C\,B_p, A\,X_b}\} = 1 + \frac{p_0 x_2}{x_0 p_2}, \qquad \{\mathbf{B\,A, C_p\,X_c}\} = -\frac{p_0 x_1}{x_0 p_1}.$$

Hence the line **p** has the equation

$$\frac{x_0}{p_0} + \frac{x_1}{p_1} + \frac{x_2}{p_2} = 0.$$

Thus every point, and likewise every line, is represented by an ordered set of three real numbers: a point by its coordinates (or "point coordinates") x_0, x_1, x_2, and a line by its *tangential coordinates* (or "line coordinates") X_0, X_1, X_2, which are the coefficients of x_0, x_1, x_2 in its equation. Any three real numbers, not all zero, are the coordinates of a definite point, and of a definite line; but the same point (x_0, x_1, x_2) or line $[X_0, X_1, X_2]$ is represented when the three coordinates are all multiplied by the same non-vanishing number.

If (x_0, x_1, x_2) is the trilinear pole of $[X_0, X_1, X_2]$, we have

$$x_0 X_0 = x_1 X_1 = x_2 X_2.$$

On the other hand, the condition for (x_0, x_1, x_2) and $[X_0, X_1, X_2]$ to be *incident* is

$$x_0X_0+x_1X_1+x_2X_2=0.$$

The expression on the left is conveniently abbreviated to $\{xX\}$, and we say that the condition for the point (x) and line $[X]$ to be incident is $\{xX\}=0$. If we regard $[X]$ as a fixed line and (x) as a variable point, this relation is the equation of the line, being the condition satisfied by all points which lie on it. Dually, if we regard (x) as a fixed point and $[X]$ as a variable line, the same relation is the *tangential equation* of the point, being the condition satisfied by all lines which pass through it. Thus the coordinates of a point are the coefficients of X_0, X_1, X_2 in its tangential equation. In particular, the vertices (1, 0, 0), (0, 1, 0), (0, 0, 1) of the triangle of reference have equations $X_\mu=0$ $(\mu=0, 1, 2)$. Dually, the sides, having equations $x_\mu=0$, are [1, 0, 0], [0, 1, 0], [0, 0, 1].

To find the condition for three points (x), (y), (z) to be collinear, we eliminate X_0, X_1, X_2 between the equations $\{xX\}=0$, $\{yX\}=0$, $\{zX\}=0$, obtaining

$$\begin{vmatrix} x_0 & x_1 & x_2 \\ y_0 & y_1 & y_2 \\ z_0 & z_1 & z_2 \end{vmatrix} = 0.$$

Thus the line joining (y) and (z) is

$$[y_1z_2-y_2z_1,\ y_2z_0-y_0z_2,\ y_0z_1-y_1z_0].$$

Dually, the condition for three lines $[X]$, $[Y]$, $[Z]$ to be concurrent is

$$\begin{vmatrix} X_0 & X_1 & X_2 \\ Y_0 & Y_1 & Y_2 \\ Z_0 & Z_1 & Z_2 \end{vmatrix} = 0,$$

and the point of intersection of $[Y]$ and $[Z]$ is

$$(Y_1Z_2-Y_2Z_1,\ Y_2Z_0-Y_0Z_2,\ Y_0Z_1-Y_1Z_0).$$

It is easily seen that any point collinear with (y) and (z) may be expressed in the form

$$(my_0+nz_0,\ my_1+nz_1,\ my_2+nz_2),$$

or briefly $(my+nz)$, and that its harmonic conjugate with respect to (y) and (z) is $(my-nz)$.

The triangle of reference divides the plane into four regions (§2.6), any one of which may be chosen as the *interior* region by taking the unit point in that region. (The fact that we intuitively designate a particular region of a triangle as interior serves to emphasize the difference between projective and Euclidean geometry.) In the interior region all three coordinates of a point have the same sign, but in the other three regions one coordinate differs in sign from the other two. The same distinction can be made as to the signs of the tangential coordinates of a line. If X_0, X_1, X_2 are all of the same sign, and $\{xX\}=0$, then x_0, x_1, x_2 are certainly not all of the same sign. Hence the lines whose coordinates are all of the same sign are *exterior* to the triangle of reference. The trilinear pole of an exterior line is an interior point.

4.4. Collineations and coordinate transformations. Before considering the general collineation, let us take an important special case: the *harmonic homology* (§3.1) with centre (u) and axis $[U]$. Let (t) be the point where the line joining any point (x) to (u) meets $[U]$. Since (x) is collinear with (t) and (u), we may write

$$x_\mu = mt_\mu + nu_\mu,$$

and the harmonic homologue is (x') where

$$x'_\mu = mt_\mu - nu_\mu = x_\mu - 2nu_\mu.$$

But, since (t) lies on $[U]$,

$$\{xU\} = m\{tU\} + n\{uU\} = n\{uU\}.$$

Hence the harmonic homology is given by

4.41. $$x'_\mu = x_\mu - 2u_\mu\{xU\}/\{uU\}.$$

In terms of homogeneous coordinates, the one-dimensional projectivity 4.25 takes the form

$$\frac{x_1'}{x_0'} = \frac{px_1 + qx_0}{rx_1 + sx_0} \qquad (ps - qr \neq 0)$$

or

$$\begin{cases} \rho x_0' = c_{00}x_0 + c_{01}x_1, \\ \rho x_1' = c_{10}x_0 + c_{11}x_1 \end{cases} \qquad (c_{00}c_{11} - c_{10}c_{01} \neq 0).$$

Analogously in two dimensions,

4.42. *The general collineation is given by the linear transformation*

$$\rho x_\mu' = c_{\mu 0}x_0 + c_{\mu 1}x_1 + c_{\mu 2}x_2 \qquad (\mu = 0, 1, 2),$$

the coefficients being any nine real numbers whose determinant

$$\gamma = \begin{vmatrix} c_{00} & c_{01} & c_{02} \\ c_{10} & c_{11} & c_{12} \\ c_{20} & c_{21} & c_{22} \end{vmatrix}$$

is not zero.

PROOF. Whenever there is no danger of confusion, the factor of proportionality ρ will be omitted or "absorbed." Since $\gamma \neq 0$, we can solve the equations

$$c_{\mu 0}x_0 + c_{\mu 1}x_1 + c_{\mu 2}x_2 = x_\mu' \qquad (\mu = 0, 1, 2),$$

obtaining

$$x_\nu = C_{0\nu}x_0' + C_{1\nu}x_1' + C_{2\nu}x_2' \qquad (\nu = 0, 1, 2),$$

where the numbers $C_{\mu\nu}$ are proportional to the co-factors of $c_{\mu\nu}$ in the determinant γ. Hence the transformation is a point-to-point correspondence; and since any linear function of the x_ν is a linear function of the x_μ', this correspondence is a collineation. Conversely, every collineation is expressible in this form. For, in order to transform the four particular points

4.43. $(1, 0, 0), \ (0, 1, 0), \ (0, 0, 1), \ (1, 1, 1)$

into the four general points

4.44. $$(x_{0\nu}, x_{1\nu}, x_{2\nu}) \qquad (\nu=0, 1, 2, 3),$$

we merely have to take $c_{\mu\nu}=x_{\mu\nu}w_\nu$, where w_0, w_1, w_2 are given by the simultaneous equations

$$x_{\mu 0}w_0+x_{\mu 1}w_1+x_{\mu 2}w_2=x_{\mu 3} \qquad (\mu=0, 1, 2),$$

which have a non-vanishing solution provided no three of the four points 4.44 are collinear. Thus any matrix $(c_{\mu\nu})$ for which $\gamma\neq 0$ determines a collineation, and conversely the collineation determines the matrix (apart from a scalar factor).

The above proof shows also that (x') is the point whose coordinates are x_ν when referred to the triangle and unit point 4.44 instead of 4.43. Thus the transformation plays a double role: it can be regarded either as a collineation which relates one point to another, or as a *transformation of coordinates*, enabling us to use a new triangle of reference and unit point.*

It is convenient to borrow from tensor calculus the convention whereby *any term involving a repeated suffix is understood to be summed for the possible values of that suffix* (in the present case, the values 0, 1, 2). In this notation the transformation 4.42 takes the form

4.45. $$x'_\mu=c_{\mu\nu}x_\nu \qquad (\mu=0, 1, 2)$$

or

4.46. $$x_\nu=C_{\mu\nu}x'_\mu \qquad (\nu=0, 1, 2).$$

The connection between these two equations becomes clearer when we use the "Kronecker delta" $\delta_{\lambda\mu}$, which is equal to 1 or 0 according as the two suffixes are the same or different. In fact, if we define $C_{\mu\nu}$ so that the co-factor of $c_{\mu\nu}$ in γ is precisely $\gamma C_{\mu\nu}$, the theory of determinants shows that

$$c_{\lambda\nu}C_{\mu\nu}=\delta_{\lambda\mu}.$$

Hence 4.46 implies $c_{\lambda\nu}x_\nu=c_{\lambda\nu}C_{\mu\nu}x'_\mu=\delta_{\lambda\mu}x'_\mu=x'_\lambda$, which is 4.45; and similarly 4.45 implies 4.46.

---

*Möbius [1], p. 304.

The corresponding transformation of tangential coordinates must be such that the incidence-condition $\{xX\} = 0$ is equivalent to $\{x'X'\} = 0$. Disregarding a factor of proportionality, we equate these expressions, obtaining

$$x_\nu X_\nu = x'_\mu X'_\mu = c_{\mu\nu} x_\nu X'_\mu.$$

Since this must hold for every point (x), we equate coefficients of x_ν, and find

4.47. $$X_\nu = c_{\mu\nu} X'_\mu \qquad (\nu = 0, 1, 2).$$

Similarly, since $x'_\mu X'_\mu = x_\nu X_\nu = C_{\mu\nu} x'_\mu X_\nu$, we have

4.48. $$X'_\mu = C_{\mu\nu} X_\nu \qquad (\mu = 0, 1, 2).$$

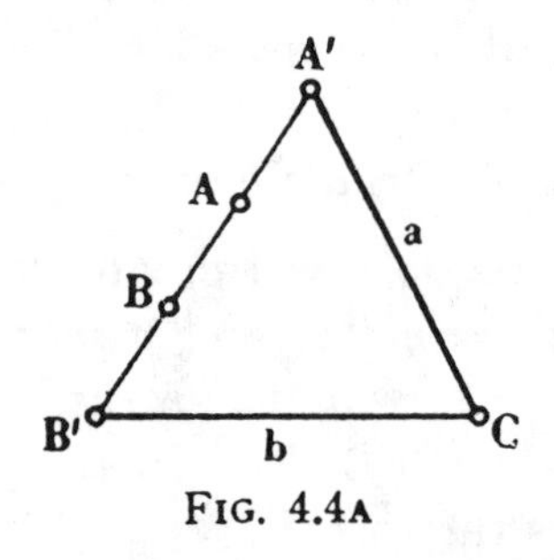

FIG. 4.4A

The advantage of being able to transform the coordinates to any triangle of reference is well illustrated by the following self-dual definition of *cross ratio*. Let **A**, **B**, **a**, **b**, or (x), (y), $[X]$, $[Y]$, be any two points and two lines in one plane. (See Fig. 4.4A.) Defining their cross ratio by the formula*

4.49. $$\{\mathbf{AB}, \mathbf{ab}\} = \frac{\{xX\}\{yY\}}{\{xY\}\{yX\}}$$

we observe that this expression does not depend on the arbitrary factors in the coordinates of the points and lines, nor on the choice of the coordinate system. Accordingly, we may suppose the triangle of reference to be formed by the lines **b**, **a**, **AB**, so that $[X]$ is $[0, 1, 0]$, $[Y]$ is $[1, 0, 0]$, and

*Heffter and Koehler [1], I, pp. 120, 136.

$$\{\mathbf{AB}, \mathbf{ab}\} = x_1 y_0 / x_0 y_1.$$

Denoting by $\mathbf{A}'$ and $\mathbf{B}'$ the points where $\mathbf{a}$ and $\mathbf{b}$ meet $\mathbf{AB}$, we see that this agrees with the value of $\{\mathbf{AB}, \mathbf{A}'\mathbf{B}'\}$ as given by 4.33.

4.5. Polarities. Since the product of any two correlations is a collineation, and the inverse of a correlation is a correlation, we see that every correlation can be regarded as the product of any particular correlation and a suitable collineation. The simplest particular correlation is clearly that which relates each point (x) to the line $[x]$. (Incidentally, this also relates the line $[X]$ to the point (X), and so is a polarity.) We have seen that the general collineation is given by

$$x'_\mu = c_{\mu\nu} x_\nu, \qquad X'_\mu = C_{\mu\nu} X_\nu \qquad (\mu = 0, 1, 2),$$
$$x_\nu = C_{\mu\nu} x'_\mu, \qquad X_\nu = c_{\mu\nu} X'_\mu \qquad (\nu = 0, 1, 2).$$

Hence

4.51. *The general correlation is given by*

$$X'_\mu = c_{\mu\nu} x_\nu, \qquad x'_\mu = C_{\mu\nu} X_\nu \qquad (\mu = 0, 1, 2),$$
$$x_\nu = C_{\mu\nu} X'_\mu, \qquad X_\nu = c_{\mu\nu} x'_\mu \qquad (\nu = 0, 1, 2).$$

Interchanging x and x', X and X', μ and ν, we find that, in the *inverse* correlation, $X'_\mu = c_{\mu\nu} x_\nu$. The correlation is a polarity if it is the same as its inverse, i.e. if $c_{\mu\nu} = m c_{\mu\nu}$. Since this implies $c_{\mu\nu} = m c_{\nu\mu}$, we must have $m^2 = 1$. But $m = -1$ would imply

$$\gamma = \begin{vmatrix} 0 & c_{01} & -c_{20} \\ -c_{01} & 0 & c_{12} \\ c_{20} & -c_{12} & 0 \end{vmatrix} = 0.$$

Therefore $m = 1$, and $c_{\nu\mu} = c_{\mu\nu}$. Moreover, no confusion can now be caused by writing x for x' and X for X'. Hence

4.52. *The general polarity is given by*

$$X_\mu = c_{\mu\nu}, \qquad x_\mu = C_{\mu\nu} X_\nu \qquad (\mu = 0, 1, 2),$$

where $c_{\nu\mu} = c_{\mu\nu}$ *and* $c_{\lambda\nu} C_{\mu\nu} = \delta_{\lambda\mu}$.

This last condition implies that the co-factor of $c_{\mu\nu}$ in the determinant $\gamma = \det(c_{\mu\nu})$ is $\gamma C_{\mu\nu}$, that the co-factor of $C_{\mu\nu}$ in $\Gamma = \det(C_{\mu\nu})$ is $\Gamma c_{\mu\nu}$, and that $\gamma\Gamma = 1$. Clearly, the condition for points (x) and (y) to be conjugate is

$$c_{\mu\nu} x_\mu y_\nu = 0,$$

and the condition for lines $[X]$ and $[Y]$ to be conjugate is

$$C_{\mu\nu} X_\mu Y_\nu = 0.$$

By transforming to a self-polar triangle of reference, we obtain the polarity in a remarkably simple form. In fact, since each pair of vertices is now a pair of conjugate points, we have $c_{12} = c_{20} = c_{01} = 0$, and the polarity is given by

4.53. $$X_0 = c_{00} x_0, \quad X_1 = c_{11} x_1, \quad X_2 = c_{22} x_2,$$

where $c_{00} c_{11} c_{22} \neq 0$. Since the order in which the new coordinates are numbered 0, 1, 2 is arbitrary, and since the polarity is not altered when the coefficients are all multiplied by the same constant (which may be negative), there is no loss of generality in assuming that c_{11} and c_{22} are positive.

A "change of scale," which preserves the triangle of reference while changing the unit point, is achieved by writing x_0/c_0, x_1/c_1, x_2/c_2 for x_0, x_1, x_2, and $c_0 X_0$, $c_1 X_1$, $c_2 X_2$ for X_0, X_1, X_2. The relations 4.53 then become

$$c_0^2 X_0 = c_{00} x_0, \quad c_1^2 X_1 = c_{11} x_1, \quad c_2^2 X_2 = c_{22} x_2.$$

Taking $c_0 = \sqrt{|c_{00}|}$, $c_1 = \sqrt{c_{11}}$, $c_2 = \sqrt{c_{22}}$, we thus reduce the polarity to its *canonical form*

4.54. $$X_0 = \pm x_0, \quad X_1 = x_1, \quad X_2 = x_2.$$

Since this relates the interior point $(1, 1, 1)$ to the exterior or interior line $[\pm 1, 1, 1]$, the polarity is elliptic or hyperbolic according as the upper or lower sign is taken. In 4.52, the condition for the point (x) to be self-conjugate is $c_{\mu\nu} x_\mu x_\nu = 0$; hence the polarity is elliptic or hyperbolic according as $c_{\mu\nu} x_\mu x_\nu$ is a definite or indefinite quadratic form, i.e. according as the three numbers C_{00}, c_{11}, γ are or are not all of the same sign.*

*Veblen and Young [2], II, p. 205.

In the hyperbolic case, the locus of self-conjugate points, and the envelope of self-conjugate lines, is the *conic*

4.55. $$c_{\mu\nu}x_\mu x_\nu=0,\quad C_{\mu\nu}X_\mu X_\nu=0,$$

or, in canonical form,

4.56. $$-x_0^2+x_1^2+x_2^2=0,\quad -X_0^2+X_1^2+X_2^2=0.$$

The line [1, 0, 0] or $x_0=0$ does not meet the conic 4.56. Hence, by considerations of continuity, all lines $[X]$ for which

$$-X_0^2+X_1^2+X_2^2<0$$

are exterior to this conic, while those for which

$$-X_0^2+X_1^2+X_2^2>0$$

are secants. Applying the polarity 4.54, we deduce that the point (x) is interior or exterior according as $-x_0^2+x_1^2+x_2^2$ is negative or positive.

4.6. Coordinates in three dimensions. The above results are easily extended to three-dimensional space.* The point (x) now has a fourth coordinate x_3, and the tangential coordinates $[X]$ or $[X_0,\ X_1,\ X_2,\ X_3]$ represent a *plane*. The condition for the point (x) and plane $[X]$ to be incident is $\{xX\}=0$ or $x_\mu X_\mu=0$, which now means

$$x_0X_0+x_1X_1+x_2X_2+x_3X_3=0.$$

This can be interpreted either as the ordinary equation of a fixed plane $[X]$, or as the tangential equation of a fixed point (x). In particular, the *tetrahedron of reference* has faces $x_\mu=0$ $(\mu=0,\ 1,\ 2,\ 3)$ and vertices $X_\mu=0$.

Since the relations $x_\mu X_\mu=y_\mu X_\mu=0$ imply $(mx_\mu+ny_\mu)X_\mu=0$, every plane through (x) and (y) contains the point

$$(mx_0+ny_0,\ mx_1+ny_1,\ mx_2+ny_2,\ mx_3+ny_3),$$

or briefly $(mx+ny)$. This point is therefore collinear with (x) and (y). The plane through three non-collinear points (x), (y), (z) is $[X]$, where

*Instead of the conic **ABCIX** we now use the twisted cubic **ABCDIX**. Cf. Robinson [1], pp. 104-106; Plücker [1], [2].

$$X_0 = \begin{vmatrix} x_1 & x_2 & x_3 \\ y_1 & y_2 & y_3 \\ z_1 & z_2 & z_3 \end{vmatrix}, \quad X_1 = -\begin{vmatrix} x_0 & x_2 & x_3 \\ y_0 & y_2 & y_3 \\ z_0 & z_2 & z_3 \end{vmatrix},$$

$$X_2 = \begin{vmatrix} x_0 & x_1 & x_3 \\ y_0 & y_1 & y_3 \\ z_0 & z_1 & z_3 \end{vmatrix}, \quad X_3 = -\begin{vmatrix} x_0 & x_1 & x_2 \\ y_0 & y_1 & y_2 \\ z_0 & z_1 & z_2 \end{vmatrix}.$$

The point of intersection of three planes is given by the same formulae with small and capital letters interchanged.

Plücker's coordinates for a *line* are defined by closely analogous formulae:

The line joining points (x) and (y) is $\{p\}$ or

$$\{p_{01}, p_{02}, p_{03}, p_{23}, p_{31}, p_{12}\},$$

where

$$p_{\mu\nu} = \begin{vmatrix} x_\mu & x_\nu \\ y_\mu & y_\nu \end{vmatrix},$$

so that

$$p_{\nu\mu} = -p_{\mu\nu}$$

and $p_{00} = p_{11} = p_{22} = p_{33} = 0$.

The line of intersection of planes $[X]$ and $[Y]$ is

$$\{P_{23}, P_{31}, P_{12}, P_{01}, P_{02}, P_{03}\},$$

where

$$P_{\mu\nu} = \begin{vmatrix} X_\mu & X_\nu \\ Y_\mu & Y_\nu \end{vmatrix},$$

so that

$$P_{\nu\mu} = -P_{\mu\nu}$$

and $P_{00} = P_{11} = P_{22} = P_{33} = 0$.

In order to identify these two sets of six coordinates, we observe that, if both the points (x) and (y) lie in both the planes $[X]$ and $[Y]$, so that $\{xX\} = \{yX\} = \{xY\} = \{yY\} = 0$, then

$$\begin{aligned} p_{\lambda\nu}P_{\mu\nu} &= (x_\lambda y_\nu - y_\lambda x_\nu)(X_\mu Y_\nu - Y_\mu X_\nu) \\ &= x_\lambda X_\mu \{yY\} - x_\lambda Y_\mu \{yX\} - y_\lambda X_\mu \{xY\} + y_\lambda Y_\mu \{xX\} = 0. \end{aligned}$$

If $\lambda \neq \mu$, two of the four terms implied by the expression $p_{\lambda\nu}P_{\mu\nu}$ vanish identically. We thus have twelve relations such as $p_{01}P_{31} + p_{02}P_{32} = 0$, which are equivalent to

$$\frac{P_{23}}{p_{01}} = \frac{P_{31}}{p_{02}} = \frac{P_{12}}{p_{03}} = \frac{P_{01}}{p_{23}} = \frac{P_{02}}{p_{31}} = \frac{P_{03}}{p_{12}}.$$

Since the $p_{\mu\nu}$ and $P_{\mu\nu}$ are *homogeneous* coordinates, we may write simply

4.61.
$$P_{23} = p_{01}, \quad P_{31} = p_{02}, \quad P_{12} = p_{03},$$
$$P_{01} = p_{23}, \quad P_{02} = p_{31}, \quad P_{03} = p_{12}.$$

To find the condition which six numbers p_{01}, p_{02}, p_{03}, p_{23}, p_{31}, p_{12} must satisfy in order to be the coordinates of a line, we put $\lambda = \mu = 0$ in the relation $p_{\lambda\nu}P_{\mu\nu} = 0$, obtaining $p_{0\nu}P_{0\nu} = 0$ or

4.62.
$$p_{01}p_{23} + p_{02}p_{31} + p_{03}p_{12} = 0.$$

To find the condition for two lines $\{p\}$ and $\{p'\}$ to intersect,* let (x) and (y) be two points on the former, $[X]$ and $[Y]$ two planes through the latter, so that

$$\begin{aligned} p_{\mu\nu}P'_{\mu\nu} &= (x_\mu y_\nu - y_\mu x_\nu)(X_\mu Y_\nu - Y_\mu X_\nu) \\ &= 2\{xX\}\{yY\} - 2\{xY\}\{yX\}. \end{aligned}$$

If the lines intersect at (x), we have $\{xX\} = \{xY\} = 0$, and therefore $p_{\mu\nu}P'_{\mu\nu} = 0$. Conversely, if this relation holds, we can choose $[Y]$ so that (x) lies on it and thus $\{xY\} = 0$. Then the relation $p_{\mu\nu}P'_{\mu\nu} = 0$ yields either $\{xX\} = 0$ or $\{yY\} = 0$. In the former case, (x) lies on both lines; in the latter, $[Y]$ contains both lines. Thus the necessary and sufficient condition for $\{p\}$ and $\{p'\}$ to intersect is

4.63.
$$p_{01}p'_{23} + p_{02}p'_{31} + p_{03}p'_{12} + p_{23}p'_{01} + p_{31}p'_{02} + p_{12}p'_{03} = 0.$$

To find conditions for the line $\{p\}$ to lie in the plane $[X]$, let $\{p\}$ be given as the join of points (x) and (y), so that, for any plane $[X]$,

$$p_{\mu\nu}X_\nu = (x_\mu y_\nu - y_\mu x_\nu)X_\nu = x_\mu\{yX\} - y_\mu\{xX\}.$$

Then the four relations

4.64.
$$p_{\mu\nu}X_\nu = 0 \qquad (\mu = 0, 1, 2, 3)$$

hold whenever $[X]$ passes through $\{p\}$. For each value of μ,

*Cartan [1], p. 120.

the relation 4.64, being linear in the X_ν, is the tangential equation of a point, viz. the point $(x_\mu y - y_\mu x)$, collinear with (x) and (y). Since there is no term in X_μ, this is the point in which the line meets the plane $x_\mu = 0$. Thus the four conditions 4.64 are sufficient to make $\{p\}$ lie in $[X]$. The dual argument shows that the four relations

4.65. $$P_{\mu\nu} x_\nu = 0 \qquad (\mu = 0, 1, 2, 3)$$

(which are the equations of the planes joining the line to the points $X_\mu = 0$) are necessary and sufficient conditions for the line $\{p\}$ to pass through the point (x).

4.7. Three-dimensional projectivities. The results of §4.4 continue to hold in three dimensions if we make the obvious changes, replacing lines by planes, triangles by tetrahedra, and extending the range of the suffixes to 0, 1, 2, 3. Thus a linear transformation of coordinates will enable us to take the tetrahedron of reference (and the unit point) in any convenient position. For any two points and two planes $\mathbf{A}, \mathbf{B}, \alpha, \beta$, or $(x), (y), [X], [Y]$, we define the *cross ratio**

4.71. $$\{\mathbf{AB}, \alpha\beta\} = \frac{\{xX\}\{yY\}}{\{xY\}\{yX\}}.$$

The convenient tetrahedron of reference is formed by β, α, and any two planes through $\mathbf{AB}$. Then $[X]$ is $[0, 1, 0, 0]$, $[Y]$ is $[1, 0, 0, 0]$, and

$$\{\mathbf{AB}, \alpha\beta\} = x_1 y_0 / x_0 y_1 = \{\mathbf{AB}, \mathbf{A'B'}\},$$

$\mathbf{A'}$ and $\mathbf{B'}$ being the points where α and β meet $\mathbf{AB}$.

The general *collineation* is given by

$$x'_\mu = c_{\mu\nu} x_\nu, \quad X'_\mu = C_{\mu\nu} X_\nu \qquad (\mu = 0, 1, 2, 3)$$
$$x_\nu = C_{\mu\nu} x'_\mu, \quad X_\nu = c_{\mu\nu} X'_\mu \qquad (\nu = 0, 1, 2, 3).$$

Combining this with the special polarity which relates the point

*Heffter and Koehler [1], II, pp. 76, 95.

(x) to the plane [x], and the plane [X] to the point (X), we obtain the general *correlation*

$$X'_\mu = c_{\mu\nu} x_\nu, \quad x'_\mu = C_{\mu\nu} X_\nu \qquad (\mu = 0, 1, 2, 3),$$
$$x_\nu = C_{\mu\nu} X'_\mu, \quad X_\nu = c_{\mu\nu} x'_\mu \qquad (\nu = 0, 1, 2, 3).$$

The condition for this to be a *polarity* is again $c_{\nu\mu} = mc_{\mu\nu}$, where $m = \pm 1$; but now both values of m are available. Hence

4.72. *The general polarity is given by*

$$X_\mu = c_{\mu\nu} x_\nu, \qquad x_\mu = C_{\mu\nu} X_\nu \qquad (\mu = 0, 1, 2, 3),$$

where $c_{\nu\mu} = \pm c_{\mu\nu}$ *and* $c_{\lambda\nu} C_{\mu\nu} = \delta_{\lambda\mu}$.

The implications of this last condition are the same as in the case of 4.52, save that now the determinants γ and Γ each have four rows and columns instead of three.

When $c_{\nu\mu} = c_{\mu\nu}$, so that the determinants are symmetric, we have an ordinary polarity, which admits non-self-conjugate points. By referring the coordinates to a self-polar tetrahedron,* we can reduce this to the canonical form

$$X_0 = c_0 x_0, \quad X_1 = c_1 x_1, \quad X_2 = c_2 x_2, \quad X_3 = c_3 x_3.$$

(A proper choice of the unit point would enable us to put $c_\nu = \pm 1$; but it is convenient to postpone this last step.) The condition for a point (x) to be self-conjugate is now $c_\nu x_\nu^2 = 0$. Hence the polarity is *uniform*, i.e. of type (6, 0), if the c_ν all have the same sign.

If a line {p} joins points (x) and (y), its polar line is the intersection of the polar planes

$$[c_0 x_0, c_1 x_1, c_2 x_2, c_3 x_3] \quad \text{and} \quad [c_0 y_0, c_1 y_1, c_2 y_2, c_3 y_3],$$

namely

4.73. $\{c_2 c_3 p_{23}, c_3 c_1 p_{31}, c_1 c_2 p_{12}, c_0 c_1 p_{01}, c_0 c_2 p_{02}, c_0 c_3 p_{03}\}.$

In particular, the polar line of {p} in the uniform polarity $X_\nu = x_\nu$ is {P}, or

*Veblen and Young [2], II, p. 263.

4.74. $\{p_{23}, p_{31}, p_{12}, p_{01}, p_{02}, p_{03}\}.$

A line $\{p\}$ is self-polar if it is the same as 4.73, i.e. if

$$\frac{c_2c_3p_{23}}{p_{01}} = \frac{c_3c_1p_{31}}{p_{02}} = \frac{c_1c_2p_{12}}{p_{03}} = \frac{c_0c_1p_{01}}{p_{23}} = \frac{c_0c_2p_{02}}{p_{31}} = \frac{c_0c_3p_{03}}{p_{12}} = m,$$

where $m^2 = c_0c_1c_2c_3$. This cannot happen if one of the c_ν differs in sign from the other three, but it can if two differ in sign from the other two. Hence, by 3.83, these distributions of sign give polarities of types (3, 3) and (2, 4), respectively. In particular, the quadric

$$-x_0^2 \pm x_1^2 + x_2^2 + x_3^2 = 0, \quad -X_0^2 \pm X_1^2 + X_2^2 + X_3^2 = 0$$

is *oval* or *ruled* according as the upper or lower sign is taken.

On the other hand, when $c_{\nu\mu} = -c_{\mu\nu}$, so that the determinants are skew-symmetric, we have a *null* polarity. For, since

$$c_{\mu\nu}x_\mu x_\nu = c_{\nu\mu}x_\nu x_\mu = \tfrac{1}{2}(c_{\mu\nu} + c_{\nu\mu})x_\mu x_\nu = 0,$$

every point in space is self-conjugate. Two distinct points, (x) and (y), are conjugate if $c_{\mu\nu}x_\mu y_\nu = 0$. Since this relation is equivalent to $c_{\mu\nu}x_\mu y_\nu + c_{\nu\mu}x_\nu y_\mu = 0$ or $c_{\mu\nu}(x_\mu y_\nu - x_\nu y_\mu) = 0$, the condition for a line $\{p\}$ to join two conjugate points is

4.75. $$c_{\mu\nu}p_{\mu\nu} = 0.$$

But the join of two conjugate points, being the line of intersection of their polar planes, is self-polar. Hence 4.75 is the equation of a *linear complex*.

When $c_{\nu\mu} = -c_{\mu\nu}$, we easily verify that

$$\gamma = (c_{01}c_{23} + c_{02}c_{31} + c_{03}c_{12})^2.$$

Thus if $c_{01}c_{23} + c_{02}c_{31} + c_{03}c_{12} = 0$, the lines which satisfy 4.75 are not the self-polar lines in a polarity. Instead, they are all the lines that meet the particular line $\{c_{23}, c_{31}, c_{12}, c_{01}, c_{02}, c_{03}\}$. They form what is called a *special* linear complex.

The common lines of two linear complexes, say $b_{\mu\nu}p_{\mu\nu} = 0$ and $c_{\mu\nu}p_{\mu\nu} = 0$, are said to form a *linear congruence*. Clearly, they belong to the linear complex $(nb_{\mu\nu} + c_{\mu\nu})p_{\mu\nu} = 0$, for all values of n.

This is a special linear complex if n satisfies the quadratic equation

$$(b_{01}n+c_{01})(b_{23}n+c_{23})+(b_{02}n+c_{02})(b_{31}n+c_{31}) \\ +(b_{03}n+c_{03})(b_{12}n+c_{12})=0.$$

If the roots are real and distinct, they provide two lines

$$\{nb_{23}+c_{23},\ nb_{31}+c_{31},\ nb_{12}+c_{12}, \\ nb_{01}+c_{01},\ nb_{02}+c_{02},\ nb_{03}+c_{03}\},$$

called *directrices*, and the congruence consists of all lines which meet these two. If the directrices are skew, the congruence is said to be *hyperbolic*, in contrast with the *elliptic* congruence* for which the above quadratic equation has no real root. By solving the equations $b_{\mu\nu}p_{\mu\nu}=c_{\mu\nu}p_{\mu\nu}=0$, along with three of the equations 4.65, we obtain a definite line of the congruence through a given point (x) of general position. For a hyperbolic congruence, this line is simply the transversal from (x) to the two directrices, and so is no longer definite when (x) lies on either of the directrices. But an elliptic congruence has no directrices, and in that case the line is definite for *every* point (x).

4.8. Line coordinates for the generators of a quadric. Consider the quadric $c_\nu x_\nu^2=0$. If $c_0c_1c_2c_3>0$, we can simplify 4.73 by the following device, due to P. W. Wood. Multiplying the c_ν by the constant $(c_0/c_1c_2c_3)^{\frac{1}{2}}$, and changing the notation so as to call the new coefficients c_ν again, we obtain the same quadric in a "normalized" form, with $c_0=c_1c_2c_3$. Then the polar line of $\{p\}$ is

$$\{p_{23}/c_1,\ p_{31}/c_2,\ p_{12}/c_3,\ c_1p_{01},\ c_2p_{02},\ c_3p_{03}\}.$$

If one of c_1, c_2, c_3 differs in sign from the other two, the quadric is ruled. If, further,

*Veblen and Young [2], I, p. 315.

4.81. $$p_{23}=c_1p_{01},\quad p_{31}=c_2p_{02},\quad p_{12}=c_3p_{03},$$

the line $\{p\}$ is self-polar. Thus a typical generator of one regulus of the quadric

$$c_1c_2c_3x_0^2+c_1x_1^2+c_2x_2^2+c_3x_3^2=0$$

is

$$\{p_1,\ p_2,\ p_3,\ c_1p_1,\ c_2p_2,\ c_3p_3\},$$

where, by 4.62,

$$c_1p_1^2+c_2p_2^2+c_3p_3^2=0.$$

A typical generator of the other regulus is obtained by reversing the signs of c_1, c_2, c_3.

4.9. Complex projective geometry. An alternative procedure (of great theoretical importance, since it identifies the consistency of the axioms with the consistency of the real number field) is to *define* a point as an ordered set of four real numbers, a line as an ordered set of six real numbers satisfying 4.62, and so on. (See p. 25.) Then every axiom about points and lines is replaced by a theorem about numbers, and most of these theorems are quite easy to prove.

Two possible generalizations immediately present themselves. First, instead of *four* numbers, we may take $n+1$, so as to obtain n-dimensional geometry.* Secondly, instead of *real* numbers, we may take elements of any field, e.g. the field of *complex* numbers.† Defining a projectivity as a linear transformation, we find, in this last case, that every polarity has self-conjugate points, and that every quadric has generators.

Axiomatically, complex geometry is derived from real geometry by denying 2.124, while retaining all the other axioms and adding an "axiom of closure" (to rule out more extensive fields).‡

---

*Schoute [1], pp. 187-219; Sommerville [3], pp. 59-72.

†Cartan [1].

‡Veblen and Young [2], II, p. 30: "Assumption I." See also Young [1].

CHAPTER V

ELLIPTIC GEOMETRY IN ONE DIMENSION

5.1. Elliptic geometry in general. As we saw in §1.7, Klein's elliptic geometry is a metrical geometry in which two coplanar lines always have a single point of intersection. One method of approaching this geometry is to introduce an undefined relation of congruence, satisfying certain axioms such as the following:

5.11. *From any point* **D** *on a given line, we can lay off two segments,* **CD** *and* **DE**, *each congruent to a given segment* **AB**.

We can then develop a chain of propositions similar to Euclid I, 1-15, and conclude* that all lines are finite and equal. A line being finite, any two points determine two "supplementary" segments. If these are equal, each is a "right" segment, and the two points are said to be "conjugate." All the lines perpendicular to a given plane are found to concur at a definite point, conjugate to every point in the plane. Conversely, the locus of points conjugate to a given point is a plane. There is thus a definite correspondence between points and planes, of the kind that was called a *uniform polarity* in §3.8.

This brings us to the alternative treatment which is followed in the present book. Observing that every axiom of real projective geometry is valid in elliptic geometry, we simply adopt the axioms of §2.1, and agree to specialize a uniform polarity. This *absolute polarity*, having once been arbitrarily chosen, is kept fixed, and a *congruent transformation* is defined as a collineation which transforms this polarity into itself. All the axioms of congruence then become theorems which can be

*Sommerville [2], p. 88.

proved. The gain in simplicity is considerable, especially as those axioms are more complicated than in Euclidean geometry (where a segment is uniquely determined by its two ends). Moreover, all the theorems of real projective geometry (Chapters II, III, IV) necessarily hold also in elliptic geometry, and form a substantial basis for the deduction of further theorems. (Another name for "congruent transformation" is *isometry*.)

If we wish to develop elliptic geometry in two dimensions only, we begin with the real projective plane, and specialize an elliptic polarity (§3.2). This will be done in Chapter VI; but in the present chapter we are still less ambitious, being content to consider the geometry of a single line. Instead of an absolute polarity we now have an *absolute involution*. This is initially any elliptic involution in the real projective line; but when once chosen it is retained throughout the whole discussion. By definition, its pairs are *conjugate points*, which divide the line into *right segments*.

Analogously, we can derive Euclidean from "affine" geometry* by specializing an elliptic polarity (or involution) in the plane at infinity (or line at infinity). In fact, the geometry of points and lines in the plane at infinity of affine space is projective, whereas the geometry of points and lines in the plane at infinity of Euclidean space is elliptic.

5.2. Models. In §1.7 we illustrated two-dimensional elliptic geometry by a model consisting of a bundle of lines and planes in Euclidean space. The analogous model for one-dimensional elliptic geometry consists of a flat pencil in the Euclidean plane, conjugate points being represented by perpendicular lines. But in this one-dimensional case, exceptionally, another simpler model is available. In fact, the geometry of the elliptic line is identical with the geometry of an ordinary circle, points being "conjugate" when they are diametrically opposite.

*Robinson [1], Chapter IV.

The existence of this model may be regarded as a consequence of the possibility of defining projectivities on a conic (§3.4). For, a circle is a conic, and the lines joining pairs of any involution on a conic are concurrent, like the diameters of a circle. There is no analogous theory for a quadric, and consequently no one-to-one representation of the elliptic plane on a sphere.

To emphasize the abstract nature of the elliptic geometry itself, we may place the two models side by side, as in the following dictionary:

The elliptic line	The Euclidean plane in the neighbourhood of a fixed point **O**	The circumference of a circle
Point	Line through **O**	Point on the circle
Segment	Angle	Arc
Supplementary segments	Supplementary angles	Major and minor arcs
Right segment	Right angle	Semicircle
Translation	Rotation about **O**	Rotation about the centre
Reflection in a point	Reflection in a line through **O**	Reflection in a diameter

It must be clearly understood that models are not part of the logical development of the subject, but merely suggestive aids, like diagrams. (Both models are used in Fig. 5.3A, the second alone in Figs. 5.4A and 5.4B.) There is no longer any question of consistency; for, since no fresh assumptions have been made since §2.1, elliptic geometry is as consistent as real projective geometry.

5.3. Reflections and translations. We saw, in 2.99, that every projectivity permutable with an elliptic involution Ω (other than Ω itself) is either elliptic but not involutory or else a hyperbolic involution. When Ω is the absolute involution (as we shall suppose from now on), we call this hyperbolic involution a *reflection*, and the elliptic projectivity a *trans-*

lation. The same theorem tells us that, for any two points **A** and **B**, there is a unique reflection ${}_{A}\Phi_{B}$, and a unique translation ${}_{A}\Psi_{B}$, each taking **A** to **B**. In the special case when **A** and **B** are conjugate, ${}_{A}\Psi_{B} = \Omega$.

The remarkable power of Theorem 2.99 is apparent in the ease with which the following important results can be deduced. The product of two projectivities, each permutable with Ω, is itself permutable with Ω. (In detail, $\Omega\Theta\Theta' = \Theta\Omega\Theta' = \Theta\Theta'\Omega$.) Hence, by 2.96, *the product of two reflections, or of two translations, is a translation; but the product of a reflection and a translation is a reflection*. In symbols, for any three points **A**, **B**, **C** on the line, we have

$$ {}_{A}\Phi_{B}\,{}_{B}\Phi_{C} = {}_{A}\Psi_{C} = {}_{A}\Psi_{B}\,{}_{B}\Psi_{C}, $$
$$ {}_{A}\Phi_{B}\,{}_{B}\Psi_{C} = {}_{A}\Phi_{C} = {}_{A}\Psi_{B}\,{}_{B}\Phi_{C}. $$

It is therefore natural to extend the notation so as to include

$$ {}_{A}\Psi_{A} = 1, $$

in agreement with the obvious fact that ${}_{A}\Psi_{B}{}^{-1} = {}_{B}\Psi_{A}$. But ${}_{A}\Phi_{A}$ is the involution whose double points are **A** and **A**Ω; we call this the *reflection in* **A** (or in **A**Ω). Since ${}_{A}\Phi_{B}$ *interchanges* **A** and **B**, it is the same as ${}_{B}\Phi_{A}$.

If Φ is any reflection, and Ψ any translation, we have $\Psi = \Phi \cdot \Phi\Psi$, where $\Phi\Psi$ is a reflection. Hence

5.31. *Every translation is the product of two reflections, one of which may be chosen arbitrarily.*

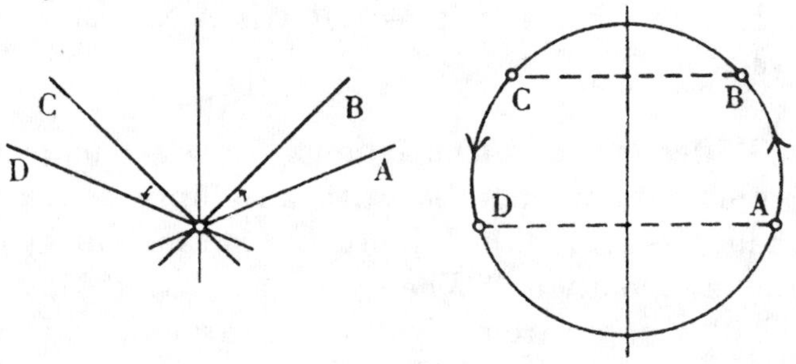

FIG. 5.3A

Again, since $\Phi\Psi$, being a reflection, is of period two, we have

5.32. $$\Psi\Phi\Psi=\Phi.$$

Thus, if $\mathbf{A}\Psi=\mathbf{B}$ and $\mathbf{B}\Phi=\mathbf{C}$, then $\mathbf{A}\Phi=\mathbf{A}\Psi\Phi\Psi=\mathbf{B}\Phi\Psi=\mathbf{C}\Psi$. Calling this last point **D**, as in Fig. 5.3A, we deduce the following theorem (which justifies the definition that we shall adopt for congruent segments):

5.33. *The relations* ${}_{\mathrm{A}}\Psi_{\mathrm{B}}={}_{\mathrm{C}}\Psi_{\mathrm{D}}$ *and* ${}_{\mathrm{B}}\Phi_{\mathrm{C}}={}_{\mathrm{A}}\Phi_{\mathrm{D}}$ *are equivalent.*

Since the second relation is symmetrical between **B** and **C**, another equivalent relation is ${}_{\mathrm{A}}\Psi_{\mathrm{C}}={}_{\mathrm{B}}\Psi_{\mathrm{D}}$.

Our use of the words "reflection" and "translation" is explained by the fact that the above theorems hold also in Euclidean geometry; but the method of proof is quite different, since the Euclidean line admits no absolute involution, no conjugate points. In Euclidean geometry, any two translations are permutable, but no two reflections are permutable (in one dimension). We now investigate the corresponding results in elliptic geometry.

Let Ψ and Ψ' be two translations, and **A** any point. Define $\mathbf{B}=\mathbf{A}\Psi$, $\mathbf{C}=\mathbf{A}\Psi'$, $\mathbf{D}=\mathbf{C}\Psi$. Then, since the relation ${}_{\mathrm{A}}\Psi_{\mathrm{B}}={}_{\mathrm{C}}\Psi_{\mathrm{D}}$ $(=\Psi)$ is equivalent to ${}_{\mathrm{B}}\Psi_{\mathrm{D}}={}_{\mathrm{A}}\Psi_{\mathrm{C}}(=\Psi')$, we have $\mathbf{B}\Psi'=\mathbf{D}$. Thus
$$\mathbf{A}\Psi\Psi'=\mathbf{D}=\mathbf{A}\Psi'\Psi,$$
and therefore $\Psi\Psi'={}_{\mathrm{A}}\Psi_{\mathrm{D}}=\Psi'\Psi$. Hence

5.34. *Any two translations are permutable.*

On the other hand, if reflections ${}_{\mathrm{A}}\Phi_{\mathrm{B}}$ and ${}_{\mathrm{B}}\Phi_{\mathrm{C}}$ are permutable, we have
$${}_{\mathrm{A}}\Psi_{\mathrm{C}}={}_{\mathrm{A}}\Phi_{\mathrm{B}}\ {}_{\mathrm{B}}\Phi_{\mathrm{C}}={}_{\mathrm{B}}\Phi_{\mathrm{C}}\ {}_{\mathrm{A}}\Phi_{\mathrm{B}}={}_{\mathrm{C}}\Phi_{\mathrm{B}}\ {}_{\mathrm{B}}\Phi_{\mathrm{A}}={}_{\mathrm{C}}\Psi_{\mathrm{A}}.$$
By 5.33 (with **C** for **B**, and **A** for **D**), this implies ${}_{\mathrm{C}}\Phi_{\mathrm{C}}={}_{\mathrm{A}}\Phi_{\mathrm{A}}$. Thus, if **A** and **C** are distinct, they are conjugate, and ${}_{\mathrm{A}}\Psi_{\mathrm{C}}=\Omega$. Hence

5.35. *If two reflections are permutable, their product is the absolute involution.*

5.4. Congruence. Since we are considering a single line, we are free to make a convention whereby one of the two sense-classes is distinguished as "positive" or "left to right." If the triad **ABC** belongs to the positive sense-class, the segment **AB/C** is denoted by **AB**, and the supplementary segment by **BA**. When **A** and **B** are conjugate, **AB** and **BA** are *right* segments.

Two segments, **AB** and **CD**, are said to be *congruent* if ${}_{A}\Psi_{B} = {}_{C}\Psi_{D}$ (or ${}_{B}\Phi_{C} = {}_{A}\Phi_{D}$, or ${}_{A}\Psi_{C} = {}_{B}\Psi_{D}$); and we write **AB≡CD**. We see at once that this implies **AC≡BD**.

The relation of congruence is reflexive, symmetric, and transitive. It is also "additive," in the following sense. If **AB≡DE** and **BC≡EF**, as in Fig. 5.4A, we have ${}_{A}\Psi_{D} = {}_{B}\Psi_{E} = {}_{C}\Psi_{F}$, and therefore **AC≡DF**.

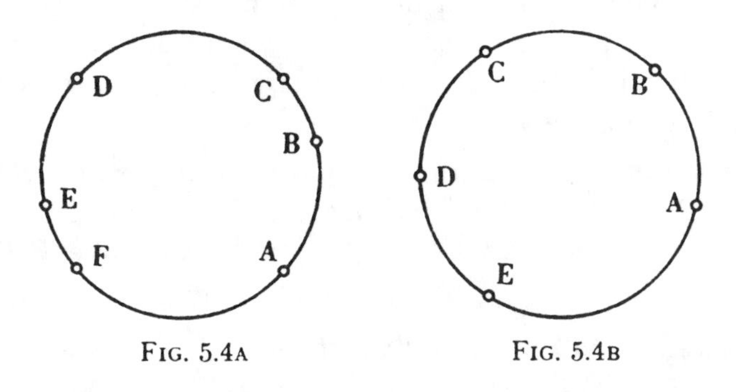

FIG. 5.4A FIG. 5.4B

To construct the segments **CD** and **DE** of 5.11, we merely have to apply ${}_{B}\Psi_{D}$ and ${}_{A}\Psi_{D}$, in turn. (See Fig. 5.4B.)

If **AB** and **CD** are right segments, ${}_{A}\Psi_{B} = \Omega = {}_{C}\Psi_{D}$. Thus *all right segments are congruent.*

We define the *mid-points* of the segments **AB** and **BA** as the double points of the hyperbolic involution ${}_{A}\Phi_{B}$. Thus ${}_{A}\Phi_{B}$ may be described as the reflection in the mid-point of **AB** (or of **BA**). Since the two double points are a pair in Ω, *the mid-points of supplementary segments are conjugate.*

Let **O** be the mid-point of **AB**. Then ${}_{A}\Phi_{B} = {}_{O}\Phi_{O}$, and therefore

$$\mathbf{AO} \equiv \mathbf{OB}.$$

The translation ${}_{A}\Psi_{B}$ is the product of reflections in **A** and **O**; for,

$${}_{A}\Phi_{A}\ {}_{O}\Phi_{O} = {}_{A}\Phi_{A}\ {}_{A}\Phi_{B} = {}_{A}\Psi_{B}.$$

5.5. Continuous translation. The following considerations enable us to define any *power* of a translation. For any $n+1$ points **A, B, C, . . . , G, H**, we have

$${}_{A}\Psi_{H} = {}_{A}\Psi_{B}\ {}_{B}\Psi_{C} \cdots {}_{G}\Psi_{H}\,.$$

In particular, if $\mathbf{AB} \equiv \mathbf{BC} \equiv \ldots \equiv \mathbf{GH}$, so that ${}_{A}\Psi_{B} = {}_{B}\Psi_{C} = \ldots = {}_{G}\Psi_{H} = \Psi$, say, we obtain

$${}_{A}\Psi_{H} = \Psi^{n}.$$

This notion of a power of Ψ can be extended to fractional values of n, as follows. If **O** is the mid-point of **AB**, we write

$${}_{A}\Psi_{O} = \Psi^{1/2},$$

observing that $(\Psi^{1/2})^2 = {}_{A}\Psi_{O}\ {}_{O}\Psi_{B} = {}_{A}\Psi_{B} = \Psi$. By repeated bisection* we define $\Psi^{1/4}$, $\Psi^{1/8}$, . . . , and deduce Ψ^{n} where n is any terminating "decimal" in the scale of 2, e.g.

$$\Psi^{0\cdot 1011} = \Psi^{1/2}\Psi^{1/8}\Psi^{1/16}.$$

Now, if n is any number between 0 and 1, the 1's in its binary expansion determine a sequence of points $\mathbf{A}_r$; e.g., $n = 0.1011\ldots$ gives

$$\mathbf{A}_1 = \mathbf{A}\Psi^{1/2},\quad \mathbf{A}_2 = \mathbf{A}_1\Psi^{1/8},\quad \mathbf{A}_3 = \mathbf{A}_2\Psi^{1/16}, \ldots;$$

and we can apply axiom 2.13 to the segment **AB**, taking as one set all points which lie in at least one segment $\mathbf{AA}_r$, and as the other the remaining points of **AB**. If **M** is the point determined by this "Dedekind cut," we write

$$\Psi^{n} = {}_{A}\Psi_{M}.$$

---

*Veblen and Young [2], II, pp. 151-154; Enriques [1], p. 393.

The extension to values of n which are negative or greater than 1 presents no difficulty.

We proceed to prove that, conversely, any translation can be expressed as a power of any other. In particular,

5.51. *Every translation is a power of* Ω.

PROOF. We shall see that, for any segment **AB**, there is a number n, between 0 and 2, such that ${}_{\mathbf{A}}\Psi_{\mathbf{B}} = \Omega^n$. In fact, the following process leads to a binary expansion

$$n = a_0.a_1a_2a_3 \ldots = a_0 + \frac{a_1}{2} + \frac{a_2}{2^2} + \frac{a_3}{2^3} + \ldots .$$

We put $a_0 = 0$ or 1 according as **B** lies in the segment **AA′** or **A′A** (where $\mathbf{A}' = \mathbf{A}\Omega$). We then bisect the segment thus selected, and put $a_1 = 0$ or 1 according as **B** lies in the left or right half. And so on. If at any stage a point of bisection coincides with **B**, we write $a_r = 1$, and the "decimal" terminates. In any case, a definite number n is obtained; and this power of Ω, interpreted as above, is precisely ${}_{\mathbf{A}}\Psi_{\mathbf{B}}$. We can include the case when **B** coincides with **A** by defining $\Omega^0 = 1$.

To express any translation as a power of any other, we merely have to observe that $\Omega^m = (\Omega^n)^{m/n}$.

As defined in §5.3, a translation is a "sudden" transformation, relating every point to another point. But since the translation Ω^n can be obtained by a gradual increase of the exponent of Ω from 0 to n, there is another aspect of it as a "deformation," gradually displacing every point to a new position. According to the first aspect, we have $\Omega^2 = 1$; but according to the second, we can distinguish Ω^2 as a displacement over the whole line in the positive sense. Thus any segment **AB** determines an infinity of translations Ω^{n+2r}, r being an arbitrary integer (positive, zero, or negative). We can select a definite one of these by making the exponent lie between 0 and 2.

A reflection, on the other hand, is essentially "sudden."

In the language of the theory of groups, the class of translations is a continuous abelian group, and is a subgroup of index two (by 5.31) in the non-abelian group of translations and reflections.

5.6. The length of a segment. To define the length of a segment, we seek a property which is invariant under congruent transformation, and additive for juxtaposed segments, i.e.

$$\mathbf{AB}+\mathbf{BC}=\mathbf{AC}$$

whenever **B** lies in the segment **AC**. Since ${}_{\mathrm{A}}\Psi_{\mathrm{B}}\,{}_{\mathrm{B}}\Psi_{\mathrm{C}}={}_{\mathrm{A}}\Psi_{\mathrm{C}}$, we consider the power to which the translation Ω must be raised in order to transform **A** into **B**. To be precise, if ${}_{\mathrm{A}}\Psi_{\mathrm{B}}=\Omega^{n}$ $(0<n<2)$, the segment **AB** is said to be of length $n\lambda$, where λ is an arbitrary constant depending on the *unit* of length; and we write

$$\mathbf{AB}=n\lambda.$$

(No confusion need result from using the same symbol **AB** for a segment and its length. The context always shows which meaning is intended.) It follows immediately that two segments are congruent if and only if they have the same length.

If **A** and **B** are conjugate, we have $\mathbf{AB}=\lambda$, and the length of the whole line is obtained by juxtaposing two such segments:

$$\mathbf{AB}+\mathbf{BA}=2\lambda.$$

Of course $\mathbf{AB}+\mathbf{BA}$ is the whole line even if **A** and **B** are not conjugate, in which case the shorter segment is said to be *acute*, and the longer *obtuse*. The length of the acute segment is called the *distance* between **A** and **B**.

The choice of a unit has not the same kind of arbitrariness here that it has in Euclidean geometry, since the distance λ plays a special role. In fact, we shall find that a judicious choice of λ greatly simplifies the appearance of certain formulae.

5.7. Distance in terms of cross ratio. We saw, in §4.1, that the points of the line can be represented continuously by the real numbers and ∞, these being the values of a non-homogeneous coordinate or *abscissa*, x. By 4.27, the simplest form for an elliptic involution is

5.71. $$xx'+1=0.$$

Let the fundamental points $\mathbf{P}_0$, $\mathbf{P}_1$, $\mathbf{P}_\infty$ be so chosen that this is the absolute involution Ω.

Then the abscissae of a typical pair of conjugate points are t and $-t^{-1}$. By 4.28, the involution which has these for double points is

5.72. $$xx'-\tfrac{1}{2}(t-t^{-1})(x+x')-1=0.$$

This, therefore, is the general *reflection*; but, by 4.29, the reflection in $\mathbf{P}_0$ is

$$x+x'=0.$$

Combining these two reflections, by putting $-x$ for x in 5.72, we obtain the general *translation* in the form

5.73. $$xx'-\tfrac{1}{2}(t-t^{-1})(x-x')+1=0$$

or

$$(x'-x)/(1+xx')=2t/(1-t^2).$$

This can be simplified by a trigonometrical substitution. In fact, putting $x=\tan\,\xi$, $x'=\tan\,\xi'$, $t=\tan\,\tfrac{1}{2}\theta$, we obtain $\tan\,(\xi'-\xi)=\tan\,\theta$, whence

5.74. $$\xi'\equiv\xi+\theta\ (\text{mod}\ \pi).$$

It need hardly be pointed out that the tangent function, though commonly defined in terms of Euclidean geometry, can just as well be defined analytically, say by means of the differential equation $\dfrac{dy}{dx}=1+y^2$ (with $y=0$ when $x=0$).

For any number ξ, with $0\leqq\xi<\pi$, there is a definite point (ξ) whose abscissa is $\tan\,\xi$. As ξ increases steadily from 0 through $\tfrac{1}{2}\pi$ to π, $\tan\,\xi$ increases from 0 to ∞ and then through negative values back to 0. Meanwhile, the point (ξ) describes the line by means of the continuous translation Ω^n, where n increases steadily from 0 to 2. If Ω^n and Ω^m translate the origin (0) to (ξ) and (θ) respectively, the combination Ω^{n+m} translates (0) to $(\xi+\theta)$, in accordance with 5.74. Hence ξ and n increase proportionately, and Ω^n translates (0) to $(\frac{1}{2}\pi n)$. In other words, the length of the segment from (0) to (ξ), or of the equal segment from (θ) to $(\xi+\theta)$, is proportional to ξ; in the notation of §5.6, it is precisely $2\xi\lambda/\pi$.

In terms of θ, the translation 5.73 takes the more familiar form

$$xx'+(x-x')\cot\theta+1=0$$

or

5.75.
$$x'=\frac{x\cos\theta+\sin\theta}{-x\sin\theta+\cos\theta}.$$

To obtain a formula for length which is independent of the choice of fundamental points, we calculate the cross ratio that the ends of a given segment make with their respective conjugates. Consider a segment **AB**, whose ends are (ξ) and (η). The conjugate points **A′** and **B′** are $(\xi\pm\frac{1}{2}\pi)$ and $(\eta\pm\frac{1}{2}\pi)$. The abscissae of these four points are $\tan\xi$, $\tan\eta$, $-\cot\xi$, $-\cot\eta$. Hence, by 4.31,

$$\{\mathbf{AA'},\ \mathbf{BB'}\}=\frac{(\tan\xi-\tan\eta)(-\cot\xi+\cot\eta)}{(\tan\xi+\cot\eta)(-\cot\xi-\tan\eta)}=$$
$$-\frac{(\tan\eta-\tan\xi)^2}{(1+\tan\eta\tan\xi)^2}=-\tan^2(\eta-\xi).$$

In other words, if **AB** is a segment of length $2\theta\lambda/\pi$, we have

$$\{\mathbf{AA'},\ \mathbf{BB'}\}=-\tan^2\theta,$$

and therefore, by 4.32,

$$\{\mathbf{AA'},\ \mathbf{B'B}\}=-\cot^2\theta,\qquad \{\mathbf{AB'},\ \mathbf{A'B}\}=\operatorname{cosec}^2\theta,$$
$$\{\mathbf{AB'},\ \mathbf{BA'}\}=\sin^2\theta,\qquad \{\mathbf{AB},\ \mathbf{B'A'}\}=\cos^2\theta,$$
$$\{\mathbf{AB},\ \mathbf{A'B'}\}=\sec^2\theta.$$

Selecting a convenient one of these cross ratios, we express the length itself as

$$\text{5.76.} \qquad \mathbf{AB} = \frac{2\lambda}{\pi} \text{ arc cos } (\pm\sqrt{\{\mathbf{AB},\ \mathbf{B'A'}\}}),$$

where the ambiguous sign is determined by the sense of the triad **ABA′**.

Our measure of length agrees with the radian measure of angle or arc (§5.2) if we take $\lambda = \frac{1}{2}\pi$ in the first model, and $\lambda = \pi$ in the second. Then the length of the whole line is π or 2π, respectively.

5.8. Alternative treatment using the complex line. Many of the above considerations are formally simplified if we regard the real projective line as a subspace of the *complex* projective line (§4.9). Then the involution Ω has two conjugate imaginary double points, say **M** and **N**. A projectivity permutable with Ω must either interchange **M** and **N**, in which case it is an involution, or leave both **M** and **N** invariant, in which case it cannot be an involution unless it is Ω itself. Hence the reflection ${}_{\mathbf{A}}\Phi_{\mathbf{B}}$ is the involution (**AB**)(**MN**), and the translation ${}_{\mathbf{A}}\Psi_{\mathbf{B}}$ is the projectivity in which $\mathbf{A\,M\,N} \barwedge \mathbf{B\,M\,N}$.

It is thus possible to define the elliptic line as the complex projective line with two conjugate imaginary *absolute points* specialized, or rather as the real part of such a complex line. From this aspect, *conjugate* points are harmonic conjugates with respect to **M** and **N**, and two segments **AB** and **CD** are *congruent* if

$$\mathbf{A\,B\,M\,N} \barwedge \mathbf{C\,D\,M\,N}.$$

When the elliptic line is regarded as the line at infinity of the Euclidean plane, **M** and **N** are recognized as Poncelet's circular points at infinity. For, as Chasles observed in 1850, two lines in the Euclidean plane are perpendicular if (and only if) their points at infinity are harmonic conjugates with respect to the circular points.* Extending this result, Laguerre showed in 1853 that *the angle between any two lines is proportional to the logarithm of the cross ratio which their points at infinity form with the circular points.*

The recognition of this as a theorem in one-dimensional elliptic geometry is due to Klein.† He saw that it can be derived from the relation

$$\{\mathbf{AB}, \mathbf{MN}\}\ \{\mathbf{BC}, \mathbf{MN}\} = \{\mathbf{AC}, \mathbf{MN}\},$$

which holds for any five collinear points. (See 4.34.) If **M** and **N** are the absolute points, the cross ratio $\{\mathbf{AB}, \mathbf{MN}\}$ is the same for **AB** as for any congruent segment; it is therefore a function of the length **AB**, say $f(\mathbf{AB})$, and satisfies the functional equation

$$f(\mathbf{AB}) \cdot f(\mathbf{BC}) = f(\mathbf{AB}+\mathbf{BC})$$

or $f(u) \cdot f(v) = f(u+v)$. Since $f(v)$ tends to $\{\mathbf{BB}, \mathbf{MN}\} = 1$ as v tends to zero, $f(u)$ is continuous. Hence

$$f(u) = e^{\kappa u},$$

for some constant κ. But $f(u)$ must be periodic, with period 2λ; therefore

5.81. $$\kappa = \pi i/\lambda.$$

Representing each point of the complex line by a complex abscissa x, we observe that **M** and **N**, being the double points of the absolute involution 5.71, have abscissae $\pm i$. Let x and y be the abscissae of **A** and **B**. Then

5.82. $$e^{\kappa \mathbf{AB}} = \{\mathbf{AB}, \mathbf{MN}\} = \frac{(x-i)\,(y+i)}{(x+i)\,(y-i)}.$$

Formula 5.76 may be verified (in the case when $\mathbf{AB} \leqq \lambda$) as follows:

---

*Poncelet [1], I, p. 48; Chasles [1], p. 425; Laguerre [1], p. 64.
†Klein [3], pp. 146, 164.

$$
\begin{aligned}
\cos(\pi \mathbf{AB}/2\lambda) &= \cos(\kappa \mathbf{AB}/2i) \\
&= \cosh(\kappa \mathbf{AB}/2) \\
&= \tfrac{1}{2}\left(e^{\kappa \mathbf{AB}/2} + e^{-\kappa \mathbf{AB}/2}\right) \\
&= \tfrac{1}{2}\left(\sqrt{\frac{(x-i)(y+i)}{(x+i)(y-i)}} + \sqrt{\frac{(x+i)(y-i)}{(x-i)(y+i)}}\right) \\
&= \frac{(x-i)(y+i)+(x+i)(y-i)}{2\sqrt{(x+i)}\sqrt{(x-i)}\sqrt{(y+i)}\sqrt{(y-i)}} \\
&= \frac{xy+1}{\sqrt{(x^2+1)}\sqrt{(y^2+1)}} \\
&= \sqrt{\frac{(x+y^{-1})(y+x^{-1})}{(x+x^{-1})(y+y^{-1})}} \\
&= \sqrt{\{\mathbf{AB}, \mathbf{B'A'}\}}.
\end{aligned}
$$

CHAPTER VI

ELLIPTIC GEOMETRY IN TWO DIMENSIONS

6.1. Spherical and elliptic geometry. As we saw in §1.7, a convenient model for the elliptic plane can be obtained by abstractly identifying every pair of antipodal points on an ordinary sphere. The reflections and rotations which we shall define in §§6.2 and 6.3 are represented on the sphere by reflections in diametral planes and rotations about diameters.

In elliptic plane geometry, every reflection is a rotation. This rather startling result is a consequence of the fact that the product of the reflection in any diametral plane and the rotation through π about the perpendicular diameter is the central inversion (or point-reflection in the centre of the sphere), which interchanges antipodal points and so corresponds to the identity in elliptic geometry. In any *orientable* space (§2.5), a reflection reverses sense, whereas a rotation preserves sense. Thus the above remarks are closely associated with the non-orientability of the real projective plane (in which elliptic geometry operates). On a sphere, corresponding rotations about antipodal points have opposite senses; so the identification of such points abolishes the distinction of sense.

It was shown by Euler in 1776 that every displacement of a sphere (keeping the centre fixed) can be obtained as a rotation. In elliptic geometry we have the stronger statement that every *congruent transformation* can be obtained as a rotation. The problem of combining two rotations of a sphere was investigated between 1840 and 1850 by Rodrigues, Cayley, Sylvester, Hamilton, and Donkin. Their results are immediately applicable to elliptic geometry, which, in turn, provides alternative proofs for them (§§6.6, 6.8).

Every diametral plane of the sphere determines a perpendicular diameter. In other words, every great circle has a pair of poles, which are antipodal points. This correspondence represents the *absolute polarity* of elliptic geometry. For the reasons given in §5.1, instead of basing our investigation on the notion of congruence, we shall take over the axioms of real projective geometry, and define all the metrical concepts in terms of this absolute polarity, which is an arbitrarily chosen elliptic polarity.

If we chose to single out a hyperbolic polarity instead, the result would be hyperbolic geometry (with "ideal elements"; see §8.1).

6.2. Reflection. Two points, or two lines, will be said to be *perpendicular* if they are conjugate in the absolute polarity. Thus any line through a point **A** is perpendicular to the absolute polar of **A**, and every point on a line **a** is perpendicular to the absolute pole of **a**. Two perpendicular lines, and the absolute polar of their point of intersection, form a *self-polar triangle*, of which every two sides are perpendicular, and likewise every two vertices. The geometry of points on any line is the one-dimensional elliptic geometry described in Chapter v, its absolute involution being induced by the absolute polarity. We now speak of perpendicular points rather than conjugate points, because the dual concept of perpendicular lines is closely analogous to the concept so named in Euclidean geometry. For the same reason, we fix the unit of length by writing $\lambda = \frac{1}{2}\pi$, so that the length of a right segment is $\frac{1}{2}\pi$, and the length of a whole line is π. (When more than one line is involved, we naturally have to abandon the convention whereby the two segments terminated by **A** and **B** were distinguished as **AB** and **BA**.)

Let **o** be any line, and **O** its absolute pole. If we define the *reflection* in **o** (or in **O**) as the harmonic homology with axis **o** and centre **O** (§3.1), we find that it has all the properties commonly associated with the idea of reflecting in a line (or in a point). This transformation is involutory; and, being invariant with respect to the absolute polarity, it relates perpendicular lines (or points) to perpendicular lines (or points).

The fact that a reflection is permutable with the absolute polarity may be seen in greater detail as follows. Let **A** be a general point of the plane, **B** its reflected image, **O′** the point (**o**, **AB**), and **a**, **b**, **o′** the absolute polars of **A**, **B**, **O′**, as in Fig. 6.2A (where the lines are drawn as arcs of circles, so as to allow angles to be represented without distortion). Then H(**AB**, **OO′**); therefore H(**ab**, **oo′**), and **b** is the reflected image of **a**.

By 3.13, the product of reflections in the three sides of a self-polar triangle is the identity. Hence

6.21. *The product of reflections in two perpendicular lines is the reflection in the common perpendicular of the two lines.*

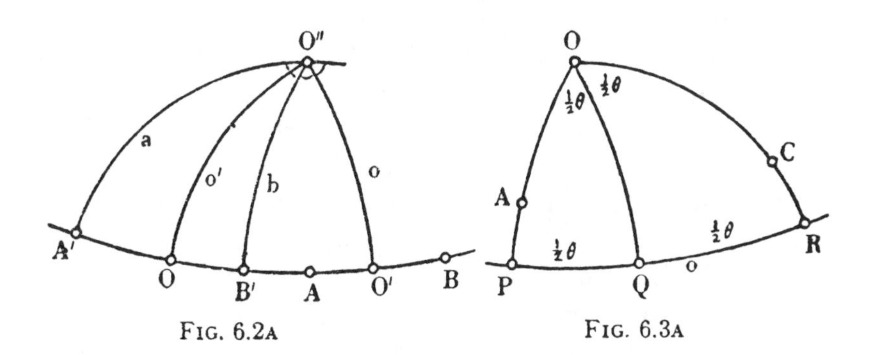

FIG. 6.2A FIG. 6.3A

6.3. **Rotations and angles.** A *rotation* about a point **O** may be defined as the product of reflections in two lines through **O**. The point **O** is transformed into itself; so also is its absolute polar, **o**. With every segment **PQ** of **o**, we associate the product of reflections in **OP** and **OQ**. (See Fig. 6.3A.) This rotation induces in **o** the one-dimensional translation

$$_{\mathbf{P}}\Phi_{\mathbf{P}}\ _{\mathbf{Q}}\Phi_{\mathbf{Q}} = {}_{\mathbf{P}}\Psi_{\mathbf{R}},$$

where **R** is the image of **P** by reflection in **Q**. When **Q** is perpendicular to **P** (so that **PQ** is a right segment), the rotation reduces to the reflection in **o**, by 6.21. In this case **R** coincides with **P**, but the two supplementary segments **OP** are interchanged. We denote this involutory rotation by Ω, regarding it as the absolute involution of the one-dimensional elliptic geometry whose "points" are right segments emanating from **O**, and whose "translations" are rotations about **O**. We define Ω^n for all real numbers n, as in §5.5. In order to agree with the ordinary radian measure of angles, we take the unit of length in this one-dimensional geometry to be such that $\lambda = \pi$. Then segments **OP** and **OR** are said to form an angle $\angle$**POR** $=\theta$ if **OR** is derived from **OP** by applying the rotation $\Omega^{\theta/\pi}$, which we describe as a rotation through angle θ. (Cf. §5.6.) This rotation induces in **o** a translation of length **PR** $=\theta$, but the "associated" segment **PQ** is of length $\frac{1}{2}\theta$. Thus the total angle at a point is 2π, although the total length of a line is only π. (If we wish to extend the principle of duality to metrical concepts, we must take, as the dual of length, the measure of a "directed angle" or "cross,"* instead of an ordinary angle.)

A convention of "positive" sense enables us to distinguish two angles $\angle$**POR** and $\angle$**ROP**, whose sum is 2π. Since **OP** may mean either of two supplementary right segments, we make this distinction precise by taking an arbitrary point **A** in the segment under discussion, and similarly a point **C** in the segment **OR**, and writing $\angle$**AOC** for the angle between the directed (or oriented) lines **OAP** and **OCR**.

*Johnson [1], [2]; Picken [1]; Forder [1], p. 120.

If **A** and **A′** lie respectively in supplementary segments **OP**, we have $\angle$**AOA′** = $\angle$**A′OA** = π (a "straight angle"). (Thus 6.21 asserts that *reflection in* **o** *is equivalent to rotation through* π *about* **O**.) Two angles $\angle$**AOB** and $\angle$**BOA′**, whose sum is π, are said to be supplementary. If **OA** and **OB** are perpendicular, so that these angles are equal, each is a *right angle* ($=\frac{1}{2}\pi$). Most of the one-dimensional theorems about segments can immediately be interpreted as theorems about angles at one point. For instance, if $\angle$**AOB** = $\angle$**COD**, so that the rotation which takes **OA** to **OB** also takes **OC** to **OD**, then $\angle$**AOC** = $\angle$**BOD**. This happens, in particular, when $\angle$**AOB** and $\angle$**COD** are (positive) right angles.

The convention which distinguishes $\angle$**BOA** from $\angle$**AOB** can to a limited extent be applied to angles at distinct points, by continuous variation within such a region as the interior of a triangle. But it cannot be applied universally, since the projective plane is not orientable. In general, therefore, the symbol $\angle$**AOB** may mean either of two angles whose sum is 2π, just as **AB** may mean either of two segments whose sum is π. If the context does not indicate otherwise, we shall take the smaller value. For instance, we say that any two lines determine two supplementary angles, respectively equal to the lengths of the two supplementary segments formed by their absolute poles.

6.4. Congruence. In §§5.4 and 5.6 we defined congruence in terms of translation, and showed that congruent segments have equal lengths. In two-dimensional geometry we can take advantage of the one-dimensional determination of length, and define a *congruent transformation* as a point-to-point correspondence which preserves length. Since a line is the locus of points distant $\frac{1}{2}\pi$ from its absolute pole, such a correspondence is a collineation. Since it preserves the relation of pole and polar, it is permutable with the absolute polarity. A congruent transformation could alternatively be *defined* as a collineation which is permutable with the absolute polarity, the preservation of length being a consequence of the one-dimensional theory.

Thus a reflection is a congruent transformation (and therefore, so also is a rotation). Conversely,

6.41. *Any congruent transformation which preserves two distinct points is a reflection (if it is not the identity).*

PROOF. If the two points, say **O** and **A** (Fig. 6.2A), are not perpendicular, the congruent transformation preserves also the absolute polar **o**, and the point **O′** where **o** meets **OA**. Hence, by 2.84, it preserves every point on **OA**. The projectivity induced in **o** is either the identity or the one-dimensional reflection in **O′**. In the former case, the whole transformation is the identity, by 3.11. In the latter, it is the reflection in **OA**; for, its product with that reflection preserves every point on **o**, as well as every point on **OA**.

If, on the other hand, the two given fixed points are **O** and **O′**, it *may* happen that every point on **OO′** is preserved, as before; but another possibility is that the projectivity induced in **OO′** is the one-dimensional reflection in **O**. In that case the whole transformation is the reflection in either **o** or **o′**; for, its product with the reflection in **o** preserves every point on **OO′**, and so is either the identity or the reflection in **OO′**.

We are now ready for the important theorem

6.42. *Every congruent transformation (of the plane) is a rotation.*

PROOF. By 3.52, any collineation preserves at least one point. In view of 6.41 and 6.21, it will suffice to consider a congruent transformation which preserves *only* one point, say **O**. This induces, in the absolute polar **o**, an elliptic projectivity permutable with the absolute involution, i.e., a translation. This translation, of length θ, say, could have been induced by a rotation through angle θ about $\mathbf{O}$. Consider the product of the given transformation with the inverse of this rotation. This product, preserving every point on $\mathbf{o}$, must be either the identity or the reflection in $\mathbf{o}$, which is the rotation through angle π about $\mathbf{O}$. Hence the given transformation is a rotation through either angle θ or $\theta+\pi$ about $\mathbf{O}$.

Two figures are said to be *congruent* if one can be derived from the other by a congruent transformation, i.e., by a rotation.

This definition is easily seen to satisfy the familiar properties of congruence. Consider, for instance, the theorem (Euclid I, 4) that two triangles $\mathbf{ABC}$ and $\mathbf{A'B'C'}$ are congruent if two sides and the included angle of the one are respectively equal to two sides and the included angle of the other. By a suitable rotation about the absolute pole of $\mathbf{AA'}$, we obtain a triangle $\mathbf{AB_1C_1}$, congruent to $\mathbf{A'B'C'}$. If we are given $\mathbf{AB}=\mathbf{A'B'}$, a suitable rotation about $\mathbf{A}$ carries $\mathbf{AB_1C_1}$ to $\mathbf{ABC_2}$. If we are also given $\angle\mathbf{BAC}=\angle\mathbf{B'A'C'}$ and $\mathbf{AC}=\mathbf{A'C'}$, the point $\mathbf{C_2}$ either coincides with $\mathbf{C}$ or is its image by reflection in $\mathbf{AB}$. (This is essentially Euclid's own proof, with the superposition of a triangle replaced by a congruent transformation of the whole plane.) By 6.42, the various transformations used could be combined into a single rotation.

On page 17 we mentioned the standpoint of Klein's *Erlanger Programm*, where the various geometries are distinguished by the groups of transformations under which their properties are invariant. We conclude from the results of Chapter III that the group for projective geometry consists of all collineations and all correlations. We see now that the group for elliptic geometry is a subgroup of this, consisting of those collineations and correlations which are permutable with a given uniform polarity. In the two-dimensional case, these collineations are just the rotations, and the correlations are derived from them by applying the absolute polarity itself.*

6.5. Circles. A *circle* with centre $\mathbf{O}$ may be defined as

*Coxeter [7], p. 6.

the locus of a point under continuous rotation about **O**, or as the locus of a point whose distance from **O** is constant, say R, the *radius*. Since no distance can be greater than $\frac{1}{2}\pi$, we have $R \leqslant \frac{1}{2}\pi$. The rotation through π gives us two "diametrically opposite" points of the circle, which are images of one another by reflection in the absolute polar of **O**, say **o**, the *axis*. A circle of radius $\frac{1}{2}\pi$ is merely the axis itself, described twice over. In the following investigation, we suppose that $R < \frac{1}{2}\pi$.

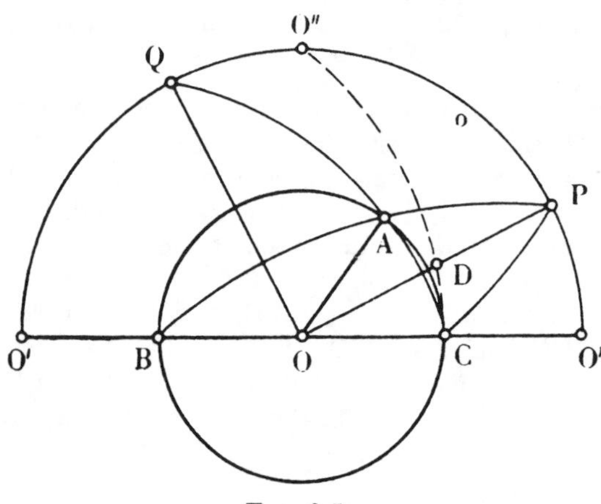

FIG. 6.5A

Given two diametrically opposite points **B** and **C**, and a variable point **A** on the circle, let **OP** (with **P** on **o**) be the perpendicular bisector of **CA**, as in Fig. 6.5A. Then the rotation which carries **C** to **A** is the product of reflections in **OC** and **OP**. The former reflection has no effect on **C**. The latter is the harmonic homology with axis **OP** and centre **Q**, the absolute pole of **OP**. If **BC** meets **o** at **O′**, we have H(**BC**, **OO′**); therefore **BP**, the reflected image of **CP**, contains **A**. Hence the circle may be constructed as the locus of (**BP**, **CQ**), where **P** and **Q** are a variable pair of perpendicular points on the axis **o**. But **P** and **Q** trace projectively related ranges on **o**. Hence, by 3.34,

6.51. *Every circle is a conic.*

It is interesting to observe that the same reasoning applies in Euclidean geometry, if we replace **o** by the line at infinity. Then $\angle$**BAC** is a right angle; but in elliptic geometry it is obtuse.

When **P** coincides with **O′**, **Q** coincides with **O″**, the absolute pole of **BC**. Thus **CO″**, perpendicular to **BC**, is the tangent at **C**. Hence

6.52. *Any tangent to a circle is perpendicular to the diameter through its point of contact.*

It follows also that **O″**, besides being the *absolute* pole of **BC**, is the pole of **BC** with respect to the circle. The axis **o**, containing no point on the circle, is an exterior line; hence the centre, its pole, is an interior point. But an arbitrary point **D** on the tangent at **C** is an exterior point. By considering the right-angled triangle **ODC**, we deduce that

6.53. *The hypotenuse of an acute-sided right-angled triangle is longer than either of the other sides.*

In other words, the shortest distance from a point to a line is along the perpendicular line. We naturally call this *the* distance from the point to the line; it is the "complement" of the distance from the point to the absolute pole of the line. Thus a circle of radius R is the envelope of a variable line distant R from the centre, and also the locus of a variable point distant $\frac{1}{2}\pi - R$ from the axis. Since $\angle$**CO″O′** = **CO′** = $\frac{1}{2}\pi - R$, the same circle is the envelope of a variable line making a constant angle with the axis.* These remarks reveal the circle as a *self-dual* figure.

Since diametrically opposite points are images of one another by reflection in **o**, we have the paradox that the circle, continuously described, remains at a constant distance from **o** "on one side" and later reappears

*Coolidge [1], p. 131.

at the same distance "on the other side." The explanation is that the two sides of a line can only be defined locally, since the whole line does not divide the projective plane into two regions. When R approaches its maximum value, $\frac{1}{2}\pi$, the circle resembles the periphery of the familiar Möbius band, which may be modelled by joining the ends of a long strip of paper after making a half-twist. The line bisecting the width of the strip represents **o**. It is now easy to see how, in the limiting case when $R=\frac{1}{2}\pi$, the circle reduces to the line **o** described twice over; so that its total length, or circumference, is then 2π, in agreement with the total angle at its centre.

6.6. Composition of rotations. Since the product of two rotations is a congruent transformation, 6.42 shows at once that

6.61. *The product of two rotations is a rotation.*

We proceed to determine the centre and angle of the rotation which thus arises from two given rotations, with centres **C** and **A**, say. By 5.31, any rotation about **C** is the product of reflections in two lines through **C**, one of which may be chosen arbitrarily. Hence there is a point **B**, such that the given rotation about **C** is the product of reflections in **CB** and **CA**, while the given rotation about **A** is the product of reflections in **AC** and **AB**. Then, since the reflection in **CA** cancels out, the product of the two rotations is the product of the reflections in **CB** and **AB**, which is a rotation about **B**.

Thus, to obtain the product of rotations through angles γ and α about **C** and **A**, we draw **CB** and **AB** so as to make $\angle \mathbf{BCA}=\frac{1}{2}\gamma$, $\angle \mathbf{CAB}=\frac{1}{2}\alpha$. The resultant rotation is then through $2\angle \mathbf{CBA}$ about **B**. The special case when $\gamma=\alpha=\pi$ and $\mathbf{CA}=\frac{1}{2}\pi$ has already been considered in 6.21. The general case may be expressed as follows:*

6.62. *The product of rotations through angles $2A, 2B, 2C$ about the three vertices of a triangle* **ABC** *is the identity.*

*Sylvester [1], p. 441.

In §6.3 we found it convenient to represent a rotation through angle θ about **O** by a segment of length $\frac{1}{2}\theta$ of the absolute polar, **o** (along which the rotation induces a translation of length θ). Since the starting-point of this directed segment is immaterial (provided it remains on **o**), we may call it a *vector* of length $\frac{1}{2}\theta$ along **o**. By considering the *polar triangle* of **ABC**, whose vertices **D**, **E**, **F** are the absolute poles of **BC**, **CA**, **AB**, we see that 6.62 is equivalent to

6.63. Donkin's Theorem.* *If* **DEF** *is any triangle, the product of rotations represented by vectors* **EF**, **FD**, **DE** *is the identity.*

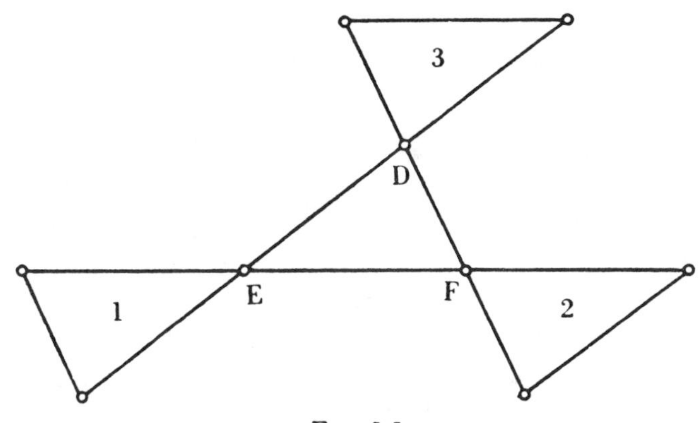

FIG. 6.6A

Donkin's own proof is delightfully perspicuous. It consists in observing the effect of the three rotations on the triangle marked 1 in Fig. 6.6A.

By applying 5.35 to the one-dimensional geometry of right segments emanating from **O**, we see that two reflections cannot be permutable unless their axes are perpendicular. By 5.34, two rotations are permutable if they have a common centre. The condition for two rotations with distinct centres to be

*Donkin [2]. See also Hamilton [4], p. 330.

permutable may be found as follows. From the construction which led to 6.62, it is clear that the product of the same rotations in the reverse order is a rotation about the image of **B** by reflection in **CA**. If this image coincides with **B**, **B** must be the absolute pole of **CA**; then **AB** and **BC** are perpendicular to **CA**, and $\gamma = \alpha = \pi$. We now have the product of reflections in **BC** and **AB**, and the product of reflections in **AB** and **BC**, which cannot be the same unless these lines are perpendicular, in which case **ABC** is a self-polar triangle. Hence*

6.64. *If two rotations about distinct points are permutable, they are rotations through π about perpendicular points.*

In §2.1, we considered the principle of transformation in general. By applying this to rotations, we see that, if Ψ is a rotation through θ about **O**, and Θ is any other rotation, then $\Theta^{-1}\Psi\Theta$ is a rotation through θ about the point $\mathbf{O}\Theta$. This result may be verified in detail by performing the rotation through θ about $\mathbf{O}\Theta$ in three stages, as follows. We first apply the rotation Θ^{-1}, which takes $\mathbf{O}\Theta$ to **O**, then the rotation Ψ about this transformed centre, and finally the rotation Θ, which restores the centre of rotation to its original position.

6.7. Formulae for distance and angle. Putting $\lambda = \frac{1}{2}\pi$ in 5.76, we see that the lengths of the two segments determined by points **A** and **B** are given by

$$\mathbf{AB} = \text{arc cos}\ (\pm\sqrt{\{\mathbf{AB}, \mathbf{B'A'}\}}),$$

A′ and **B′** being the points where the line **AB** meets the absolute polars **a** and **b**. The upper sign refers to the acute segment **AB/A′**, and gives the *distance* **AB**. There is some advantage in using the "mixed" cross ratio, of two points and two lines (4.49), so that

6.71. $$\mathbf{AB} = \text{arc cos}\ (\pm\sqrt{\{\mathbf{AB}, \mathbf{ba}\}}).$$

*Stephanos [1], p. 357.

The angles between **a** and **b**, having these same values, may be expressed in the dually corresponding form

6.72. $\angle(\mathbf{ab}) = \text{arc cos}\,(\pm\sqrt{\{\mathbf{ab}, \mathbf{BA}\}}).$

According to the definition which was justified by 6.53, the distance from the point **A** to the line **b** is

$$\mathbf{AB}' = \text{arc cos}\,\sqrt{\{\mathbf{AB}', \mathbf{BA}'\}}$$

6.73. $= \text{arc cos}\,\sqrt{\{\mathbf{Ab}, \mathbf{Ba}\}} = \text{arc sin}\,\sqrt{\{\mathbf{AB}, \mathbf{ba}\}}.$

In terms of coordinates, if **A**, **B**, **a**, **b** are (x), (y), $[X]$, $[Y]$, we have

$$\{\mathbf{AB}, \mathbf{ba}\} = \frac{\{xY\}\{yX\}}{\{xX\}\{yY\}},$$

where $\{xY\} = x_0Y_0 + x_1Y_1 + x_2Y_2$, and so on. If we take the absolute polarity in its canonical form

6.74. $X_\mu = x_\mu$

(see 4.54), this can just as well be written in any of the following ways:

$$\{\mathbf{AB}, \mathbf{ba}\} = \frac{\{xy\}^2}{\{xx\}\{yy\}} = \frac{\{xY\}^2}{\{xx\}\{YY\}} = \frac{\{XY\}^2}{\{XX\}\{YY\}},$$

where $\{xx\} = x_0^2 + x_1^2 + x_2^2$, and so on. Thus

$$\cos \mathbf{AB} = \pm\frac{\{xy\}}{\sqrt{\{xx\}}\sqrt{\{yy\}}}, \quad \cos\angle(\mathbf{ab}) = \pm\frac{\{XY\}}{\sqrt{\{XX\}}\sqrt{\{YY\}}}.$$

In particular, **A** and **B** are perpendicular points if $\{xy\} = 0$, and **a** and **b** are perpendicular lines if $\{XY\} = 0$.

If we suppose the coordinates of **B** to be derived from those of **A** by continuous variation along the intended segment **AB**, we can take the upper sign in the above expression for cos **AB**. But the choice of sign in the expression for $\cos\angle(\mathbf{ab})$ is less obvious. Of the two supplementary angles between **a** and **b**, the one usually preferred is that which contains some standard point, such as (1, 1, 1). For two lines whose coordinates differ sufficiently little, this will in general be the obtuse angle; we therefore take the lower sign.

On this understanding, the segment between points (x) and (y), the angle between lines $[X]$ and $[Y]$, and the distance from (x) to $[Y]$, are respectively

6.75. $$\text{arc cos}\ \frac{\{xy\}}{\sqrt{\{xx\}}\sqrt{\{yy\}}},\quad \text{arc cos}\ \frac{-\{XY\}}{\sqrt{\{XX\}}\sqrt{\{YY\}}},$$
$$\text{arc sin}\ \frac{|\{xY\}|}{\sqrt{\{xx\}}\sqrt{\{YY\}}}.$$

It follows that the circle of radius R with centre (z) has the ordinary equation

6.76. $$\{zx\}^2 = \{zz\}\ \{xx\}\ \cos^2 R$$

and the tangential equation

$$\{zX\}^2 = \{zz\}\ \{XX\}\ \sin^2 R.$$

It is sometimes convenient to *normalize* the coordinates x_0, x_1, x_2, replacing them by such multiples that $\{xx\}=1$. Homogeneity is then lost, although (x) and $(-x)$ are still the same point. (We have here the analytical form of the model described in §6.1.) The normalized equation of the circle is simply

6.77. $$\{zx\} = \cos R.$$

When the centre is $(1, 0, 0)$, this reduces to

6.78. $$x_0 = \cos R,$$

and the general point on the circle is $(\cos R, x_1, x_2)$, where

6.79. $$x_1^2 + x_2^2 = \sin^2 R.$$

6.8. Rotations and quaternions. Quaternions were invented by W. R. Hamilton in 1843, and were soon applied, by Cayley and others, to the analytical geometry of rotations.* The basic ideas are as follows.

*Hamilton [1], [2], [3]; Cayley [1]; Boole [1]; Donkin [1].

Quaternions are generalized numbers, consisting of ordered tetrads of real numbers, for which addition and multiplication are defined by special rules. If the quaternion consisting of the real numbers a, b, c, d is written as

$$a + b\boldsymbol{i} + c\boldsymbol{j} + d\boldsymbol{k},$$

the rules are those of ordinary arithmetic, excluding the commutative law of multiplication, together with

$$\boldsymbol{i}^2 = \boldsymbol{j}^2 = \boldsymbol{k}^2 = \boldsymbol{ijk} = -1.$$

These clearly imply $\boldsymbol{jk} = -\boldsymbol{kj} = \boldsymbol{i}$, $\boldsymbol{ki} = -\boldsymbol{ik} = \boldsymbol{j}$, $\boldsymbol{ij} = -\boldsymbol{ji} = \boldsymbol{k}$.

For a quaternion $\boldsymbol{u} = a + b\boldsymbol{i} + c\boldsymbol{j} + d\boldsymbol{k}$, we define the *conjugate* $\overline{\boldsymbol{u}} = a - b\boldsymbol{i} - c\boldsymbol{j} - d\boldsymbol{k}$, the *scalar part* $\frac{1}{2}(\boldsymbol{u} + \overline{\boldsymbol{u}}) = a$, the *norm* $\overline{\boldsymbol{u}}\boldsymbol{u} = \boldsymbol{u}\overline{\boldsymbol{u}} = a^2 + b^2 + c^2 + d^2$, and the *reciprocal* $\boldsymbol{u}^{-1} = \overline{\boldsymbol{u}}/\overline{\boldsymbol{u}}\boldsymbol{u}$. By examining the conjugates of $\boldsymbol{ii}$, $\boldsymbol{ij}$, etc., we easily verify that

$$\overline{\boldsymbol{uv}} = \overline{\boldsymbol{v}}\,\overline{\boldsymbol{u}},$$

whence the norm of a product equals the product of norms:

$$\overline{\boldsymbol{uv}}\,\boldsymbol{uv} = \overline{\boldsymbol{v}}\,\overline{\boldsymbol{u}}\boldsymbol{u}\,\boldsymbol{v} = \overline{\boldsymbol{u}}\boldsymbol{u}\,\overline{\boldsymbol{v}}\boldsymbol{v}$$

(The real number $\overline{\boldsymbol{u}}\boldsymbol{u}$ commutes with the quaternion $\overline{\boldsymbol{v}}$.)

If the scalar part vanishes the quaternion is said to be *pure*. Then $\overline{\boldsymbol{u}} = -\boldsymbol{u}$, and $\boldsymbol{u}^2 = -\overline{\boldsymbol{u}}\boldsymbol{u}$. Thus any pure quaternion of unit norm is a square root of -1. The scalar part of the product of two pure quaternions $\boldsymbol{u}$ and $\boldsymbol{v}$ is

$$\tfrac{1}{2}(\boldsymbol{uv} + \overline{\boldsymbol{uv}}) = \tfrac{1}{2}(\boldsymbol{uv} + \boldsymbol{vu}).$$

Before applying these ideas to geometry, let us obtain the analytical expression for a rotation. The collineation 4.45 or 4.48 is a rotation if it commutes with the absolute polarity 6.74, i.e., if

$$c_{\mu\nu} = C_{\mu\nu}.$$

Since $c_{\lambda\mu}C_{\lambda\nu} = \delta_{\mu\nu}$, this condition is equivalent to

6.81. $$c_{\lambda\mu}c_{\lambda\nu} = \delta_{\mu\nu},$$

which makes the transformation $x'_\mu = c_{\mu\nu}x_\nu$ "orthogonal." By this transformation we have $\{x'x'\} = x'_\lambda x'_\lambda = c_{\lambda\mu}x_\mu c_{\lambda\nu}x_\nu = c_{\lambda\mu}c_{\lambda\nu}x_\mu x_\nu$. But $\{xx\} = x_\nu x_\nu = \delta_{\mu\nu}x_\mu x_\nu$. Thus the relation 6.81 is a consequence of $\{x'x'\} = \{xx\}$, and we deduce the following theorem:

6.82. *If a linear transformation leaves the quadratic form $\{xx\}$ invariant, it represents a rotation.*

Consider the linear transformation

6.83. $$x'_0\boldsymbol{i}+x'_1\boldsymbol{j}+x'_2\boldsymbol{k} = \boldsymbol{s}^{-1}(x_0\boldsymbol{i}+x_1\boldsymbol{j}+x_2\boldsymbol{k})\boldsymbol{s},$$

where $\boldsymbol{s}$ is a quaternion. Since

$$-\{x'x'\} = (x'_0\boldsymbol{i}+x'_1\boldsymbol{j}+x'_2\boldsymbol{k})^2 = \boldsymbol{s}^{-1}(x_0\boldsymbol{i}+x_1\boldsymbol{j}+x_2\boldsymbol{k})^2\boldsymbol{s}$$
$$= \boldsymbol{s}^{-1}(-x_0^2-x_1^2-x_2^2)\boldsymbol{s} = -\{xx\},$$

this transformation represents a rotation. Clearly, there will be no loss of generality in supposing $\boldsymbol{s}$ to be a quaternion of *unit norm.*

With any line $[u]$, or $\{ux\} = 0$, normalized so that $\{uu\} = 1$, we may associate a pure quaternion $\boldsymbol{u} = u_0\boldsymbol{i} + u_1\boldsymbol{j} + u_2\boldsymbol{k}$ of unit norm. Since the scalar part of the product

$$(u_0\boldsymbol{i} + u_1\boldsymbol{j} + u_2\boldsymbol{k})\,(v_0\boldsymbol{i} + v_1\boldsymbol{j} + v_2\boldsymbol{k})$$

is $-\{uv\}$, the cosine of the angle between two such lines $[u]$ and $[v]$ is $\{uv\} = -\frac{1}{2}(\boldsymbol{uv} + \boldsymbol{vu})$. (This choice of sign makes the angle vanish when $\boldsymbol{u} = \boldsymbol{v}$. In 6.75 we employed the opposite convention.)

The reflection in the line $[u]$, being the harmonic homology with axis $[u]$ and centre (u) (see 4.41), is

$$x_\mu' = x_\mu - 2u_\mu\{ux\}.$$

As a transformation of the quaternion $\boldsymbol{x} = x_0\boldsymbol{i} + x_1\boldsymbol{j} + x_2\boldsymbol{k}$, this takes the form

$$\boldsymbol{x}' = \boldsymbol{x} - 2\boldsymbol{u}\{ux\} = \boldsymbol{x} + \boldsymbol{u}(\boldsymbol{ux} + \boldsymbol{xu})$$
$$= \boldsymbol{uxu}$$

(since $\boldsymbol{u}^2 = -1$).

Now, every transformation $\boldsymbol{x}' = \boldsymbol{s}^{-1}\boldsymbol{xs}$ is a rotation, every rotation is expressible as the product of two reflections, and the product of reflections in lines $[u]$ and $[v]$ is

$$\boldsymbol{x}' = \boldsymbol{vuxuv}.$$

Since $(\boldsymbol{uv})^{-1} = \boldsymbol{v}^{-1}\boldsymbol{u}^{-1} = (-\boldsymbol{v})(-\boldsymbol{u}) = \boldsymbol{vu}$, this transformation may be identified with 6.83 by setting $\boldsymbol{s} = -\boldsymbol{uv}$. (It would have been formally simpler to set $\boldsymbol{s} = +\boldsymbol{uv}$, but the minus sign has the advantage of making $\boldsymbol{s} = 1$ when $\boldsymbol{u} = \boldsymbol{v}$.) If this is a rotation through θ, the angle between $[u]$ and $[v]$ must be $\frac{1}{2}\theta$, so

$$\cos\tfrac{1}{2}\theta = -\tfrac{1}{2}(\boldsymbol{uv} + \boldsymbol{vu}),$$

which is the scalar part of $\boldsymbol{s}$. For a reason that will appear in a moment, let us write the non-scalar part of $\boldsymbol{s}$ as $\boldsymbol{z}\sin\frac{1}{2}\theta$, where $\boldsymbol{z} = z_0\boldsymbol{i} + z_1\boldsymbol{j} + z_2\boldsymbol{k}$. Then $\bar{\boldsymbol{s}} = \cos\frac{1}{2}\theta - \boldsymbol{z}\sin\frac{1}{2}\theta$, and $\cos^2\frac{1}{2}\theta - \boldsymbol{z}^2\sin^2\frac{1}{2}\theta = \bar{\boldsymbol{s}}\boldsymbol{s} = 1$; therefore $-\boldsymbol{z}^2 = 1$, and $\boldsymbol{z}$ is a pure quaternion of unit norm. Since $\boldsymbol{s}$ commutes with $\boldsymbol{z}$, we have $\boldsymbol{z} = \boldsymbol{s}^{-1}\boldsymbol{zs}$, so the point $(\boldsymbol{z})$ is left invariant by the rotation $\boldsymbol{x}' = \boldsymbol{s}^{-1}\boldsymbol{xs}$. We have now proved Cayley's remarkable theorem

6.85. *The rotation through θ about the point (z), where $\{zz\} = 1$, is given by the transformation*

$$x_0'\boldsymbol{i} + x_1'\boldsymbol{j} + x_2'\boldsymbol{k} = \boldsymbol{s}^{-1}(x_0\boldsymbol{i} + x_1\boldsymbol{j} + x_2\boldsymbol{k})\boldsymbol{s},$$

where $$\boldsymbol{s} = \cos\tfrac{1}{2}\theta + (z_0\boldsymbol{i} + z_1\boldsymbol{j} + z_2\boldsymbol{k})\sin\tfrac{1}{2}\theta.$$

In particular, the rotation through θ about $(1, 0, 0)$ is given by

$$\begin{aligned} & x_0'\boldsymbol{i} + x_1'\boldsymbol{j} + x_2'\boldsymbol{k} \\ &= (\cos\tfrac{1}{2}\theta - \boldsymbol{i}\sin\tfrac{1}{2}\theta)(x_0\boldsymbol{i} + x_1\boldsymbol{j} + x_2\boldsymbol{k})(\cos\tfrac{1}{2}\theta + \boldsymbol{i}\sin\tfrac{1}{2}\theta) \\ &= x_0\boldsymbol{i} + (x_1\boldsymbol{j} + x_2\boldsymbol{k})(\cos\tfrac{1}{2}\theta + \boldsymbol{i}\sin\tfrac{1}{2}\theta)^2 \\ &= x_0\boldsymbol{i} + (x_1\boldsymbol{j} + x_2\boldsymbol{k})\cos\theta + (x_2\boldsymbol{j} - x_1\boldsymbol{k})\sin\theta, \end{aligned}$$

or

6.86. $$\begin{cases} x_0' = x_0, \\ x_1' = x_1\cos\theta + x_2\sin\theta, \\ x_2' = -x_1\sin\theta + x_2\cos\theta, \end{cases}$$

which agrees with 5.75 when we put $x = x_1/x_2$.

To combine two given rotations, we merely have to multiply the corresponding quaternions. For, the effect of transforming by $\boldsymbol{s}$, and then by $\boldsymbol{t}$, is to transform by $\boldsymbol{st}$. However, the correspondence between rotations and quaternions (of unit norm) is not one-to-one but one-to-two, since s and $-s$ have the same effect.

The advantage of the quaternion notation is also seen in its conciseness. The general rotation of 6.85, when written in full (with $s = a + bi + cj + dk$), takes the far less elegant form*

$$\begin{aligned} x_0' &= (a^2 + b^2 - c^2 - d^2)x_0 + 2(bc + ad)x_1 + 2(db - ac)x_2, \\ x_1' &= 2(bc - ad)x_0 + (a^2 - b^2 + c^2 - d^2)x_1 + 2(cd + ab)x_2, \\ x_2' &= 2(db + ac)x_0 + 2(cd - ab)x_1 + (a^2 - b^2 - c^2 + d^2)x_2. \end{aligned}$$

6.9. Alternative treatment using the complex plane. The expressions for distance and angle in terms of coordinates (§6.7) were obtained by Cayley in 1859. His derivation of them will be described later (§12.1). The idea of expressing distance in terms of cross ratio is due to Klein.† Both Cayley and Klein regarded the real projective plane as a subspace of the complex projective plane, and defined the absolute polarity by means of a conic, called the Absolute. A congruent transformation is then a collineation which preserves the Absolute, and may be described (by 3.41) as a projectivity on the conic itself.

The tangents to the Absolute are self-perpendicular lines, and the points on the Absolute are self-perpendicular points. Thus the points in which any real line meets the Absolute are the "absolute points" for that line. By 5.81, our chosen unit of length is such that $\kappa = 2i$. Hence, by 5.82, if the line **AB** meets the Absolute at **M** and **N**, the two segments **AB** are given by

$$e^{2i\mathbf{AB}} = \{\mathbf{AB}, \mathbf{MN}\}^{\pm 1},$$

and the *distance* **AB** is

6.91.
$$\left| \frac{\log\{\mathbf{AB}, \mathbf{MN}\}}{2i} \right| .$$

*Klein [3], p. 102.

†Cayley [3], pp. 88-89; Klein [1], p. 574.

The acute angle between the absolute polar lines **a** and **b**, being equal to this distance, may be expressed in the dually corresponding form

$$\left|\frac{\log\{\mathbf{ab}, \mathbf{mn}\}}{2i}\right|,$$

where **m** and **n** are the tangents to the Absolute from the point $(\mathbf{a}, \mathbf{b})$.

In both these expressions, the logarithm is a "principal value." The other values would give the same distance or angle plus an arbitrary multiple of π.

By 6.52, the tangents to a circle from its centre are self-perpendicular, i.e. they are tangents also to the Absolute. Since the centre has the same polar with respect to the circle and the Absolute, the points of contact of these tangents are the same in both cases. Hence

6.92. *A circle is a conic which has double contact with the Absolute.*

This is also clear from the form of 6.76, since the Absolute has the equation $\{xx\} = 0$.

CHAPTER VII

ELLIPTIC GEOMETRY IN THREE DIMENSIONS

7.1. Congruent transformations. In the present section we describe those properties of perpendicularity, congruence, distance, angle, etc., which closely resemble their two-dimensional analogues. We mention also certain other properties which more closely resemble their *one*-dimensional analogues, because projective spaces of odd dimension are orientable. In contrast to 6.42, we find various kinds of congruent transformation: reflections, rotations, rotatory reflections, and double rotations. The last of these resembles the one-dimensional translation in leaving no point invariant. A special case of it, known as a Clifford translation,* is intimately associated with the existence of pairs of skew lines which are *parallel* in the sense of 1.11. Most of this chapter is concerned with various devices for the elucidation of this fascinating idea, which is quite different from the hyperbolic parallelism described in §1.4.

As in two dimensions, so in three, we begin with real projective geometry, and introduce metrical concepts by singling out an arbitrarily chosen uniform (or elliptic) polarity, calling it the *absolute polarity*. Then every point **A** has an absolute polar plane α, every line **AB** has an absolute polar line (α, β), and every plane α has an absolute pole **A**. Every point of α is perpendicular to **A**, and every plane through **A** is perpendicular to α. The relation between two absolute polar lines is symmetrical: every point on the one is perpendicular to every point on the other, and every plane through the one is perpendicular to every plane through the other. Two perpen-

*Clifford [1].

dicular planes through each of two such lines form a *self-polar tetrahedron*, such that any two opposite edges are absolute polar lines.

In any plane (or line), we have an elliptic geometry whose absolute polarity (or involution) is induced by the absolute polarity in space. Thus lengths can be measured as in §5.6. The *dihedral angles* formed by two planes α and β can be measured as ordinary angles by taking their section by any plane, γ, perpendicular to the line (α, β). The result is independent of our choice of γ, since any such plane contains the absolute polar line **AB**, and the angles in question are equal to the lengths of the two supplementary segments **AB**.

7.11. *The angles between two intersecting lines are equal to the angles between the respective absolute polar lines.*

PROOF. Let **l** and **m** be the lines, intersecting at **D** and lying in γ. Then the absolute polar lines, **l**′ and **m**′, lie in the absolute polar plane δ, and intersect at the absolute pole **C**. Since **CD** is perpendicular to γ, the angles between **l** and **m** are equal to the angles between the planes **Cl** and **Cm**, and so are equal to the segments formed by the absolute poles of these planes, which are the points where **l**′ and **m**′ meet γ.

The orientability of space (§2.5) enables us to associate, in either of two definite ways, the senses along each line with the senses along the absolute polar line. We can thus improve 7.11 by distinguishing the two supplementary angles formed by the two lines. The distinction is made by considerations of continuity, beginning with the case when the angle is zero. The result is as follows:

7.12. *The angle between two directed intersecting lines is equal to the angle between the correspondingly directed absolute polar lines.*

The *reflection* in any plane ω (or in its absolute pole, **O**) is defined as the harmonic homology with centre **O** and axial plane ω. There are two reflections which will interchange two given points **A** and **B**, namely those in which **O** is the mid-point of either segment **AB**. It is easy to see, as in §6.2, that a reflection is permutable with the absolute polarity. By 3.72, *the product of reflections in all four faces of a self-polar tetrahedron is the identity.*

Corresponding to the two-dimensional rotation through angle θ about a point **O** in a plane α, we have a three-dimensional rotation through the same angle about the line **AO**, perpendicular to α. When the former rotation is expressed as the product of reflections in lines **OP** and **OQ**, the latter is the product of reflections in the planes **AOP** and **AOQ**, whose line of intersection is the *axis* of the rotation. The three-dimensional rotation preserves every point on its axis, **AO**, and induces a translation of length θ in the absolute polar line. This translation reduces to the identity in the special case when $\theta = \pi$; the rotation is then the product of reflections in two perpendicular planes. Thus it is the same as the product of reflections in two other planes which form with the first two a self-polar tetrahedron, and it may therefore be regarded indifferently as a rotation through π about either of two absolute polar lines. Any line which intersects both axes is rotated into itself.

As in §6.4, we may define a *congruent transformation* either as a point-to-point correspondence preserving distance or as a collineation which is permutable with the absolute polarity. Any reflection or rotation is a congruent transformation, but there are other kinds as well. The possibilities, however, are limited by the following theorem:

7.13. *Every congruent transformation is the product of at most four reflections.*

PROOF. Let any given congruent transformation Ψ relate

A to **B**, and let Φ be one of the two reflections which interchange **A** and **B**. Then the product $\Psi\Phi$ leaves **A** invariant, and induces, in the absolute polar plane α, a certain two-dimensional congruent transformation, which, by 6.42, is a rotation about some point **O**. Let Θ denote the corresponding three-dimensional rotation about the line **AO**. Then $\Theta^{-1}\Psi\Phi$, preserving every point of α, is either the identity or the reflection in α, say Φ'. (This follows from 3.71, since the projectivity induced in **AO** is either the identity or the one-dimensional reflection in **A**.) Thus Ψ is either $\Theta\Phi$ or $\Theta\Phi'\Phi$. Since Θ is the product of two reflections, Ψ is now expressed as the product of three or four reflections.

By 5.31, when a rotation is expressed as the product of two reflections, one of the reflecting planes may be *any* plane through the axis, and so can be chosen to pass through a given point, or to be perpendicular to a given plane. Thus, in the product of *three* reflections, we can choose the first two in such a way that the plane of the second is perpendicular to that of the third. Then (if necessary) we can modify the second and third so as to make the plane of the second perpendicular to that of the first. We now have a product $\Phi\Phi'\Phi''$, where both Φ and Φ'' are permutable with Φ'. This can be expressed as $\Theta\Phi'$, where the rotation $\Theta = \Phi\Phi''$ is permutable with the reflection Φ', the axis of Θ being perpendicular to the plane of Φ'. In other words,

7.14. *The product of three reflections is a rotatory reflection.*

It follows from 7.13 that any congruent transformation which leaves no point invariant is expressible (in many ways) as the product of four reflections, or of two rotations. We shall see, in 7.49, that the axes of the two rotations can be chosen so as to be absolute polars; then the rotations are permutable, and we are justified in calling their product a *double rotation*.

By 3.73, a congruent transformation is direct or opposite

according as it is the product of an even or odd number of reflections. Hence

7.15. *A direct congruent transformation is either a rotation or a double rotation, and an opposite congruent transformation is either a reflection or a rotatory reflection.*

Two figures are again said to be *congruent* if one can be derived from the other by a congruent transformation. The congruence may be either direct or opposite (as between left and right hands).

In terms of cross ratio (4.71), the lengths of the two segments **AB** are

$$\text{arc cos}(\pm\sqrt{\{\mathbf{AB}, \beta\alpha\}}),$$

where α and β are the absolute polar planes of **A** and **B**. (Cf. §6.71.) These are the same as the angles between α and β, and as such may be written in the form

$$\text{arc cos}(\pm\sqrt{\{\alpha\beta, \mathbf{BA}\}}).$$

The distance from the point **A** to the plane β is

$$\text{arc cos}\sqrt{\{\mathbf{A}\beta, \mathbf{B}\alpha\}} = \text{arc sin}\sqrt{\{\mathbf{AB}, \beta\alpha\}}.$$

In terms of coordinates, if **A**, **B**, α, β are (x), (y), $[X]$, $[Y]$, we have

$$\{\mathbf{AB}, \beta\alpha\} = \frac{\{xY\}\{yX\}}{\{xX\}\{yY\}},$$

where now $\{xY\} = x_0Y_0 + x_1Y_1 + x_2Y_2 + x_3Y_3$, and so on. If we take the absolute polarity in its canonical form, and make the proper conventions for selecting one of the two supplementary lengths or angles, we deduce the following expressions for the segment between points (x) and (y), the angle between planes $[X]$ and $[Y]$, and the distance from the point (x) to the plane $[Y]$:

$$\text{arc cos}\frac{\{xy\}}{\sqrt{\{xx\}}\sqrt{\{yy\}}}, \quad \text{arc cos}\frac{-\{XY\}}{\sqrt{\{XX\}}\sqrt{\{YY\}}},$$

$$\text{arc sin}\frac{|\{xY\}|}{\sqrt{\{xx\}}\sqrt{\{YY\}}}.$$

7.2. Clifford parallels. Some of the properties of equidistant lines can be obtained in quite an elementary way, by means of the following chain of theorems.

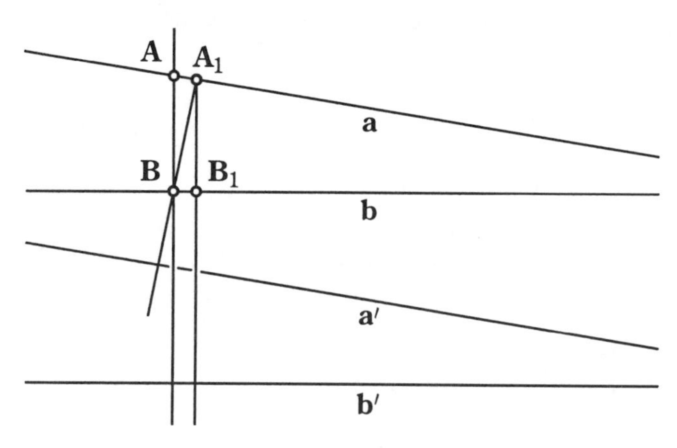

FIG. 7.2A

7.21. *The shortest segment connecting two skew lines is perpendicular to both lines.*

PROOF. Let **a** and **b** be the two skew lines. If they are absolute polars, *every* connecting segment is perpendicular to both, and of length $\frac{1}{2}\pi$. If not, then by 6.53, the shortest segment from any point **A** on **a** is the acute segment of the line perpendicular to **b** (which is the transversal from **A** to **b** and its absolute polar, **b**′). Considerations of continuity suffice to show that, among such segments, there must be one or more minima. If such a minimum segment **AB**, perpendicular to **b**, is not also perpendicular to **a**, draw $\mathbf{BA_1}$ perpendicular to **a**, and $\mathbf{A_1B_1}$ perpendicular to **b**, as in Fig. 7.2A. Then

$$\mathbf{A_1B_1} < \mathbf{A_1B} < \mathbf{AB},$$

which is absurd.

7.22. *Any two lines have at least two common perpendiculars.*

PROOF. Two intersecting lines have one common perpendicular in their plane, and one perpendicular to their plane at the point of intersection. By 7.21, two skew lines have at least one common perpendicular, **AB**; another is the absolute polar of **AB**.

7.23. *If two skew lines have more than two common perpendiculars, they have infinitely many, and the segments intercepted on these are all equal.*

PROOF. The common perpendiculars to **a** and **b** are transversals of four lines: **a**, **b**, and their absolute polars. By 3.62, if there are more than two transversals, the four lines belong to a regulus, and the sets of four points intercepted are projectively related.

Two such equidistant lines are called *Clifford parallels*, or "paratactics," and the ruled quadric generated by their common perpendiculars is called a *rectangular Clifford surface.* Clearly, any two intersecting generators are perpendicular, and any two skew generators are Clifford parallels. Thus four generators, two from each regulus, form a "skew rectangle," having opposite sides equal, and four right angles, like a rectangle in the Euclidean plane.

Theorem 3.61 provides the following sufficient condition for Clifford parallelism:*

7.24. *Two skew lines which have two common perpendiculars of equal length, are Clifford parallels.*

To construct a Clifford parallel from a point **A** to a line **b**,† draw **AB** perpendicular to **b**, and let the absolute polar of **AB** meet **b** at **B′**, as in Fig. 7.2B. On this absolute polar line, take a point **A′** so that **B′A′** = **BA**. Then, by 7.24, **AA′** is a Clifford parallel to **BB′**. By 5.11, there are two possible positions for **A′**. Hence

*Coolidge [1], p. 114. †Bonola [2], p. 204.

7.25. ***Two Clifford parallels to a given line can be drawn through a given point of general position.***

The two parallels reduce to one if the point lies on the line or on its absolute polar.

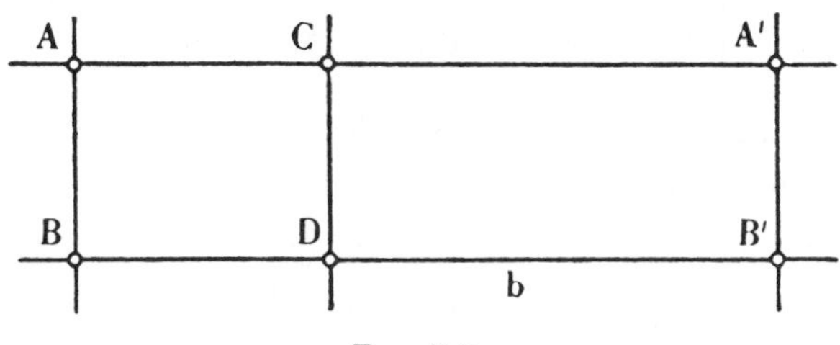

FIG. 7.2B

A *Clifford translation* may be defined as the product of rotations through equal angles θ about two absolute polar lines. Each rotation leaves one of these lines pointwise invariant, and induces a one-dimensional translation in the other; thus the product induces equal translations in the two lines. We proceed to investigate its effect on an arbitrary point **C**.

Let **AA′** be the transversal from **C** to the two lines (i.e., the perpendicular from **C** to either of the lines). Suppose the Clifford translation takes **AA′** to **BB′**, and **C** to **D**. Then the segments **AB** and **A′B′** are equal, and the line **BB′** is a Clifford parallel to **AA′**. By varying θ, we obtain an infinity of lines **BB′**, which generate a rectangular Clifford surface. The generators of the regulus that includes **AA′** are permuted by the Clifford translation; the generators of the other regulus, including **AB** and **A′B′**, are translated along themselves. Hence **D** is the point where the generator through **C** of this second regulus meets **BB′**.

Since the senses along **AB** and its absolute polar can be associated in either of two ways, there are two possible positions for **B′**, given **A**, **B**, and **A′**. Hence

7.26. *For any two points* **A** *and* **B**, *there are just two Clifford translations which will take* **A** *to* **B**.

Since the point **C** was chosen arbitrarily, we see that a Clifford translation of length (or angle) θ moves every point of space through a distance θ along a line which is preserved by being translated along itself. Conversely,

7.27. *Any congruent transformation which moves every point through the same distance, and preserves the line which joins each point to its transform, is a Clifford translation.*

PROOF. If such a congruent transformation takes **A** to **B**, let **A′** be any point on the absolute polar of **AB**, and suppose the transformation takes **A′** to **B′**, so that **A′B′** = **AB**. Then the line **AB** is invariant; its absolute polar, being likewise invariant, contains **B′**, and is **A′B′**.

The invariant lines of a Clifford translation are called its *axes*; the same translation can be regarded as the product of rotations through θ about any one of them and its absolute polar. Clearly, every power of the Clifford translation has the same axes, and any two axes are Clifford parallels. Since two Clifford parallel lines whose distance apart is $\frac{1}{2}\pi$ are absolute polars,

7.28. *A Clifford translation of length* $\frac{1}{2}\pi$ *transforms any line perpendicular to an axis into its absolute polar line.*

A Clifford translation of length π, being the product of reflections in the faces of a self-polar tetrahedron, is the identity.

7.3. The Stephanos-Cartan representation of rotations by points. A rotation about a line **l** induces, in any plane perpendicular to **l**, a rotation through the same angle about the point where this plane meets **l**. Therefore Donkin's Theorem

6.63 continues to hold in three dimensions, if a rotation through θ about **l** is represented by a vector of length $\frac{1}{2}\theta$ along the absolute polar line, **l**′. In this manner the directed sides of any triangle **DEF** represent definite rotations, whose product is the identity. The axes of these rotations concur at the absolute pole, $\mathbf{P}_1$, of the plane **DEF**.

Since the senses along pairs of absolute polar lines can be consistently associated, the rotation through θ about **l** may be represented (in a different but definite manner) by a vector of length $\frac{1}{2}\theta$ along **l** itself, instead of along **l**′. The various rotations S, about lines through $\mathbf{P}_1$, are then represented by vectors $\mathbf{P}_1\mathbf{P}_S$ along their axes, and so, finally, by points $\mathbf{P}_S$, such that the length $\mathbf{P}_1\mathbf{P}_S$ is equal to half the angle of rotation. When we keep the axis fixed, and increase θ from 0 to 2π, the point $\mathbf{P}_S$ starts at $\mathbf{P}_1$, and traverses the whole length of the axis. In particular, the point $\mathbf{P}_{S^{-1}}$, which represents the inverse rotation S^{-1} (i.e. the rotation through $-\theta$ or $2\pi-\theta$ about $\mathbf{P}_1\mathbf{P}_S$), is the image of $\mathbf{P}_S$ by reflection in $\mathbf{P}_1$. Hence*

7.31. *A point* $\mathbf{P}_1$ *and a doubly oriented line determine a definite correspondence between all the points of space and all the rotations about lines through* $\mathbf{P}_1$.

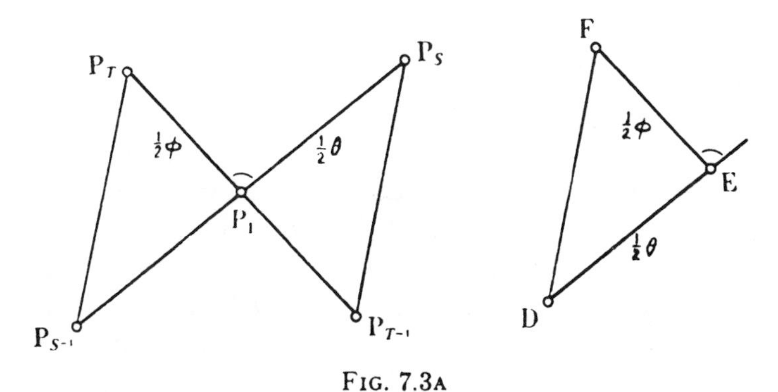

FIG. 7.3A

*Stephanos [1], pp. 344, 360; Cartan [1], p. 265; Coxeter [4], pp. 473–475.

Consider two rotations $\boldsymbol{S}$ and $\boldsymbol{T}$, through angles θ and ϕ, about lines $\mathbf{P}_1\mathbf{P}_S$ and $\mathbf{P}_1\mathbf{P}_T$. We seek the point $\mathbf{P}_{ST}$ which represents their product. Let **DE** and **EF** (Fig. 7.3A) be vectors of lengths $\frac{1}{2}\theta$ and $\frac{1}{2}\phi$ along the absolute polars of $\mathbf{P}_1\mathbf{P}_S$ and $\mathbf{P}_1\mathbf{P}_T$ (so that **E** is the absolute pole of the plane $\mathbf{P}_1\mathbf{P}_S\mathbf{P}_T$). By 7.12, the angle between the vectors $\mathbf{P}_1\mathbf{P}_S$ and $\mathbf{P}_1\mathbf{P}_T$ is equal to the angle between the vectors **DE** and **EF**, which is the supplement of the angle E of the triangle **DEF**. Thus triangles **DEF** and $\mathbf{P}_{S^{-1}}\mathbf{P}_1\mathbf{P}_T$ are congruent, the sides **DE**, **EF**, and angle E of the former being equal to the corresponding sides and angle of the latter. By Donkin's Theorem, $\mathbf{P}_{ST}$ *lies on the absolute polar of* **DF**, *at distance* **DF** *from* $\mathbf{P}_1$. Thus

$$\mathbf{P}_1\mathbf{P}_{ST} = \mathbf{P}_{S^{-1}}\mathbf{P}_T.$$

Transforming the rotation $\boldsymbol{ST}$ by $\boldsymbol{S}$, we see that $\boldsymbol{TS}$ is a rotation through the same angle. Hence

$$\mathbf{P}_1\mathbf{P}_{TS} = \mathbf{P}_1\mathbf{P}_{ST} = \mathbf{P}_{S^{-1}}\mathbf{P}_T.$$

Replacing $\boldsymbol{S}$ by $\boldsymbol{S}^{-1}$,

7.32. $$\mathbf{P}_1\mathbf{P}_{TS^{-1}} = \mathbf{P}_1\mathbf{P}_{S^{-1}T} = \mathbf{P}_S\mathbf{P}_T.$$

Thus *the length* $\mathbf{P}_S\mathbf{P}_T$ *is equal to half the angle of either of the rotations* $\boldsymbol{TS}^{-1}$, $\boldsymbol{S}^{-1}\boldsymbol{T}$.

It follows readily that both the segments $\mathbf{P}_{SR}\mathbf{P}_{TR}$ and $\mathbf{P}_{RS}\mathbf{P}_{RT}$ are congruent to $\mathbf{P}_S\mathbf{P}_T$.

7.4. Right translations and left translations. Consider the group of all rotations about lines through $\mathbf{P}_1$, and the corresponding set of all points of space. The effect of multiplying all the rotations by any one of them, say $\boldsymbol{R}$, on the right or on the left, is to permute the representative points according to a certain correspondence, Δ_R or Λ_R, so that, for every point $\mathbf{P}_S$, we have

$$\mathbf{P}_S\Delta_R = \mathbf{P}_{SR}, \quad \mathbf{P}_S\Lambda_R = \mathbf{P}_{RS}.$$

By the remark at the end of §7.3, Δ_R and Λ_R preserve the distance between any two points, and so are congruent transformations. Let us call them *right* and *left translations*. In order to identify them with the Clifford translations defined in §7.2, we shall apply the criterion 7.27, after establishing the following lemma:

7.41. *Any line may be expressed as the locus of a point* $\mathbf{P}_{U^x S}$, *or of a point* $\mathbf{P}_{SV^y}$, *where* $\mathbf{P}_S$ *is an arbitrary point on the line,* U *and* V *are rotations through arbitrary angles about definite axes, and* x *and* y *vary over a sufficient range of real numbers.*

PROOF. Since the various powers of a rotation all have the same axis, the general point on the line $\mathbf{P}_1\mathbf{P}_U$ is $\mathbf{P}_{U^x}$. Applying the right translation Δ_S, we deduce that the general point on $\mathbf{P}_S\mathbf{P}_{US}$ is $\mathbf{P}_{U^x S}$. Similarly, the general point on $\mathbf{P}_S\mathbf{P}_{SV}$ is $\mathbf{P}_{SV^y}$. Since any line $\mathbf{P}_S\mathbf{P}_T$ may be expressed as $\mathbf{P}_S\mathbf{P}_{US}$ or as $\mathbf{P}_S\mathbf{P}_{SV}$, the desired result follows.

Since $\mathbf{P}_S\mathbf{P}_{RS} = \mathbf{P}_1\mathbf{P}_R = \mathbf{P}_S\mathbf{P}_{SR}$, Λ_R and Δ_R translate every point $\mathbf{P}_S$ through the same distance $\mathbf{P}_1\mathbf{P}_R$. By 7.41, the line $\mathbf{P}_S\mathbf{P}_{RS}$ is preserved by Λ_R, and $\mathbf{P}_S\mathbf{P}_{SR}$ by Δ_R. Hence

7.42. *If* R *is a rotation through* θ *(about a line through* $\mathbf{P}_1$*),* Λ_R *and* Δ_R *are Clifford translations of length* $\frac{1}{2}\theta$, *each having* $\mathbf{P}_1\mathbf{P}_R$ *as an axis.*

Given two points, $\mathbf{P}_S$ and $\mathbf{P}_T$, there is a unique left translation $\Lambda_{TS^{-1}}$, and a unique right translation $\Delta_{S^{-1}T}$, which will take $\mathbf{P}_S$ to $\mathbf{P}_T$.* Hence, by 7.26,

7.43. *Every Clifford translation is either a left translation or a right translation.*

Since $\mathbf{P}_S\Lambda_U\Delta_V = \mathbf{P}_{USV} = \mathbf{P}_S\Delta_V\Lambda_U$ (for every point $\mathbf{P}_S$),

7.44. *Any left translation is permutable with any right translation.*

*Klein [3], p. 236.

Since $\Delta_U \Delta_V = \Delta_{UV}$ and $\Lambda_U \Lambda_V = \Lambda_{VU}$,

7.45. *The product of two right (or left) translations is a right (or left) translation.**

Thus, in the language of group-theory, the class of right (or left) translations is a continuous group, simply isomorphic with the group of rotations about lines through $\mathbf{P}_1$.

We come now to the most important theorem on Clifford translations:

7.46. *Every direct congruent transformation is uniquely expressible as the product of a left translation and a right translation.*

PROOF. A rotation through θ about any line whatever may be regarded as the product of two equal rotations through $\frac{1}{2}\theta$ about this line and two inverse rotations through $\frac{1}{2}\theta$ about the absolute polar line in opposite senses. It is then expressed as the product of a left translation and a right translation, each of length $\frac{1}{2}\theta$. Hence, by 7.44 and 7.45, the product of any two rotations, being the product of two left and two right translations, is also the product of one left and one right translation. By 7.15, this is the most general direct congruent transformation.

Finally, if this expression were not unique, we could find a left translation (other than the identity) which was also a right translation, say $\Lambda_S = \Delta_T$. This would imply $\boldsymbol{SR} = \boldsymbol{RT}$ for every $\boldsymbol{R}$, including $\boldsymbol{R} = 1$; hence $\boldsymbol{S} = \boldsymbol{T}$, and $\boldsymbol{SR} = \boldsymbol{RS}$. By 6.64, this is impossible.

The last three theorems may be summarized in group-theoretic terms, as follows:

7.47. *The group of all direct congruent transformations, being the direct product of the group of left translations and the group of right translations, is simply isomorphic with the direct square of the group of all rotations about lines through a fixed point.*

*Klein [3], p. 236.

From the above proof of 7.46, it is seen that, when a rotation through θ about any line $\mathbf{l}$ is expressed as $\Lambda_U\Delta_V$, the Clifford translations Λ_U and Δ_V are of equal length $\frac{1}{2}\theta$. Therefore, by 7.42, $\boldsymbol{U}$ and $\boldsymbol{V}$ are rotations through equal angles θ about two lines through $\mathbf{P}_1$. The effect of replacing $\boldsymbol{U}$ or Λ_U by its inverse is to change the rotation through θ about $\mathbf{l}$ into a rotation through θ about the absolute polar line. Hence

7.48. *If U and V are rotations through equal angles about two lines through $\mathbf{P}_1$, then $\Lambda_U\Delta_V$ and $\Lambda_{U^{-1}}\Delta_V$ are rotations through equal angles about two lines which are absolute polars of each other.*

In particular, since any $\boldsymbol{R}$ transforms any $\boldsymbol{S}$ into $\boldsymbol{R}^{-1}\boldsymbol{S}\boldsymbol{R}$, we have

$$\boldsymbol{R} = \Lambda_{R^{-1}}\Delta_R,$$

for a rotation through $2\mathbf{P}_1\mathbf{P}_R$ about the line $\mathbf{P}_1\mathbf{P}_R$. Putting $\boldsymbol{T}\boldsymbol{S}^{-1}$ for $\boldsymbol{R}$, and transforming by Δ_S, we obtain the expression $\Lambda_{ST^{-1}}\Delta_{S^{-1}T}$ for a rotation through $2\mathbf{P}_S\mathbf{P}_T$ about the line $\mathbf{P}_S\mathbf{P}_T$.

The rotations about absolute polar lines (in 7.48) can be re-combined to form Clifford translations, as follows:

$$\Lambda_U\Delta_V \cdot \Lambda_{U^{-1}}\Delta_V = \Delta_{V^2}, \quad \Lambda_U\Delta_V(\Lambda_{U^{-1}}\Delta_V)^{-1} = \Lambda_{U^2}.$$

More generally,

7.49. *Every direct congruent transformation is the product of rotaticns about two absolute polar lines.*

PROOF. Since any two rotations about lines through $\mathbf{P}_1$ may be expressed as $\boldsymbol{U}^x$ and $\boldsymbol{V}^y$, where $\boldsymbol{U}$ and $\boldsymbol{V}$ are rotations through equal angles, the general direct congruent transformation is

$$\Lambda_{U^x}\Delta_{V^y} = (\Lambda_U\Delta_V)^{\frac{1}{2}(x+y)}(\Lambda_{U^{-1}}\Delta_V)^{\frac{1}{2}(y-x)}.$$

7.5. Right parallels and left parallels. By 7.25, there are, in general, two Clifford parallels to a given line through a given point. These may be distinguished as follows. Two lines which occur among the axes of a right (or left) translation are said to be right (or left) parallels. It follows at once that right (or left) parallelism is transitive, and that two lines which are both right and left parallels must be absolute polars.

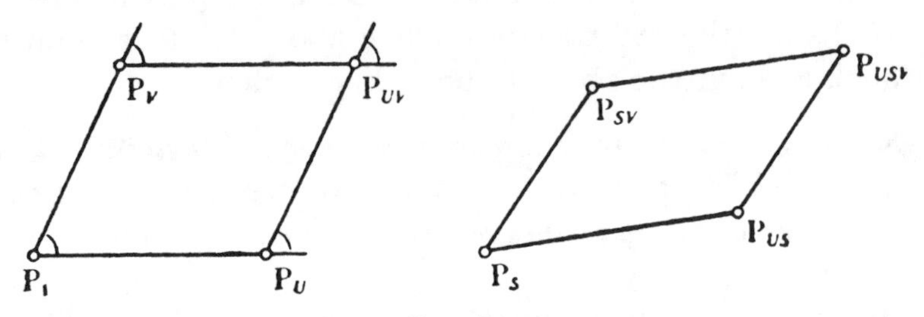

FIG. 7.5A

If U and V are any two rotations about distinct lines through $\mathbf{P}_1$, the skew quadrangle $\mathbf{P}_1\mathbf{P}_U\mathbf{P}_{UV}\mathbf{P}_V$ is called a *Clifford parallelogram.** (See Fig. 7.5A.) The opposite sides $\mathbf{P}_1\mathbf{P}_U$ and $\mathbf{P}_V\mathbf{P}_{UV}$, being axes of Λ_U, are left parallels; the opposite sides $\mathbf{P}_1\mathbf{P}_V$ and $\mathbf{P}_U\mathbf{P}_{UV}$, being axes of Δ_V, are right parallels. The former two opposite sides are both of length half the angle of the rotation U, and the latter two are both of length half the angle of V. Since Λ_U takes $\mathbf{P}_1\mathbf{P}_V$ to $\mathbf{P}_U\mathbf{P}_{UV}$, and translates $\mathbf{P}_1\mathbf{P}_U$ along itself, the external angle at $\mathbf{P}_U$ is equal to the internal angle at $\mathbf{P}_1$. Proceeding similarly with Δ_V, we see that the four internal angles are equal or supplementary, just like those of a Euclidean parallelogram; their sum is exactly 2π. For simplicity, we have taken one vertex at $\mathbf{P}_1$; but similar results evidently hold for the general Clifford parallelogram $\mathbf{P}_S\mathbf{P}_{US}\mathbf{P}_{USV}\mathbf{P}_{SV}$. Since Λ_U translates the line $\mathbf{P}_S\mathbf{P}_{SV}$ to $\mathbf{P}_{US}\mathbf{P}_{USV}$, while Δ_V translates $\mathbf{P}_S\mathbf{P}_{US}$ to $\mathbf{P}_{SV}\mathbf{P}_{USV}$, we have

7.51. *Two lines which are derived from one another by a left (or right) translation are right (or left) parallels.*

---

*Klein [3], p. 235.

It is now clear that the problem of representing the product of two given rotations (§7.3) amounts to completing a Clifford parallelogram. Fig. 7.5B indicates the representative points of three rotations $\boldsymbol{R}$, $\boldsymbol{S}$, $\boldsymbol{T}$ whose product is the identity (so that the inverse of each is the product of the other two, in cyclic order). Six Clifford parallelograms are easily picked out.

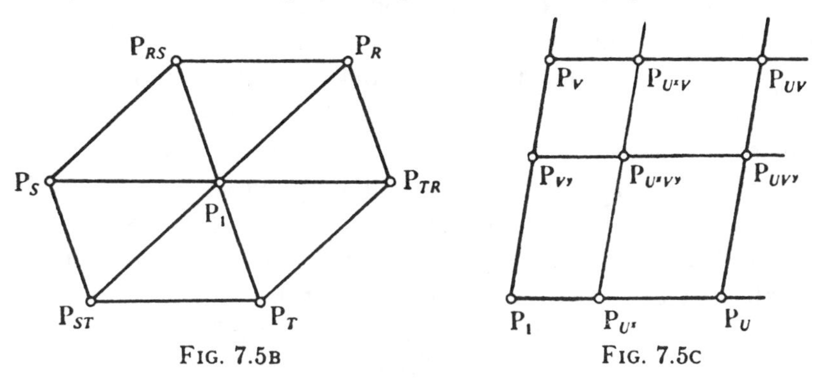

FIG. 7.5B FIG. 7.5C

The following theorem enables us to define a generalization of the surface considered in §7.2:

7.52. *Given any two intersecting lines, the left parallel to one through any point of the other meets the right parallel to the latter through any point of the former, so as to form a Clifford parallelogram.*

PROOF. Since any point $\mathbf{P}_S$ can be translated (by $\Lambda_{S^{-1}}$ or $\Delta_{S^{-1}}$) to $\mathbf{P}_1$, there is no loss of generality in taking the two intersecting lines to be $\mathbf{P}_1\mathbf{P}_U$ and $\mathbf{P}_1\mathbf{P}_V$, as in Fig. 7.5C. Then $\mathbf{P}_{U^x}$ is the general point of the former, and $\mathbf{P}_{V^y}$ of the latter. The point $\mathbf{P}_{U^xV^y}$ lies on both the left parallel to $\mathbf{P}_1\mathbf{P}_U$ through $\mathbf{P}_{V^y}$ and the right parallel to $\mathbf{P}_1\mathbf{P}_V$ through $\mathbf{P}_{U^x}$.

The quadric generated by the above systems of lines is called a *Clifford surface.** Given three parallel lines and a transversal, we can apply 7.52 and obtain

*Sommerville [2], pp. 105-114.

7.53. *The quadric determined by any three left (or right) parallels is a Clifford surface.*

Clearly, any two left parallel generators and any two right parallel generators form a Clifford parallelogram of angle $\psi = \angle \mathbf{P}_U \mathbf{P}_1 \mathbf{P}_V$. By a famous theorem of Gauss, the "total curvature" of a geodesic polygon on any surface is measured by its angular excess (as compared with a Euclidean polygon of the same number of sides). Since a Clifford surface is covered with a network of Clifford parallelograms, as small as we please, it follows that this surface has zero curvature everywhere, which means that it can be mapped without distortion on a suitable region of the Euclidean plane.* (This will be seen more clearly in §13.8.) To put the matter picturesquely, a small creature whose world was a Clifford surface would find his practical geometry to be Euclidean, until he explored so far as to discover that his "flat earth" was finite (though unbounded). In other words, after cutting the Clifford surface along any two intersecting generators, we can "develop" it on the Euclidean plane, just as we can unroll a cylinder. The Clifford parallelograms will then become ordinary parallelograms, and the whole surface will appear as the interior of a rhombus of side π and angle ψ. (Thus its total area is $\pi^2 \sin \psi$.)

The general point of the Clifford surface being $\mathbf{P}_{U^x V^y}$, the numbers x and y serve as intrinsic coordinates, which resemble affine coordinates in the Euclidean plane, save that they are periodic (since every generator is of length π). Along each of the right parallel generators, x is constant; and along each of the left parallel generators, y is constant. Linear equations involving both x and y define curves called *geodesics*, which develop into straight lines. If U and V are rotations through angles θ and ϕ, we have

$$\mathbf{P}_1 \mathbf{P}_{U^x} = \tfrac{1}{2}\theta x, \qquad \mathbf{P}_1 \mathbf{P}_{V^y} = \tfrac{1}{2}\phi y.$$

*Klein [3], pp. 243-247.

The coordinates x and y will be proportional to actual distances (like oblique Cartesian coordinates) if we choose the points $\mathbf{P}_U$ and $\mathbf{P}_V$ to be equidistant from $\mathbf{P}_1$, so that $\theta = \phi$. In this case the surface, consisting of points $\mathbf{P}_{U^x V^y}$, is preserved by every power of either of the rotations $\Lambda_U \Delta_V$ and $\Lambda_{U^{-1}} \Delta_V$. Hence

7.54. *A Clifford surface is a surface of revolution in two distinct ways.*

It follows that the geodesics $x - y = c$ are circles (of the same radius for all values of c), and likewise the geodesics $x + y = c$. When the surface is developed on the Euclidean plane, two such circles, one of each kind, appear as diagonals of the rhombus. Hence, by Euclidean trigonometry, their circumferences are $2\pi \sin \frac{1}{2}\psi$ and $2\pi \cos \frac{1}{2}\psi$. In the special case of a rectangular Clifford surface, the rhombus is a square, and the circles are equal.

If we had taken the two intersecting lines of 7.52 to be $\mathbf{P}_S \mathbf{P}_{US}$ and $\mathbf{P}_S \mathbf{P}_{SV}$, the general point of the Clifford surface would have been $\mathbf{P}_{U^x S V^y}$ (and the angle $\psi = \angle \mathbf{P}_{US} \mathbf{P}_S \mathbf{P}_{SV} = \angle \mathbf{P}_{S^{-1}US} \mathbf{P}_1 \mathbf{P}_V$). The loci of points $\mathbf{P}_{U^x S^z V^y}$ for various values of z are a family of "coaxal" Clifford surfaces. The numbers x, y, z could be used as coordinates for the whole elliptic space.*

A Clifford surface, being closed and uniform like a sphere, developable like a cylinder, and ring-shaped like a torus, is not easy to visualize. Unlike a sphere, it has zero curvature; moreover, in the rectangular case, its "inside" and "outside" are congruent. Unlike a cylinder, it is finite, and has two systems of generating lines.

---

*A treatment closely resembling the above §§7.3-7.5 was worked out independently by W. S. Morris (*The Geometry of the Rotation Group*, Princeton Junior Paper, 1936) under the direction of Professor A. W. Tucker. Morris considers the abstract space whose "points" are rotations about a point in ordinary space. He defines the "motions" of this space

7.6. Study's representation of lines by pairs of points. Any line determines a unique pair of lines through the arbitrary fixed point $\mathbf{P}_1$: one left parallel and one right parallel. These meet the absolute polar plane of $\mathbf{P}_1$ in a definite pair of "representative" points. Any other line represented by the same pair of points (in the same order) is both left and right parallel to the first line, and so can only be the absolute polar line. Clearly, point-pairs such as $(\mathbf{A}, \mathbf{B})$ and $(\mathbf{A}, \mathbf{D})$ represent *left* parallel lines, while pairs such as $(\mathbf{A}, \mathbf{D})$ and $(\mathbf{C}, \mathbf{D})$ represent *right* parallel lines. In particular, $(\mathbf{A}, \mathbf{A})$ represents the line $\mathbf{P}_1\mathbf{A}$ (and its absolute polar).

7.61. *If* U *and* V *are rotations through* π *about lines through* $\mathbf{P}_1, \Lambda_U\Delta_V$ *is the rotation through* π *about either of the lines represented by the point-pair* $(\mathbf{P}_U, \mathbf{P}_V)$.

PROOF. The two absolute polar lines represented by $(\mathbf{P}_U, \mathbf{P}_V)$, being left parallel to $\mathbf{P}_1\mathbf{P}_U$ and right parallel to $\mathbf{P}_1\mathbf{P}_V$, are axes of both the Clifford translations Λ_U and Δ_V. Therefore either of them can be regarded as the axis of the rotation $\Lambda_U\Delta_V$.

7.62. *The general direct congruent transformation* $\Lambda_S\Delta_T$ *is represented by independent rotations* S^{-1} *and* T, *applied respectively to the first and second representative points of a line.*[*]

PROOF. $\Lambda_S\Delta_T$ transforms the rotation $\Lambda_U\Delta_V$ into

$$(\Lambda_S\Delta_T)^{-1}\Lambda_U\Delta_V(\Lambda_S\Delta_T) = \Lambda_{S^{-1}}\Lambda_U\Lambda_S \cdot \Delta_{T^{-1}}\Delta_V\Delta_T = \Lambda_{SUS^{-1}}\Delta_{T^{-1}VT},$$

---

to be the transformations that we have called $\Lambda_U\Delta_V$, the "lines" to be the one-dimensional subgroups V^y (V fixed) and their cosets SV^y (cf. 7.41), the "distance" from S to T to be proportional to the angle of the rotation $S^{-1}T$ (c.f. 7.32), the "polar plane" of S to be the class of rotations T for which $S^{-1}T$ is of angle π, and a "Clifford surface" to be the product of two one-dimensional subgroups. He shows that these definitions are consistent, and that the properties of the space so constructed are those of elliptic space.

*Study [1], pp. 121-123.

and thus replaces the point-pair $(\mathbf{P}_U, \mathbf{P}_V)$ by $(\mathbf{P}_{SUS^{-1}}, \mathbf{P}_{T^{-1}VT})$.

7.63. *A necessary and sufficient condition for two lines, represented by point-pairs* $(\mathbf{A}, \mathbf{B})$ *and* $(\mathbf{C}, \mathbf{D})$, *to be either intersecting or perpendicular, is* $\mathbf{AC} = \mathbf{BD}$.

PROOF. If the two lines intersect at an angle $\psi (\leq \frac{1}{2}\pi)$, the respective left parallels through $\mathbf{P}_1$ make this same angle, and therefore $\mathbf{AC} = \psi$. Similarly $\mathbf{BD} = \psi$. If, on the other hand, the two lines are perpendicular (i.e. conjugate in the absolute polarity), one intersects the absolute polar of the other, and again $\mathbf{AC} = \mathbf{BD}$. Conversely, if we are given $\mathbf{AC} = \mathbf{BD}$, we can find a two-dimensional rotation (in the absolute polar plane of $\mathbf{P}_1$) which takes $\mathbf{A}$ to $\mathbf{B}$, and $\mathbf{C}$ to $\mathbf{D}$. Let T be the corresponding three-dimensional rotation (about a line through $\mathbf{P}_1$). By 7.62, a right translation affects the second representative point alone. Thus Δ_T transforms the point-pairs $(\mathbf{A}, \mathbf{A})$ and $(\mathbf{C}, \mathbf{C})$, which represent lines through $\mathbf{P}_1$, into $(\mathbf{A}, \mathbf{B})$ and $(\mathbf{C}, \mathbf{D})$.

In the above theorem, the word "perpendicular" is used in a wider sense than in 7.22, where the "common perpendiculars" of two lines were understood to intersect the two lines. If lines represented by $(\mathbf{A}, \mathbf{B})$ and $(\mathbf{E}, \mathbf{F})$ are perpendicular in this stricter sense, we have $\mathbf{AE} = \mathbf{BF} = \frac{1}{2}\pi$. Hence

7.64. *If two non-parallel lines are represented by point-pairs* $(\mathbf{A}, \mathbf{B})$ *and* $(\mathbf{C}, \mathbf{D})$, *their two common perpendiculars are represented together by* $(\mathbf{E}, \mathbf{F})$ *where* $\mathbf{E}$ *and* $\mathbf{F}$ *are the absolute poles of the lines* $\mathbf{AC}$ *and* $\mathbf{BD}$ *(in the plane of the representation).*

The bundle of lines through $\mathbf{P}_1$ is represented by all pairs of coincident points $(\mathbf{A}, \mathbf{A})$. Applying Δ_T, we deduce that the bundle of lines through any point $\mathbf{P}_T$ is represented by all pairs $(\mathbf{A}, \mathbf{A}T)$, where $\mathbf{A}$ is arbitrary. In other words, the point $\mathbf{P}_T$ is represented by the rotation T. In this sense, Study's representation is just the inverse of the Stephanos-Cartan representation.

By 7.54, the reguli of a Clifford surface are the loci of two intersecting lines, say **l** and **r**, by continuous rotation about either of the two lines which are left parallel to **l** and right parallel to **r**. Suppose that **l** and **r** are represented by (**A**, **B**) and (**C**, **D**), where **AC** = **BD** = ψ. Then the two axes of rotation are represented together by (**A**, **D**). Thus the Clifford surface is represented by two equal circles, with centres **D** and **A**, in the following manner. The left parallel generators are represented by point-pairs (**A**, **B**), where **A** is fixed and **B** describes the circle with centre **D** and radius ψ; and the right parallel generators by point-pairs (**C**, **D**), where **C** describes the circle with centre **A** and the same radius.

In the special case when $\psi = \frac{1}{2}\pi$, the circles reduce to lines. In every other case, a second Clifford surface, whose generators are the absolute polars of those of the first, is automatically represented at the same time. Both surfaces belong to the coaxal system which is obtained by varying ψ while keeping **A** and **D** fixed. If ψ = **AD**, one of the surfaces passes through $\mathbf{P}_1$.

Three left parallel lines, represented by (**A**, **B**), (**A**, **C**), (**A**, **D**), determine a Clifford surface which is represented by the circle **BCD** and an equal circle with centre **A**. (Cf. 7.53.)

7.7. Clifford translations and quaternions. Let us choose our canonical coordinate system so that the point $\mathbf{P}_1$ is (1, 0, 0, 0). A rotation about a line through (1, 0, 0, 0) induces, in the perpendicular plane $x_0 = 0$, a rotation through the same angle about the point where this plane meets the line. By 6.85, if this point is $(0, z_1, z_2, z_3)$, where $z_1^2 + z_2^2 + z_3^2 = 1$, such a rotation S, through angle θ, leaves x_0 unchanged, and transforms x_1, x_2, x_3 according to the formula

$$x_1'\boldsymbol{i} + x_2'\boldsymbol{j} + x_3'\boldsymbol{k} = \boldsymbol{s}^{-1}(x_1\boldsymbol{i} + x_2\boldsymbol{j} + x_3\boldsymbol{k})\,\boldsymbol{s},$$

where

7.71. $$\boldsymbol{s} = \cos\tfrac{1}{2}\theta + (z_1\boldsymbol{i} + z_2\boldsymbol{j} + z_3\boldsymbol{k})\sin\tfrac{1}{2}\theta.$$

In other words, the rotation through θ about the line

7.72.
$$\frac{x_1}{z_1} = \frac{x_2}{z_2} = \frac{x_3}{z_3}$$
is given by the transformation
$$x_0' + x_1'\boldsymbol{i} + x_2'\boldsymbol{j} + x_3'\boldsymbol{k} = \boldsymbol{s}^{-1}(x_0 + x_1\boldsymbol{i} + x_2\boldsymbol{j} + x_3\boldsymbol{k})\,\boldsymbol{s},$$
or, in brief,

7.73.
$$\boldsymbol{x}' = \boldsymbol{s}^{-1}\boldsymbol{x}\boldsymbol{s}.$$

We naturally call $\boldsymbol{s}$ *the quaternion of the rotation*, observing that quaternions combine by multiplication in the same manner as the corresponding rotations.

The Stephanos-Cartan representative point $\mathbf{P}_S$, distant $\frac{1}{2}\theta$ from (1, 0, 0, 0) along the line 7.72, is
$$(\cos \tfrac{1}{2}\theta,\ \ z_1 \sin \tfrac{1}{2}\theta,\ \ z_2 \sin \tfrac{1}{2}\theta,\ \ z_3 \sin \tfrac{1}{2}\theta).$$
These coordinates being the constituents of $\boldsymbol{s}$, we have

7.74. *The point* (s) *represents the rotation whose quaternion is* $\boldsymbol{s}$.

For a given point (s), there are *two* quaternions of unit norm: $\pm\boldsymbol{s}/\sqrt{\{ss\}}$. These correspond to values of θ which differ by 2π, and so give essentially the same rotation.

If $\boldsymbol{u}$ and $\boldsymbol{v}$ are the quaternions of rotations U and V, the effect of multiplying all rotations on the left by U, and on the right by V, is to multiply the corresponding quaternions on the left by $\boldsymbol{u}$ and on the right by $\boldsymbol{v}$. Thus the transformation $\Lambda_U\Delta_V$ is*

7.75.
$$\boldsymbol{x}' = \boldsymbol{u}\boldsymbol{x}\boldsymbol{v}.$$

In particular, the Clifford translations Λ_U and Δ_V are, respectively,

$\boldsymbol{x}' = \boldsymbol{u}\,\boldsymbol{x}$,	$\boldsymbol{x}' = \boldsymbol{x}\,\boldsymbol{v}$,
or	or
$x_0' = u_0x_0 - u_1x_1 - u_2x_2 - u_3x_3,$	$x_0' = v_0x_0 - v_1x_1 - v_2x_2 - v_3x_3,$
$x_1' = u_1x_0 + u_0x_1 - u_3x_2 + u_2x_3,$	$x_1' = v_1x_0 + v_0x_1 + v_3x_2 - v_2x_3,$
$x_2' = u_2x_0 + u_3x_1 + u_0x_2 - u_1x_3,$	$x_2' = v_2x_0 - v_3x_1 + v_0x_2 + v_1x_3,$
$x_3' = u_3x_0 - u_2x_1 + u_1x_2 + u_0x_3,$	$x_3' = v_3x_0 + v_2x_1 - v_1x_2 + v_0x_3.$

*Cayley [2], p. 812.

By 6.86 (with x_0 for x_2, and then x_3 for x_1), the product of rotations through given angles θ and ϕ about two opposite edges of the tetrahedron of reference is*

$$\begin{array}{l|l} x_0' = x_0 \cos\theta - x_1 \sin\theta, & x_2' = x_2 \cos\phi - x_3 \sin\phi, \\ x_1' = x_0 \sin\theta + x_1 \cos\theta, & x_3' = x_2 \sin\phi + x_3 \cos\phi. \end{array}$$

(Cf. 7.49.) This may be expressed in the form 7.75 by taking $\boldsymbol{U}$ and $\boldsymbol{V}$ to be rotations through respective angles $\theta \pm \phi$ about the line $x_2 = x_3 = 0$, so that

$$\boldsymbol{u}\boldsymbol{x}\boldsymbol{v} = \left(\cos\frac{\theta+\phi}{2} + \boldsymbol{i}\sin\frac{\theta+\phi}{2}\right)(x_0 + x_1\boldsymbol{i} + x_2\boldsymbol{j} + x_3\boldsymbol{k}) \left(\cos\frac{\theta-\phi}{2} + \boldsymbol{i}\sin\frac{\theta-\phi}{2}\right)$$

$$= (\cos\theta + \boldsymbol{i}\sin\theta)(x_0 + x_1\boldsymbol{i}) + (\cos\phi + \boldsymbol{i}\sin\phi)(x_2\boldsymbol{j} + x_3\boldsymbol{k})$$

$$= (x_0\cos\theta - x_1\sin\theta) + (x_0\sin\theta + x_1\cos\theta)\,\boldsymbol{i} + (x_2\cos\phi - x_3\sin\phi)\,\boldsymbol{j} + (x_2\sin\phi + x_3\cos\phi)\,\boldsymbol{k}.$$

By 7.46, 7.75 is the general direct transformation of elliptic space. To obtain the general opposite transformation, it is sufficient to combine this with a *particular* opposite transformation, such as the reflection in $x_0 = 0$, which merely reverses the sign of x_0. Hence

7.76. *The general congruent transformation (direct or opposite according as we take the upper or lower sign for x_0) is*

$$x_0' + x_1'\boldsymbol{i} + x_2'\boldsymbol{j} + x_3'\boldsymbol{k} = (u_0 + u_1\boldsymbol{i} + u_2\boldsymbol{j} + u_3\boldsymbol{k})(\pm x_0 + x_1\boldsymbol{i} + x_2\boldsymbol{j} + x_3\boldsymbol{k})(v_0 + v_1\boldsymbol{i} + v_2\boldsymbol{j} + v_3\boldsymbol{k}).$$

As in §4.6, we define the Plücker coordinates of the line joining two points (x) and (y) by the formula $p_{\mu\nu} = x_\mu y_\nu - y_\mu x_\nu$. If (y) is derived from (x) by applying Λ_U, we multiply both sides of the equation $\boldsymbol{u}\boldsymbol{x} = \boldsymbol{y}$ on the right by $x_0 - x_1\boldsymbol{i} - x_2\boldsymbol{j} - x_3\boldsymbol{k}$, obtaining

$$(u_0 + u_1\boldsymbol{i} + u_2\boldsymbol{j} + u_3\boldsymbol{k})(x_0^2 + x_1^2 + x_2^2 + x_3^2)$$

$$= (y_0 + y_1\boldsymbol{i} + y_2\boldsymbol{j} + y_3\boldsymbol{k})(x_0 - x_1\boldsymbol{i} - x_2\boldsymbol{j} - x_3\boldsymbol{k})$$

$$= (x_0y_0 + x_1y_1 + x_2y_2 + x_3y_3) + (p_{01} + p_{23})\,\boldsymbol{i} + (p_{02} + p_{31})\,\boldsymbol{j} + (p_{03} + p_{12})\,\boldsymbol{k}.$$

---

*Goursat [1], p. 36.

If, on the other hand, (y) is derived from (x) by applying Δ_V, the equation $\boldsymbol{x}\,\boldsymbol{v}=\boldsymbol{y}$ gives similarly

$$\{xx\}(v_0+v_1\boldsymbol{i}+v_2\boldsymbol{j}+v_3\boldsymbol{k}) \\ =\{xy\}+(p_{01}-p_{23})\boldsymbol{i}+(p_{02}-p_{31})\boldsymbol{j}+(p_{03}-p_{12})\boldsymbol{k}.$$

Comparing the coefficients of $\boldsymbol{i}, \boldsymbol{j}, \boldsymbol{k}$, we deduce that the axes of Λ_U and Δ_V form the linear congruences

7.77.
$$\begin{cases} \dfrac{p_{01}+p_{23}}{u_1}=\dfrac{p_{02}+p_{31}}{u_2}=\dfrac{p_{03}+p_{12}}{u_3}, \\[2ex] \dfrac{p_{01}-p_{23}}{v_1}=\dfrac{p_{02}-p_{31}}{v_2}=\dfrac{p_{03}-p_{12}}{v_3}. \end{cases}$$

Hence*

7.78. *A necessary and sufficient condition for two lines to be left (or right) parallel is that the sums (or differences) of complementary Plücker coordinates for one line be proportional to the corresponding sums (or differences) for the other.*

7.8. Study's coordinates for a line. The linear congruences 7.77 are *elliptic*, in the sense of p. 93, since there is just one axis through each point of space. If a line $\{p\}$ belongs to both of them, its left and right parallels through $(1, 0, 0, 0)$ are

$$\{u_1, u_2, u_3, 0, 0, 0\} \text{ and } \{v_1, v_2, v_3, 0, 0, 0\};$$

therefore the points of Study's representative pair are

7.81.
$$(0, u_1, u_2, u_3), \ (0, v_1, v_2, v_3).$$

Since the coordinates are homogeneous, we may write

$$2u_1=p_{01}+p_{23},\ 2u_2=p_{02}+p_{31},\ 2u_3=p_{03}+p_{12},$$
$$2v_1=p_{01}-p_{23},\ 2v_2=p_{02}-p_{31},\ 2v_3=p_{03}-p_{12},$$

whence

7.82.
$$\begin{cases} p_{01}=u_1+v_1,\ p_{02}=u_2+v_2,\ p_{03}=u_3+v_3, \\ p_{23}=u_1-v_1,\ p_{31}=u_2-v_2,\ p_{12}=u_3-v_3. \end{cases}$$

---

*Coolidge [1], p. 125.

The line is specified by its "Study coordinates"* $u_1, u_2, u_3, v_1, v_2, v_3$, just as well as by its Plücker coordinates $p_{\mu\nu}$. We then speak of "the line $\{u; v\}$." The identical relation 4.62 is replaced by

7.83. $$u_1^2+u_2^2+u_3^2=v_1^2+v_2^2+v_3^2,$$

and the condition 4.63 for two lines to intersect becomes

7.84. $$u_1u_1'+u_2u_2'+u_3u_3'=v_1v_1'+v_2v_2'+v_3v_3'.$$

By 4.74, the absolute polar of $\{u;v\}$ is $\{u; -v\}$, which clearly has the same representative point-pair. Two lines $\{u; v\}$ and $\{u';v'\}$ are perpendicular (without necessarily intersecting) if one intersects the absolute polar of the other, i.e. if

$$u_1u_1'+u_2u_2'+u_3u_3'+v_1v_1'+v_2v_2'+v_3v_3'=0.$$

These results resemble 7.63, but are slightly stronger in that they distinguish between intersection and perpendicularity.

By 7.78, two lines are left parallel if their u's are proportional, and right parallel if their v's are proportional. Since the line $\{u;v\}$ is represented by the point-pair 7.81, 7.62 implies the following theorem:

7.85. *As a transformation of Study coordinates,* $\Lambda_S\Delta_T$ *is*

$$u_1'\boldsymbol{i}+u_2'\boldsymbol{j}+u_3'\boldsymbol{k}=\boldsymbol{s}(u_1\boldsymbol{i}+u_2\boldsymbol{j}+u_3\boldsymbol{k})\boldsymbol{s}^{-1},$$
$$v_1'\boldsymbol{i}+v_2'\boldsymbol{j}+v_3'\boldsymbol{k}=\boldsymbol{t}^{-1}(v_1\boldsymbol{i}+v_2\boldsymbol{j}+v_3\boldsymbol{k})\boldsymbol{t},$$

where $\boldsymbol{s}$ *and* $\boldsymbol{t}$ *are the quaternions of the rotations* $\boldsymbol{S}$ *and* $\boldsymbol{T}$.

To illustrate the use of these coordinates, let us apply them to a Clifford surface. We saw, at the end of §7.6, that the right parallel generators of a Clifford surface of angle ψ are represented by a circle of radius ψ and centre **A**, paired with a fixed point **D**. By applying suitable left and right translations, we can make **A** and **D** coincide at (0, 1, 0, 0). By 6.79, a typical point of the circle is then

*Study [1], p. 119.

$$(0, \cos \psi, u_2, u_3), \text{ where } u_2^2+u_3^2=\sin^2 \psi.$$

Hence a typical generator (of the right parallel regulus) is

$$\{\cos \psi, u_2, u_3;\ 1, 0, 0\}.$$

By 7.82, the same line in Plücker coordinates is

$$\{2 \cos^2 \tfrac{1}{2}\psi, u_2, u_3, -2 \sin^2 \tfrac{1}{2}\psi, u_2, u_3\}.$$

Comparing this with 4.81, we see that $c_1=-\tan^2 \frac{1}{2}\psi$ and $c_2=c_3=1$. Thus the Clifford surface is

$$-(x_0^2+x_1^2) \tan^2 \tfrac{1}{2}\psi+x_2^2+x_3^2=0,$$

or

$$(x_0^2+x_1^2) \sin^2 \tfrac{1}{2}\psi=(x_2^2+x_3^2) \cos^2 \tfrac{1}{2}\psi,$$

or*

7.86. $$x_0^2+x_1^2-x_2^2-x_3^2=\{xx\} \cos \psi.$$

In particular, the rectangular Clifford surface is simply

$$x_0^2+x_1^2=x_2^2+x_3^2.$$

7.9. Complex space. When the real projective space of elliptic geometry is regarded as a subspace of *complex* projective space, the locus of self-perpendicular points and the envelope of self-perpendicular planes is a quadric, called the Absolute. Klein's formulae for distance and dihedral angle† are precisely analogous to his formulae for distance and angle in the plane (§6.9).

A congruent transformation, being a collineation which preserves the Absolute, either preserves each of the two reguli or interchanges them. Consider, in particular, a reflection. Let **L** be the point where the reflecting plane meets an arbitrary generator **l**. Since **L** reflects into itself, the reflected generator **l′** passes through **L**. Thus **l** and **l′** intersect, and belong to different reguli. Hence a reflection interchanges the two reguli, and, by 7.15,

*Klein [3], p. 241.
†Klein [1], p. 621.

7.91. *A congruent transformation is direct or opposite, according as it preserves each regulus of the Absolute, or interchanges the two reguli.*

Since there are no real points on the Absolute, its generators are imaginary lines of the *second kind,** and occur in conjugate imaginary pairs which, being skew, belong to one regulus. Any real line meets the Absolute in two conjugate imaginary points, and the two generators through one of these points are conjugate imaginary to the two generators through the other. Thus any real line which meets a particular generator also meets the conjugate imaginary generator.

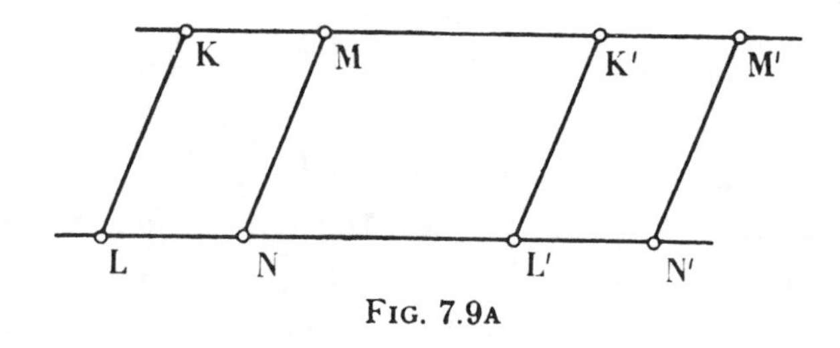

FIG. 7.9A

7.92. *Two real lines which meet the same generator of the Absolute are Clifford parallels.*†

PROOF. Let **KL** and **MN** be two real lines which meet both the conjugate imaginary generators **KM** and **LN**. Since generators are self-polar, the respective polar lines **K′L′** and **M′N′** will meet the same generators, as in Fig. 7.9A. Now, the polar planes of **K**, **M**, **K′**, **M′** are **KK′L′**, **MM′N′**, **K′KL**, **M′MN**, which meet **LN** in **L′**, **N′**, **L**, **N**, respectively. Hence

$$\mathbf{K\,M\,K'M'} \barwedge \mathbf{L'N'L\,N} \barwedge \mathbf{L\,N\,L'N'},$$

and, by 3.61 and 7.23, **KL** and **MN** are Clifford parallels.

*Klein [3], p. 79; Robinson [1], p. 147.

†Klein [3], p. 234; Study [1], p. 133.

This theorem enables us to recognize each pair of conjugate imaginary generators as the directrices of a linear congruence of left or right parallels. Accordingly, Study distinguishes the two reguli as consisting of left and right generators, respectively. The opposite convention would perhaps be more natural, since the *left* generators (which intersect all right generators) belong to every congruence of *right* parallels, and vice versa.

7.93. *A real collineation which preserves each right (or left) generator is a left (or right) translation.*

PROOF. Since such a collineation preserves the Absolute, it is a congruent transformation, in fact (by 7.91) a direct congruent transformation. It transforms the left (or right) generators according to a certain projectivity having two self-corresponding elements. There are thus two conjugate imaginary left (or right) generators of which every point is invariant. The Clifford parallels which are transversals of these two generators are consequently invariant, and the transformation is a left (or right) translation having these transversals for axes.

Theorem 7.92 suggests the possibility of defining a Clifford surface without mentioning parallels. This can be done by means of 7.53. The result is as follows:*

7.94. *A real quadric which has four generators in common with the Absolute is a Clifford surface.*

In other words, a Clifford surface may be defined as a real quadric which has quadruple contact with the Absolute. Comparison with 6.92 emphasizes the analogy between a circle and a Clifford surface, though a more obvious analogue of the circle is a *sphere* (which can be shown to have "ring contact" with the Absolute).

*Klein [3], p. 241.

To close this chapter, let us express some of the above results in terms of coordinates.

The Absolute, being the locus of self-perpendicular points, has the equation $\{xx\}=0$, or

$$x_0^2+x_1^2+x_2^2+x_3^2=0.$$

To obtain the generators, it is easiest to use Study coordinates. Since the polar line of $\{u;v\}$ is $\{u;-v\}$, or $\{-u;v\}$, a self-polar line has either vanishing v's or vanishing u's. Thus the generators are

$$\{p_1, p_2, p_3; 0, 0, 0\} \text{ and } \{0, 0, 0; p_1, p_2, p_3\},$$

where, by 7.83,

$$p_1^2+p_2^2+p_3^2=0.$$

Since $\{p; 0\}$ belongs to every congruence of right parallels, and $\{0; p\}$ to every congruence of left parallels, these are left and right generators, respectively. The same lines in Plücker coordinates are

$$\{p_1, p_2, p_3, \pm p_1, \pm p_2, \pm p_3\},$$

in agreement with §4.8 (which is valid in complex space without any restriction on the signs of c_1, c_2, c_3).

The Clifford surface 7.86 meets the Absolute where

$$x_0^2+x_1^2=x_2^2+x_3^2=0.$$

This locus, being the intersection of the planes $x_0 \pm ix_1=0$ with the planes $x_2 \pm ix_3=0$, consists of four lines

$$\{0, 1, \pm i, 0, 1, \pm i\}, \{0, 1, \pm i, 0, -1, \mp i\},$$

in agreement with 7.94.

CHAPTER VIII

DESCRIPTIVE GEOMETRY

8.1. Klein's projective model for hyperbolic geometry. The two chief ways of approaching non-Euclidean geometry are that of Gauss, Lobatschewsky, Bolyai, and Riemann, who began with Euclidean geometry and modified the postulates, and that of Cayley and Klein, who began with projective geometry and singled out a polarity.

In Klein's treatment, two lines are *perpendicular* if they are conjugate in the absolute polarity, and the geometry is elliptic or hyperbolic according to the nature of this polarity. We have considered the elliptic case exhaustively in the preceding three chapters; the null polarity is easily seen to be unsuitable. Setting these aside, we are left with a polarity of the kind that determines a conic or quadric: "the Absolute." If we accept Postulate IV (§1.1), which rules out the possibility of self-perpendicular lines, we find that the Absolute cannot be a *ruled* quadric, and we are led to consider points interior to a conic in the plane, or to an oval quadric in space.*

Defining a congruent transformation as a collineation which preserves the Absolute, Klein showed, as in §5.8, that if a line **AB** meets the Absolute at **M** and **N**, then the *length* **AB** is given by the formula

$$e^{\kappa \mathbf{AB}} = \{\mathbf{AB}, \mathbf{MN}\} ,$$

where κ is a real constant depending on the unit of length. Taking κ to be positive, we find that this formula makes **AB** positive or negative according as **AM** || **BN** or **AN** || **BM**. In either case the *distance* **AB** is

*Veblen and Young [2], II, pp. 350-370.

$$\frac{|\log \{\mathbf{AB}, \mathbf{MN}\}|}{\kappa}.$$

In particular, $\mathbf{AM} = \infty$; so we speak of the points on the Absolute as *points at infinity*.

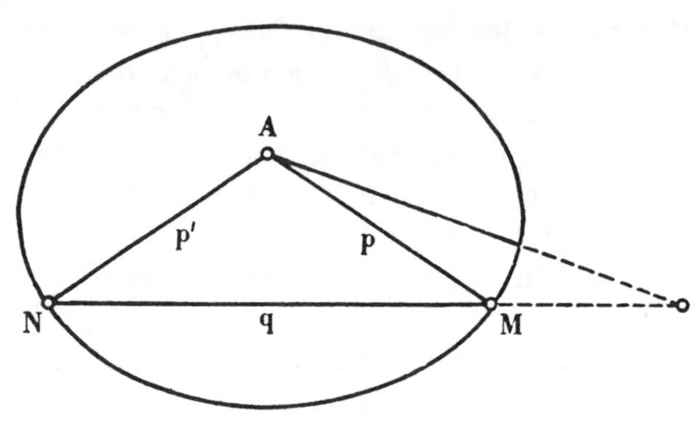

FIG. 8.1A

We easily see that this geometry of the interior of a conic or quadric has all the properties of Lobatschewsky's "imaginary geometry." Consider, for instance, a flat pencil with centre **A**, and a line **q** in the same plane, as in Figs. 1.2A and 8.1A. The parallels **p** and **p'** join **A** to the points at infinity on **q**, and divide the pencil into two parts. The "second" part consists of lines which meet **q** outside the Absolute, so that, from the standpoint of hyperbolic geometry, they do not meet **q** at all.

The *ideal* points, on and exterior to the Absolute, will be found very useful; e.g. the polar of an exterior point is the common perpendicular of any two lines through it. But since all the ordinary points of hyperbolic geometry are *interior* to the absolute, it is by no means obvious that these other points can still be defined when we approach hyperbolic geometry in the classical manner, as a modification of Euclidean geometry. Klein indicated a method for achieving this extension of hyperbolic space. The details were worked out by Pasch. The adjunction of points at infinity to affine or Euclidean space appears as a special case.

It is natural enough that metrical notions are irrelevant. What is more startling is that nothing need be said as to the number of lines, parallel to a given line, that can be drawn through a given point. We are thus dealing with a more general geometry, which includes both Euclidean and hyperbolic.

8.2. Geometry in a convex region. When we described real projective geometry as "What can be done with an ungraduated straight edge," we ignored the practical difficulty that a complicated construction is apt to involve pairs of lines whose desired point of intersection is outside the sheet of paper we are using.* We can take account of this familiar trouble by asking what becomes of real projective geometry when we restrict consideration to a *convex region* (i.e. to a region which does not include the whole of any line but includes the whole of one of the two segments determined by any two points within it). The only axioms that are violated are 2.115 and 2.117. (Two lines, or two planes, are said not to meet at all if they do not meet within the chosen region.) The resulting *descriptive* geometry is of theoretical as well as practical interest, since the loss of the principle of duality is compensated by the fact that two points now determine a unique segment. Consequently the relations of incidence and separation, instead of being undefined, are both expressible in terms of the single relation of "three point" order, or *intermediacy*. In fact, the collinearity of three points is tested by seeing whether one of them lies *between* the other two, just as in practical surveying.

The name "descriptive geometry," which commonly refers to something quite different (namely, the technique of repre-

*Robinson [1], p. 68.

senting a solid figure by projections on planes), was adopted by Russell and Whitehead as a convenient abbreviation for "the geometry of serial order." Possibly a better name for it would be *ordered geometry* (as in Coxeter [9]).

A set of axioms for descriptive geometry will be given in §8.3. The geometry of a convex region then serves as a model. But the variety of possible convex regions shows that descriptive geometry is not categorical; in other words, it is strictly not one geometry but a family of geometries. The two most useful regions (in the two-dimensional case) are the interior of a conic, and the whole projective plane with the exception of one line; these give hyperbolic geometry and affine geometry, respectively. But there are other possibilities. For instance, we might use a triangular region and the corresponding trilinear polarity; however, the metrical properties are then quite bizarre, as perpendicularity is no longer a symmetric relation.

Descriptive geometry, being high school geometry with congruence and parallelism left out, is more familiar than projective geometry. It is therefore interesting that the former provides a model for the latter, as well as *vice versa*. In fact, this happens in two distinct ways. We shall see in §8.5 that the lines and planes through a point in descriptive space form a model for the points and lines in the real projective plane. More generally, the lines and planes through a point in n-dimensional descriptive space form a model for the points and lines in real projective $(n-1)$-space. Secondly, following Klein's suggestion as elaborated by Pasch, we shall see in §§8.6-8.8 that certain classes of lines and planes in descriptive space form a model for the points and lines in real projective space of the *same* number of dimensions. In effect this means that, given a descriptive space, we can construct a projective space of which the descriptive space is a part, namely a convex region.

8.3. Veblen's axioms of order. After giving a sufficient set of axioms for descriptive geometry, we shall save space by omitting most of the consequent elementary theorems, as they are intuitively obvious and have been elegantly proved elsewhere.*

In Veblen's treatment there is one undefined entity, a *point*, and one undefined relation, *intermediacy*. Following Forder, we use the symbol [**ABC**] to express this relation in the form "**B** lies between **A** and **C**" or "the three points **A**, **B**, **C** are in the order **ABC**."

AXIOMS FOR DESCRIPTIVE GEOMETRY

8.311. *There are at least two points.*

8.312. *If* **A** *and* **B** *are two points, there is at least one point* **C** *such that* [**ABC**].

8.313. *If* [**ABC**], *then* **A** *and* **C** *are distinct.*

8.314. *If* [**ABC**], *then* [**CBA**] *but not* [**BCA**].

If **A** and **B** are any two points, the *segment* **AB** is the class of points **X** such that [**AXB**], the half-line or *ray* **A/B** is the class of points **Y** such that [**BAY**], the *interval* $\overline{\mathbf{AB}}$ consists of the segment **AB** together with its end-points **A** and **B**, and the *line* **AB** consists of the interval $\overline{\mathbf{AB}}$ together with the rays **A/B** and **B/A**, as in Fig. 8.3A. The ray **A/B** is said to *emanate* from **A**. Points are said to lie *on* a segment, ray, interval, or line, if they belong to the respective class. Several other words, such as *collinear*, will be used in the same sense as in §2.1.

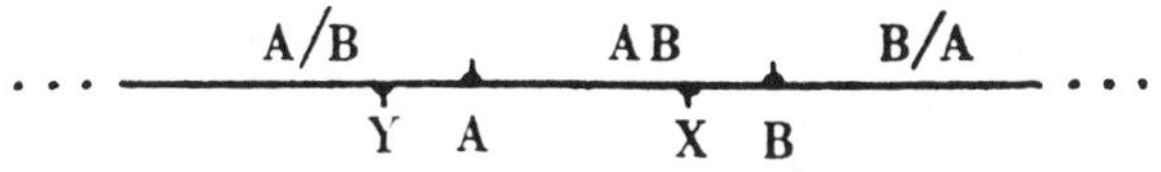

A/B AB B/A

Y A X B

FIG. 8.3A

8.315. *If* **C** *and* **D** *are two points on the line* **AB**, *then* **A** *is on the line* **CD**.

*Veblen [1], pp. 353-370.

8.316. *There is at least one point not on the line* **AB**.

8.317. *If* **A**, **B**, **C** *are three non-collinear points, and* **D** *and* **E** *are such that* [**BCD**] *and* [**CEA**], *then there is a point* **F** *on the line* **DE** *with* [**AFB**]. (See Fig. 8.8A.)

If **A**, **B**, **C** are non-collinear, the *plane* **ABC** is the class of points collinear with pairs of points on the intervals $\overline{\mathbf{BC}}$, $\overline{\mathbf{CA}}$, $\overline{\mathbf{AB}}$. It can be deduced* that the plane contains the line joining any two of its points.

8.318. *There is at least one point not in the plane* **ABC**.

8.319. *Two planes which have one common point have another.*

8.32. *For every partition of all the points of a line into two non-vacuous sets, such that no point of either lies between two points of the other, there is a point of one set which lies between every other point of that set and every point of the other set.*

This last axiom† differs from 2.13 only in the substitution of "line" for "segment." There would be no harm in using 2.13 itself, although we could not conversely use 8.32 in projective geometry (where it would be meaningless).

8.4. Order in a pencil. A convex region is slightly easier to define in descriptive geometry than in projective; in fact it is simply a set of points which includes the whole of the segment bounded by any two points within it. In particular, segments and rays are convex regions.

For any three collinear points **A**, **B**, **C**, we have either [**ABC**] or [**BCA**] or [**CAB**]. In the first two cases we say that **B** and **C** are on the *same side* of **A** (on the line considered), and in the third case that they are on *opposite* sides of **A**. Thus the points on either side of **A** form a ray emanating from **A**. Two such rays (which are "halves of one line") are said to be *supple-*

*Veblen [1], p. 360; Forder [1], p. 59.

†Dedekind [1], §3.

mentary. Analogously,* if **a** is any line through **A**, and α any plane through **a**, then **a** divides the rest of α into two convex regions called supplementary *half-planes*. Points **B** and **C** are said to be on the same side of **a** (in α) if they belong to the same half-plane, i.e. if the segment **BC** contains no point of **a**. Again the plane α divides the rest of space into two convex regions called supplementary half-spaces, which determine whether two points are on the same side of α or on opposite sides.

Two rays a_1, b_1, emanating from a point O, along with O itself, are said to form the *angle* a_1b_1. Another ray x_1, emanating from O in the same plane, is said to lie *within* the angle a_1b_1, or *between* a_1 and b_1, if and only if there are points **A**, **B**, **X**, on a_1, b_1, x_1, with [**AXB**]. There is a precisely analogous condition for a half-plane to lie within the *dihedral angle* $\alpha_1\beta_1$, where α_1 and β_1 are two half-planes bounded by a common line.

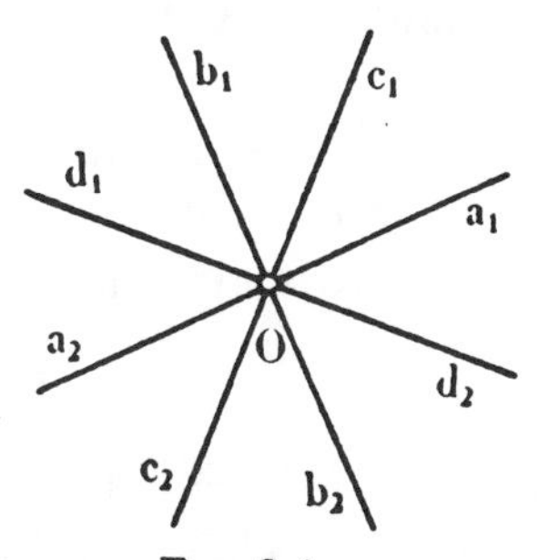

FIG. 8.4A

Let a_1 and a_2 denote the two rays into which a point O divides a line **a** (through O). Then two lines **a** and **b** through O divide the rest of the plane into four convex regions† bounded by the angles a_1b_1, b_1a_2, a_2b_2, b_2a_1. A ray c_1 within the angle a_1b_1 divides it into two angles a_1c_1, c_1b_1, and the supplementary

*Forder [1], p. 68.

†Veblen [1], p. 365; Forder [1], p. 82.

ray c_2 effects a corresponding division of a_2b_2. In this manner we see that n coplanar lines through **O** divide the rest of the plane into $2n$ convex regions. In particular, the notation **a, b, c, d** can be assigned to four concurrent and coplanar lines in such a way that the eight regions are bounded by the angles

$$a_1c_1, \quad c_1b_1, \quad b_1d_1, \quad d_1a_2, \quad a_2c_2, \quad c_2b_2, \quad b_2d_2, \quad d_2a_1,$$

as in Fig. 8.4A. We then say that **a** and **b** *separate* **c** and **d**, writing

$$\mathbf{ab} \parallel \mathbf{cd}.$$

The separation $\alpha\beta \parallel \gamma\delta$ for four coaxial planes can be defined in an analogous manner, since n coaxial planes divide the rest of space into $2n$ convex regions bounded by dihedral angles.

8.5. The geometry of lines and planes through a fixed point. The following definition will be needed soon. A *trihedron* consists of three concurrent but non-coplanar lines **a, b, c** (its *edges*) and their joining planes **bc, ca, ab** (its *faces*).

In §2.2 we asserted that the "third model" for real projective geometry can be set up in Euclidean or non-Euclidean space. More generally, it can be set up in descriptive space. This is proved in detail by taking the axioms of real projective geometry in two dimensions, namely 2.111-2.114, 2.31, 2.32, 2.121-2.126, 2.13, and translating them into provable theorems in the geometry of lines and planes through a fixed point. For instance, the translations of 2.32, 2.126, 2.13 are as follows:

8.51. *If the edges of two covertical trihedra correspond in such a way that the planes joining corresponding edges are coaxial, then the lines of intersection of corresponding faces are coplanar.*

8.52. *If* **abcd** *and* **a′b′c′d′** *are two plane sections of a set of four coaxial planes, and* **ab** ∥ **cd**, *then also* **a′b′** ∥ **c′d′**.

8.53. *For every partition of all the rays within an angle into two non-vacuous sets, such that no ray of either lies between two rays of* *the other, there is a ray of one set which lies between every other ray of that set and every ray of the other set.*

The deduction of these theorems from Axioms 8.311-8.32 is left to the reader. The particularly significant theorem 8.51 is due to Reyes y Prósper.* The final result is as follows:

8.54. *The descriptive geometry of lines and planes through a point can be identified with the real projective geometry of points and lines in a plane.*

In other words, the properties of a bundle are the same in descriptive geometry as in real projective geometry. In particular, there is a principle of duality between lines and planes of a bundle. Using this, or applying 8.54 to 2.33, we obtain the bundle-dual of 8.51, which is also its converse:

8.55. *If the faces of two covertical trihedra correspond in such a way that the lines of intersection of corresponding faces are coplanar, then the planes joining corresponding edges are coaxial.*

8.6. Generalized bundles and pencils. Given two coplanar lines **a** and **b**, which fail to meet within a certain convex region of projective space (or as drawn on an ordinary sheet of paper), the problem of constructing another line through their inaccessible point of intersection can be solved by means of 2.33. But it can be solved more elegantly as follows: Join **a** and **b** by planes to a point **E** outside **ab**. Then lines of the desired kind are coplanar with the line (**Ea**, **Eb**). (It will be proved in 8.61 that the result is independent of the choice of **E**.) Essentially, instead of the inaccessible point (**a**, **b**) we are using the bundle of lines and planes through it. From the standpoint of the convex region, or of the corresponding descriptive geometry, we cannot call this a bundle without

*Reyes y Prósper [1]; Robinson [1], pp. 59-60.

extending the meaning of that word. Let us call it an *improper* bundle, in contrast to the *proper* bundle whose centre is an accessible or *ordinary* point. The following definition applies equally well to either kind.*

If **a** and **b** are any two coplanar lines, we define the *bundle* [**a**, **b**] as a class of lines and planes, consisting of all lines of intersection of planes through **a** and planes through **b**, together with all planes through every such line, and all lines of intersection of such planes with the plane **ab**. If **a** and **b** intersect in a point **O**, we have a proper bundle, consisting of all the lines and planes through the centre **O**. We shall see that an improper bundle has many of the same properties.

The common planes of two bundles are said to form a *pencil*.† If the two bundles contain a common line **o**, we have a *proper* (axial) pencil, consisting of all the planes through the axis **o**. The essential properties of these generalized bundles and pencils are developed in the following theorems.

8.61. *There is just one line of a given bundle through any point (other than the centre, in case the bundle is proper).*

PROOF. Let [**a**, **b**] be the bundle, and **C** the point. When **C** is outside the plane **ab**, the unique line through it is (**Ca**, **Cb**). The theorem is again obvious when **C** lies on **a** or **b**. Suppose, then, that **C** lies in **ab**, but not on **a** or **b**. Let **d** and **e** be any two lines of the bundle outside **ab**. What we have to prove is that the line **c** = (**Cd**, **ab**) could just as well be constructed as (**Ce**, **ab**), or that **c** and **e** are coplanar. Since this is obvious when **a** and **b** intersect, we shall suppose that they do not.

Let **A**, **A′** be two points on **a**, as in Fig. 8.6A. Take a point **N**, in the plane **ab**, on the other side of **b**, and construct **B** = (**AN**, **b**), **B′** = (**A′N**, **b**). Take a point **C′** on **c**, on the

*Pasch and Dehn [1], pp. 33-36; Schur [1], pp. 16-18; Whitehead [2], pp. 18-22; Baker [1], pp. 110-114; Robinson [1], pp. 61-67.

†Veblen [1], p. 372.

opposite side of **BC** from **B′**, and construct **L**=(**BC**, **B′C′**). By 8.317, **NL** meets **CA** in a point **M**. Since **a**, **b**, **c** are each coplanar with **d**, we can apply 8.51 to the trihedra which join any point on **d** to the triangles **ABC**, **A′B′C′**, and conclude that **C′A′** passes through **M**. Since **L**, **M**, **N** are collinear, we can apply 8.55 to the trihedra which join any point **E** on **e** to the same triangles, and conclude that the planes **Ea**, **Eb**, **Ec** are coaxial, i.e. that **Ec** passes through (**Ea**, **Eb**), which is **e**.

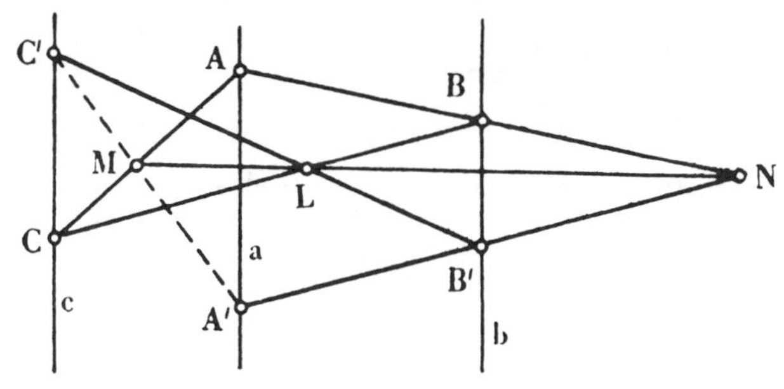

FIG. 8.6A

It follows that the bundle [**a**, **b**] is the same as [**a**, **c**]. Also any line **d** and any plane ρ determine a bundle, provided ρ does not contain **d**; the bundle so determined is naturally denoted by [**d**, ρ].

8.62. *Any two lines of a bundle are coplanar.*

PROOF. Let **d** and **e** be two lines of the bundle [**a**, **b**]. If **d** lies in the plane **ab**, we know that **d** and **e** are coplanar. The same conclusion holds if **d** and **e** lie on opposite sides of **ab**; for, the plane joining **d** to any point on **e** meets **ab** in a line of the bundle, and so contains **e** entirely. Suppose, then, that **d** and **e** lie on the same side of **ab**, and let **g** be any line of the bundle on the other side of **ab**, as in Fig. 8.6B. Let **D** and **E** be any points on **d** and **e**, respectively, and **F** any point between them. Since the lines **g** and **c**=(**Fg**, **ab**) are each coplanar with **d**, and similarly with **e**, we see that **d** and **e** belong to the bundle [**g**, **c**]. But **d** and **e** lie on opposite sides of the plane **gc**. Hence **d** and **e** are coplanar.

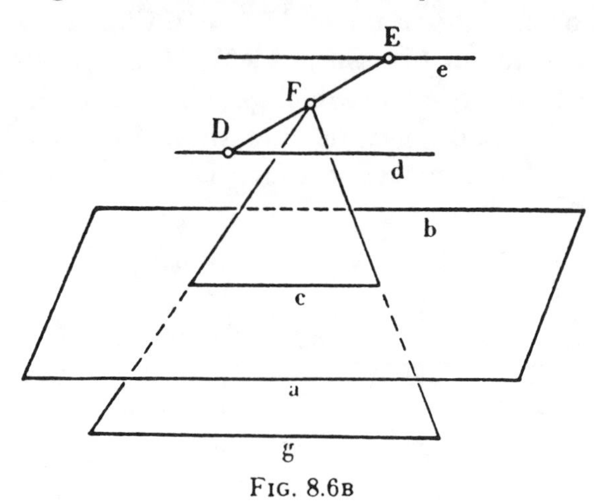

FIG. 8.6B

8.63. *A bundle is determined by any two of its lines.*

PROOF. Let **d** and **e** be two lines of [**a**, **b**]. It follows from 8.62 that **a** and **b** are lines of the bundle [**d**, **e**]. Let **F** be any point in the plane **de**; then the line through **F** of the bundle [**a**, **b**] is coplanar with **a**, and so belongs to [**d**, **e**]. Let **G** be any point outside the plane **de**; then the line through **G** of [**a**, **b**] is coplanar with each of **d**, **e**, and so belongs to [**d**, **e**]. Thus the bundles [**a**, **b**] and [**d**, **e**] are identical.

8.64. *If ρ is any plane of a given bundle, and* **C** *any point in ρ, then the line through* **C** *of the bundle lies in ρ.*

PROOF. Otherwise, any plane through that line would meet ρ in another line through **C** belonging to the bundle. This contradicts 8.61.

8.65. *There is just one plane of a given bundle through any line not belonging to the bundle.*

PROOF. Let **AB** be the given line, and **BC** the line through **B** of the bundle. Then **ABC** is the required plane. It is unique since, if two planes of a bundle intersect, their common line belongs to the bundle.

8.66. *There is just one plane of a given pencil through any point (not on the axis, in case the pencil is proper).*

PROOF. Let **l** and **m** be the lines, through the given point **O**, of the two bundles which determine the pencil. Then **lm**, belonging to both bundles, is the required plane. Any other plane through **O** of the pencil would meet **lm** in a line belonging to both bundles; the pencil would then consist of all the planes through this line, its axis.

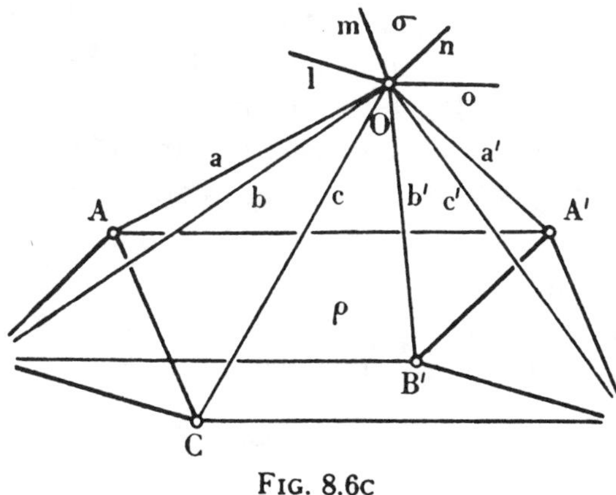

FIG. 8.6c

8.67. *Any two planes belong to a unique pencil.*

PROOF. A pencil containing two given planes ρ and σ is determined by two bundles [**l**, ρ] and [**m**, ρ], where **l** and **m** are two intersecting lines in σ whose common point **O** does not lie in ρ. To establish the uniqueness of this pencil, we shall show that any other bundle which contains ρ and σ also contains every common plane of [**l**, ρ] and [**m**, ρ].

Let [**n**, ρ] be such a bundle, **n** being its line through **O**. By

8.64, $\mathbf{n}$ lies in σ. Take any two points $\mathbf{A}$, $\mathbf{A}'$ in ρ, any point $\mathbf{C}$ on the line $(\mathbf{Am}, \rho)$, any point $\mathbf{B}'$ on $(\mathbf{A}'\mathbf{n}, \rho)$, and construct

$$\mathbf{a} = \mathbf{OA}, \quad \mathbf{b} = (\mathbf{An}, \mathbf{Cl}), \quad \mathbf{c} = \mathbf{OC},$$
$$\mathbf{a}' = \mathbf{OA}', \quad \mathbf{b}' = \mathbf{OB}', \quad \mathbf{c}' = (\mathbf{A}'\mathbf{m}, \mathbf{B}'\mathbf{l}),$$

as in Fig. 8.6c. By 8.55, since $\mathbf{l}$, $\mathbf{m}$, $\mathbf{n}$ are coplanar, the three planes $\mathbf{aa}'$, $\mathbf{bb}'$, $\mathbf{cc}'$ pass through one line $\mathbf{o}$. Let $\mathbf{O}_1$ be any point on neither of the planes ρ, σ. Let $\mathbf{l}_1$, $\mathbf{m}_1$, $\mathbf{n}_1$, $\mathbf{o}_1$ be the lines through $\mathbf{O}_1$ of the respective bundles $[\mathbf{l}, \rho]$, $[\mathbf{m}, \rho]$, $[\mathbf{n}, \rho]$, $[\mathbf{o}, \rho]$, and construct

$$\mathbf{a}_1 = \mathbf{O}_1\mathbf{A}, \quad \mathbf{b}_1 = (\mathbf{An}_1, \mathbf{Cl}_1), \quad \mathbf{c}_1 = \mathbf{O}_1\mathbf{C},$$
$$\mathbf{a}_1' = \mathbf{O}_1\mathbf{A}', \quad \mathbf{b}_1' = \mathbf{O}_1\mathbf{B}', \quad \mathbf{c}_1' = (\mathbf{A}'\mathbf{m}_1, \mathbf{B}'\mathbf{l}_1).$$

Now, the bundle $[\mathbf{b}, \rho]$ contains the planes $\mathbf{An}$, $\mathbf{Cl}$, which meet ρ in the lines through $\mathbf{A}$, $\mathbf{C}$ of the respective bundles $[\mathbf{n}, \rho]$, $[\mathbf{l}, \rho]$. But $[\mathbf{b}_1, \rho]$, containing the same lines through $\mathbf{A}$ and $\mathbf{C}$, is the same bundle. Hence the planes $\mathbf{bb}'$ and $\mathbf{b}_1\mathbf{b}_1'$ (through $\mathbf{B}'$) are coaxial with ρ. Similarly, $\mathbf{cc}'$ and $\mathbf{c}_1\mathbf{c}_1'$ (through $\mathbf{C}$) are coaxial with ρ. Thus the three planes $\mathbf{a}_1\mathbf{a}_1'$, $\mathbf{b}_1\mathbf{b}_1'$, $\mathbf{c}_1\mathbf{c}_1'$ belong to the bundle $[\mathbf{o}, \rho]$, and pass through the line $\mathbf{o}_1$. By 8.51, applied to the trihedra $\mathbf{a}_1\mathbf{b}_1\mathbf{c}_1$ and $\mathbf{a}_1'\mathbf{b}_1'\mathbf{c}_1'$, the three lines $\mathbf{l}_1$, $\mathbf{m}_1$, $\mathbf{n}_1$ are coplanar. Thus any plane belonging to both bundles $[\mathbf{l}, \rho]$ and $[\mathbf{m}, \rho]$ belongs also to $[\mathbf{n}, \rho]$.

8.68. *Any three planes, not belonging to a pencil, belong to a unique bundle.*

PROOF. Let ρ, σ, τ be the three planes, and $\mathbf{O}$ any point in τ. Let ω be the plane through $\mathbf{O}$ of the pencil determined by ρ and σ. Then ω meets τ in a line through $\mathbf{O}$, say $\mathbf{l}$. The bundle $[\mathbf{l}, \rho]$, containing the planes ω and ρ of the pencil, contains σ also, by 8.67. This bundle is unique, since two such would determine a pencil.

8.69. *If two pencils contain a common plane, they belong to one bundle.*

PROOF. Let each pencil be determined by the common plane and one other plane, and apply 8.68.

8.7. Ideal points and lines. In this section and the next, we shall establish the validity of the "first model" for real projective geometry, by translating the above theorems according to the dictionary on page 24, and deducing the axioms 2.111-2.13. *Proper* and *improper* bundles (or pencils) will be translated as *ordinary* and *ideal* points (or lines). It is natural to identify the ordinary points and lines with the centres and axes of the proper bundles and pencils, and to think of the ideal points and lines as centres and axes of improper bundles and pencils.*

An analogous change of meaning occurs several times in arithmetic.† For instance, we derive the field of rational numbers from the ring of integers by defining a rational number as the class of "equivalent" pairs of integers n/d, the criterion for equivalence of such pairs being $nd' = dn'$. We then observe that certain classes of pairs (namely those in which d divides n) are isomorphic to the integers themselves. Therefore we agree to include the integers among the rational numbers, identifying the integer n with the class of pairs equivalent to $n/1$. It is to be clearly understood, however, that the use of such devices in mathematics is psychological rather than logical. They aid our thinking in much the same way as diagrams do in the discovery of geometrical theorems.

Returning to geometry, we say that a point and a line (ordinary or ideal) are *incident* if the bundle contains the pencil. In particular, an ordinary line passes through an ideal point if it belongs to the corresponding improper bundle. Similarly, we say that a given plane passes through an ideal point or line

*Pasch and Dehn [1], pp. 40-49; Schur [1], pp. 18-20; Bonola [1], pp. 110-116; Veblen [1], pp. 373-376; Whitehead [2], pp. 22-29.

†Robinson [1], pp. 77, 86.

if it belongs to the corresponding improper bundle or pencil. Thus Theorems 8.61-8.69 are translated as follows:

An ideal point is joined to an ordinary point by an ordinary line.

Any two ordinary lines through an ideal point are coplanar.

If two ordinary lines lie in a plane, they meet in a point (ordinary or ideal).

Any plane containing an ideal point and an ordinary point contains their join.

An ideal point (or line) is joined to an ordinary line (or point) by a plane.

Any two planes meet in a line (ordinary or ideal).

Any three non-coaxial planes meet in a point (ordinary or ideal).

Any two coplanar lines (ordinary or ideal) meet in a point.

It is easily deduced that an ideal line meets a plane (not containing it) in an ideal point. (Naturally, every point on an ideal line is ideal.) Such results become quite obvious when we think of the ordinary points and lines as interior points and "secants" of a convex region, such as a sphere. This model forms a suggestive guide, but we have not yet completed its justification.

8.8. Verifying the projective axioms. Clearly, Axioms 2.111-2.114 present no difficulty. Before considering the next, it is useful to remark that we can prove Desargues' Theorem 2.32, and its converse 2.33, by joining the vertices and sides of the given triangles to an ordinary point outside their plane, and applying 8.51 or 8.55 to the consequent trihedra.

Axiom 2.115 (Fig. 8.8A) is easily verified whenever at least one of **A**, **B**, **C** is an ordinary point; but it is no longer obvious when all these points are ideal. To establish it we construct, in any plane through **BC**, two triangles whose points of intersection of corresponding sides are **D**, **B**, **C**. This can be done* in such a way that each triangle has at least two ordinary vertices, while no two corresponding vertices are both ideal.

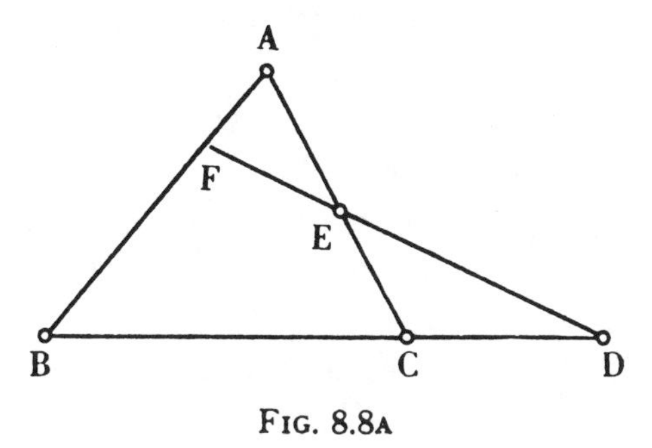

FIG. 8.8A

Then all three sides of each triangle are ordinary lines, and so also are the three joins of corresponding vertices, which, by 2.33, are concurrent. We now join these nine ordinary lines to **A** by nine planes, which intersect any plane through **DE** in a new set of nine lines (ordinary or ideal). These again form two triangles with their corresponding vertices joined by concurrent lines. Hence, by 2.32, the points of intersection of corresponding sides are collinear. These three points, lying on **AD**, **AB**, **AC**, respectively, are **D**, **E**, and the desired point **F**.

We may now consider the class of points lying on lines which join **A** to the points of **BC** as a *plane*: *ordinary* if it contains any ordinary point, *ideal* otherwise. (Even when both **A** and **BC** are ideal, the plane is not necessarily ideal; for, **A** and **BC** might be an ideal point and line in a given ordinary plane.) Since an ideal plane contains no ordinary point, the verification of 2.116 is trivial.

It remains to be shown that an ideal plane ρ meets any plane ω in a line. Let **a** and **b** be two arbitrary lines in ρ. If ω is ordinary, it meets ρ in the line $(\mathbf{a}, \omega)(\mathbf{b}, \omega)$. The same

*Robinson [1], p. 66. (His A and A' may be taken to coincide with our **D**.) For an alternative procedure see Pasch and Dehn [1], pp. 52-53.

conclusion holds when ω is ideal, provided there is a point of intersection ($\mathbf{a}$, ω) for any line $\mathbf{a}$. This can be constructed as the point of intersection of $\mathbf{a}$ with the line ($\mathbf{Oa}$, ω), where $\mathbf{O}$ is any ordinary point. Thus 2.117 is verified.

To verify the Axioms of Separation and Continuity, we define the separation of four collinear points in a way that will apply to ideal points just as well as to ordinary points. This is done by joining the four points to an ordinary point $\mathbf{O}$ by lines $\mathbf{a}$, $\mathbf{b}$, $\mathbf{c}$, $\mathbf{d}$, and taking the definition of $\mathbf{ab} \parallel \mathbf{cd}$ from §8.4. To see that this way of defining separation is independent of the choice of $\mathbf{O}$, we join the same four points to another point $\mathbf{O'}$ by lines $\mathbf{a'}$, $\mathbf{b'}$, $\mathbf{c'}$, $\mathbf{d'}$, and apply 8.52 to the coaxial planes $\mathbf{aa'}$, $\mathbf{bb'}$, $\mathbf{cc'}$, $\mathbf{dd'}$. Thus Axioms 2.121-2.13 follow from the corresponding results in the geometry of a proper bundle (§8.5).

The desired extension of descriptive space is now accomplished. We have found that the descriptive geometry of bundles and pencils is "isomorphic" with the projective geometry of points and lines. But it is obvious that the geometry of *proper* bundles and pencils is isomorphic with the *descriptive* geometry of points and lines. In this sense, therefore, descriptive space is a part of projective space, and the use of ideal elements is entirely justified.

We saw, in 8.54, that the geometry of a proper bundle is projective. It can be proved similarly that the geometry of an improper bundle is descriptive. In other words,

8.81. *The geometry of ordinary lines and planes through an ideal point can be identified with the geometry of ordinary points and lines in a plane.*

8.9. Parallelism. It is interesting to observe that, if Axiom 8.32 were omitted, we could have a "descriptive" geometry* in which two coplanar lines always intersect; so

*Veblen [1], p. 348.

that all bundles (and pencils) are proper, and the projective axioms (except 2.13) can be deduced without extending the space. The manner in which continuity provides non-intersecting lines will be seen in the proof of the following theorem:*

8.91. *Every line contains at least one ideal point.*

PROOF. Suppose, if possible, that a certain line **q** contains no ideal point, so that it is a projective line whose points are all ordinary. Take points **B, C, D** on **q**, such that [**BCD**], and apply 2.13 to the partition of the segment **BD/C** into the two rays **B/C** and **D/C**. The dividing point **E** belongs to one of these rays, say the latter. By 8.312, this ray contains a point **F** such that [**DEF**]. But by 2.13, such a point **F** belongs to the *other* ray; so we have a contradiction, and our theorem is proved.

In order to define parallelism, we consider once more a flat pencil with centre **A**, and a line **BC** or **q** in the same plane, as in Fig. 1.2A. The pencil includes all the lines that join **A** to ordinary points on **q**. By 8.91, it contains at least one more line, say **s**. If (**q, s**) is the *only* ideal point on **q**, we say (with Euclid) that **s** is *parallel* to **q**.

But if **q** contains *more than one* ideal point, let **R** and **S** be two, such that **BR** || **CS**. Then, by 2.13 applied to the segment **CS/B**, the ray **C/B** contains a *first* ideal point **M**, such that all points between **C** and **M** are ordinary, while all points "beyond" **M** are ideal. Similarly, the supplementary ray **C/D** contains a first ideal point **N**. We now say (with Gauss and Lobatschewsky) that both **AM** and **AN** are parallel to **q**.

Without making a further assumption, we cannot say whether **q** contains just one ideal point or more than one. In other words, we cannot say whether *one* or *two* lines parallel

*Veblen [1], pp. 369-370.

to **q** can be drawn through **A**. We may combine these alternatives by using rays instead of lines, thus:

A ray p_1, emanating from **A**, is said to be parallel to a ray q_1 if it joins **A** to the first ideal point on q_1. It is then also said to be parallel to the line **q** which contains q_1.

According to this definition, there are always two rays from **A** parallel to **q**. If **q** contains only one ideal point, these two rays are supplementary (and the word "first" in the definition is superfluous). The rays are again supplementary if **A** lies on **q**, any ray being parallel to itself. When **A** does not lie on **q**, the parallels from **A** separate every ray that meets **q** (in an ordinary point) from every other ray that fails to do so.

If the ray p_1 is parallel to **q**, so also is any ray proceeding in the same sense along the line **p** which contains p_1. Thus the position of **A** on **p** is immaterial,* and we can define the parallelism of two *lines* as follows:

A line **p** is said to be parallel to a line **q** if it contains the first ideal point on **q** in either sense. Hence, if there is only one ideal point on **q**, there is only one line parallel to **q** through *any* point **A**; but if there are two (and so infinitely many) ideal points on **q**, then there are two lines parallel to **q** through any point **A** not lying on **q**.

The relation of parallelism, as defined above, is obviously reflexive. We shall now prove that it is also symmetric and transitive.†

8.92. *If* p_1 *is parallel* to q_1, *then* q_1 *is parallel to* p_1.

PROOF. Take any points **A**, **B**, on **p**, **q**, respectively, and **A′** on **A/B** (or "**BA** produced"), as in Fig. 8.9A. From **A′**,

---

*Gauss [1], p. 203.

†Gauss gave a metrical proof of 8.92, and a descriptive proof of 8.93. Lobatschewsky's treatment is somewhat simpler, but still metrical. (See Carslaw [1], p. 45.) Our descriptive proof of 8.92 was suggested by Gauss's proof of 8.93.

draw r_1 parallel to q_1. Then any ray from **A′** within the angle between **A′B** and r_1 meets q_1, say at **C**. It is a consequence of Axioms 8.311-8.317 that* if a line coplanar with **A′**, **B**, **C** meets the segment **A′B** and does not pass through **C**, it meets one of the segments **A′C**, **BC**. Hence **p**, not meeting **BC**, must meet **A′C**. Since this property of **A′C** applies to *every* ray between **A′B** and r_1, r_1 is parallel to p_1. Let **M** be the first ideal point on q_1, so that p_1 is **AM** and r_1 is **A′M**. Then, since r_1 is parallel to p_1, **M** is the first ideal point on p_1, and q_1 (or **BM**) is likewise parallel to p_1.

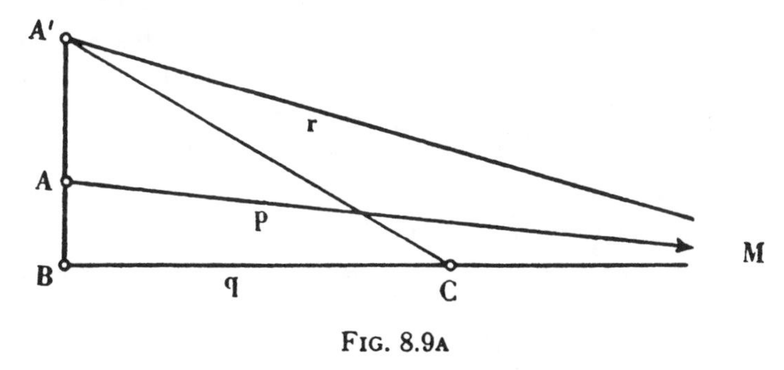

FIG. 8.9A

Thus parallelism is symmetric, and we may say that two rays are parallel (to one another) if and only if they have the same first ideal point. Hence

8.93. *Two rays which are parallel to the same ray are parallel to one another*, and

8.94. *The lines parallel to a given ray form an improper bundle.*

We naturally call this a *bundle of parallels*, and the corresponding ideal point a *point at infinity*. Moreover, any line and plane of the bundle are said to be parallel (to one another). Thus two lines, or a line and a plane, are parallel if and only if they have a common point at infinity.

*Forder [1], p. 58.

The following theorem will enable us to distinguish two kinds of point at infinity:

8.95. *If there is a line containing only one ideal point, then every parallel line contains only one ideal point.*

PROOF. Let **q** be such a line, and **p** any parallel line. Choose **A′** as in Fig. 8.9A. Since **q** contains only one ideal point, the rays from **A** parallel to **q**, and likewise those from **A′**, are supplementary. But the latter pair of rays are parallel also to **p**. Hence **p** contains only one ideal point.

It follows that every point at infinity is of one of two kinds. Every ordinary line through a point of the first kind contains no other ideal point; but every line through a point of the second kind contains an infinity of ideal points.

If every point at infinity is of the first kind, then every ordinary plane contains just one ideal line (its *line at infinity*), and there is just one ideal plane (the *plane at infinity*). This geometry is *affine* (see §2.1), and can be metricized by singling out an elliptic or hyperbolic polarity in the plane at infinity. We thus obtain three-dimensional Euclidean or Minkowskian geometry, respectively.

In hyperbolic geometry, which may now be considered as the geometry of points interior to an oval quadric, every point at infinity is of the second kind. But this is not the only such case: the locus of points at infinity may be any convex surface, not necessarily a quadric. The geometry can then be considered as a "distorted" hyperbolic geometry.

Between these extremes we have many geometries in which points at infinity of the two kinds coexist. The most obvious case is a Euclidean half-space,* wherein any line parallel to the bounding plane contains only one ideal point, but any other line contains infinitely many (in the supplementary half-space). Another instance is the interior of a cone (in real projective space); here the vertex alone is of the first kind.

*Forder [1], p. 303.

CHAPTER IX

EUCLIDEAN AND HYPERBOLIC GEOMETRY

9.1. The introduction of congruence. In Chapters V-VII we introduced the elliptic metric into real projective geometry by means of the "absolute polarity," and observed the equivalence of two alternative definitions for a congruent transformation: a point-to-point transformation preserving distance, and a collineation permutable with the absolute polarity. It is quite easy to introduce the hyperbolic metric similarly (see §8.1). But in order to follow the historical development more closely, we prefer to reverse the process, introducing congruence into descriptive geometry as a second undefined relation, and stating its properties in the form of axioms. The propositions of Bolyai's "absolute geometry" can then be deduced in a straightforward manner. After imbedding the descriptive space in a real projective space by the method of Chapter VIII, we shall find a definite polarity which is permutable with every congruent transformation.

The relation of congruence applies initially to point-pairs, and we write $\mathbf{AB} \equiv \mathbf{CD}$ to mean that the point-pair **AB** is congruent to the point-pair **CD**. But since every point-pair determines a unique segment, no confusion will be caused by reading the same formula as "the segment **AB** is congruent to the segment **CD**."

The idea of introducing congruence axiomatically is due to Pasch. His axioms were simplified by Hilbert and R. L. Moore.* For the sake of reducing the number of axioms to a minimum, it should be noticed that 9.11 makes 8.312 superfluous. Accordingly, we begin with the single undefined

*Pasch and Dehn [1], pp. 92-101; Hilbert [1], pp. 9-19; Moore [1].

entity, a *point*, the two undefined relations of *intermediacy* and *congruence*, the "descriptive" axioms 8.311, 8.313-8.32, and the following five

AXIOMS OF CONGRUENCE

9.11. *If* **A** *and* **B** *are distinct points, then on any ray* **C/E** *there is just one point* **D** *such that* **AB≡CD**.

9.12. *If* **AB≡CD** *and* **CD≡EF**, *then* **AB≡EF**.

9.13. **AB≡BA**. (Therefore **AB≡AB**.)

9.14. *If* [**ABC**] *and* [**A′B′C′**] *and* **AB≡A′B′** *and* **BC≡B′C′**, *then* **AC≡A′C′**.

9.15. *If* **ABC** *and* **A′B′C′** *are two triads of non-collinear points, with* **BC≡B′C′**, **CA≡C′A′**, **AB≡A′B′**, *while* **D** *and* **D′** *are two further points, such that* [**BCD**], [**B′C′D′**], *and* **BD≡B′D′**, *then* **AD≡A′D′**.

It is easily deduced that the relation of congruence is not only reflexive and transitive but also symmetric.* Moreover, this relation is readily extended from point-pairs or segments to figures of any kind. The only really complicated axiom is 9.15, which may be roughly described as ensuring the rigidity of "a triangle with a tail."

The detailed deduction of elementary theorems would take too much space, so we shall be content to mention some of the most important steps. Axioms 9.11-9.14 enable us to define the *length* of a segment (or of an interval), and then 8.32 shows that lengths form a continuous set of magnitudes. In particular, every interval has a *mid-point*. After defining congruent angles in the natural manner, we can deduce† that *if* p_1q_1 *is any angle, and* a_1 *any ray, there are not more than two rays* b_1, *in a plane through* a_1, *such that* $a_1b_1 \equiv p_1q_1$.

If a_1 and a_2 are supplementary rays, and b_1 emanates from

*Forder [1], p. 92.

†Forder [1], p. 132. Hilbert took this as an axiom, instead of 9.15.

the same vertex, the angles a_1b_1 and a_2b_1 are said to be supplementary. If $a_1b_1 \equiv a_2b_1$, then the rays a_1 and b_1, or the lines **a** and **b** which contain them, are said to be *perpendicular*, and the angle a_1b_1 is called a *right angle*.

The statement $\mathbf{AB} \equiv \mathbf{CD}$ for segments is clearly equivalent to the statement $\mathbf{AB} = \mathbf{CD}$ for lengths, so no confusion arises from using the same symbol for a segment and its length. A similar remark applies to angles, although there the situation is slightly more complicated.*

The *circle* with centre **O** and radius **OP** is defined as the class of points **X**, in a plane through **O**, such that $\mathbf{OX} \equiv \mathbf{OP}$ (or $\mathbf{OX} = \mathbf{OP}$). A point **Q** such that $\mathbf{OQ} > \mathbf{OP}$ is said to be *outside* the circle. Points neither on nor outside the circle are said to be *inside*. It can be proved† that if a circle with centre **A** has a point inside and a point outside a circle with centre **C**, in the same plane, then the two circles meet in at least one point on each side of **AC**. It then follows that *any two right angles are congruent*, and we have reached Euclid's starting-point (with Postulates I-IV). Accordingly, we accept Euclid's propositions I, 1-28, with the word "parallel" replaced by "non-intersecting." We shall also adopt the customary notation $\angle\mathbf{BAC}$ for the angle formed by the rays from **A** that contain **B** and **C**, respectively.

9.2. Perpendicular lines and planes. In Chapter VIII we saw how to extend descriptive space by defining ideal elements, and found that the result is real projective space. The method used has the advantage that no new axioms are needed, as the ideal elements are sets of things already defined. Our purpose in the present chapter is to show that the effect of introducing congruence into the descriptive space is to single out an *absolute polarity*, either in one ideal plane or in the whole pro-

*Forder [1], pp. 113-115.
†Forder [1], pp. 308, 131, 133.

jective space. In other words, the treatment of Cayley and Klein will be derived from that of Bolyai and Lobatschewsky. As a first step we shall obtain, in §9.3, a metrical description for the bundles and pencils which determine the various kinds of ideal element, beginning with the following chain of theorems about perpendicularity. (Until §9.4, all points, lines, and planes considered are ordinary.)

9.21. *A line which is perpendicular to each of two intersecting lines at their common point* **A**, *is perpendicular to every line through* **A** *in the plane of the two lines.*

Euclid's proof (XI, 4) is valid. Such a line and plane are said to be perpendicular (to one another).

9.22. *There is just one line through any given point, perpendicular to any given plane.*

Euclid's constructions (XI, 11 and 12) give such a line. (But his proof of the former has to be modified* to avoid using parallels.) The uniqueness follows (as in Euclid XI, 13) by considering the plane that two such lines would determine.

9.23. *If a plane* σ *contains a line* **r** *perpendicular to a plane* ρ, *then* ρ *contains a line* **s** *perpendicular to* σ.

This follows easily from 9.21. Two planes such as ρ and σ are said to be perpendicular. (Cf. Euclid XI, 18.)

9.24. *If* ρ *and* σ *are perpendicular planes, any line in* σ *which is perpendicular to the line* (ρ, σ) *is perpendicular to* ρ.

PROOF. Let σ contain the line **BC** perpendicular to ρ, and ρ contain the line **BD** perpendicular to σ. (See Fig. 9.2A.) Let **AE** be any line in σ perpendicular to (ρ, σ) or **BA**. On **A/E**, take **F** so that **AF** $\equiv$ **AE**. By considering pairs of congruent triangles (as in Euclid's proof of XI, 6), we deduce in turn that **BE** $\equiv$ **BF**, that **DE** $\equiv$ **DF**, and that **DA** is perpendicular to **EA**.

*Forder [1], p. 123.

Hence **EA**, being perpendicular to both **AB** and **AD**, is perpendicular to ρ.

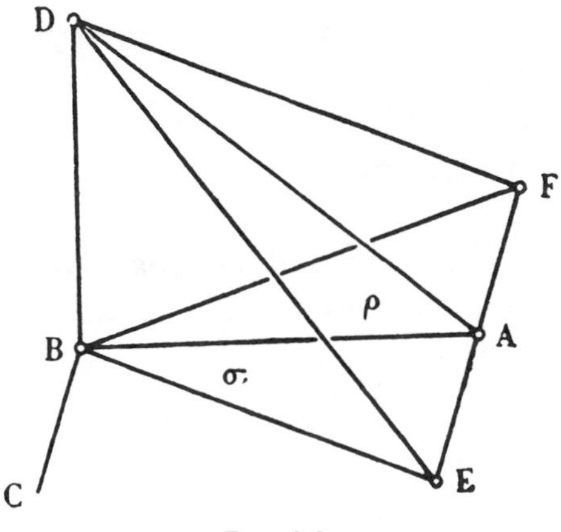

FIG. 9.2A

9.25. *Two lines which are perpendicular to the same plane are coplanar.*

PROOF. Let **BC** and $\mathbf{AE_1}$ be any two lines perpendicular to ρ (at **B** and **A**). By 9.24, the line **AE** (perpendicular to **AB** in the plane **ABC**) is perpendicular to ρ. By 9.22, $\mathbf{AE_1}$ coincides with **AE**, and so is coplanar with **BC**.

9.26. *A plane which is perpendicular to each of two intersecting planes is perpendicular to their common line.*

PROOF. Let ρ be perpendicular to the two intersecting planes σ, τ. From any point **C** on (σ, τ) draw lines perpendicular to (ρ, σ) and (ρ, τ), respectively. By 9.24, each of these lines is perpendicular to ρ. But by 9.22, there is only one line through **C** perpendicular to ρ. This therefore lies in both planes σ, τ, and is (σ, τ) itself.

9.27. *There is just one plane through any given point, perpendicular to any given line.*

PROOF. In the plane of the given point **A** and given line **r** (or, if **r** contains **A**, in *any* plane through **r**) draw the line **t** perpendicular to **r** through **A**. Let **s** be the line through **A** perpendicular to the plane **rt**. Then **st** is the required plane (by 9.24 with $\rho = \mathbf{st}$, $\sigma = \mathbf{rt}$). The uniqueness can be established (as in Euclid XI, 14) by considering the section which two such planes would make on **rt**.

9.3. Improper bundles and pencils. We now consider two kinds of bundle and two kinds of pencil, without saying that they are necessarily different.

9.31. *The lines and planes perpendicular to a given plane form an improper bundle.*

PROOF. Let **a** and **b** be any two lines perpendicular to the given plane ρ. By 9.25, these are coplanar. All the planes through **a** or **b**, all the lines of intersection of pairs of these planes, all planes through each of these lines, and all lines of intersection of such planes with **ab**, are perpendicular to ρ (by 9.23 and 9.26). These lines and planes form the bundle [**a**, **b**], which is improper by 9.22.

9.32. *The planes perpendicular to a given line form an improper pencil.*

PROOF. Take any two planes ρ, σ, through the given line, and consider the bundles perpendicular to ρ and σ, respectively. The common planes of these bundles form a pencil. Being perpendicular to both ρ and σ, they are perpendicular to the given line. Conversely, every plane perpendicular to the given line is perpendicular to both ρ and σ, and so belongs to the pencil. The pencil is improper, by 9.27.

By 8.94, there is a "bundle of parallels" consisting of all lines and planes parallel to a given ray. Let ρ be any plane of such a bundle. Then those planes of the bundle which are perpendicular to ρ, being the common planes of two bundles, form a pencil, called a *pencil of parallels*. (This pencil is improper. For otherwise, by 9.26, its axis would intersect ρ; but it would also be parallel to ρ.) Any two planes of such a pencil are said to be parallel (to one another).

9.4. The absolute polarity. The ideal point **R**, defined by the bundle perpendicular to a plane ρ, is called the *absolute pole* of ρ; it lies on every line or plane perpendicular to ρ. The ideal line **RS**, defined by the pencil of planes perpendicular to a line (ρ, σ), is called the *absolute polar line* of (ρ, σ); it lies on every plane perpendicular to (ρ, σ).

9.41. *The absolute poles of the planes through an ordinary line* **l** *are the points of its absolute polar line* **l′**.

PROOF. Let ω and ω_1 be any two planes perpendicular to **l**, so that **l′** is (ω, ω_1). Since all planes through **l** are perpendicular to ω and ω_1, their absolute poles lie on **l′**. Conversely, if **R** is any given point on **l′**, there is a plane through **l** whose absolute pole is **R**, namely the plane through **l** perpendicular to **Rl**.

9.42. *The absolute poles of the planes through an ordinary point* **A** *are the points of an ideal plane* α.

PROOF. Let **l** be any line through **A**, and ρ any plane through **A** but not through **l**. Let **l′** be the absolute polar line of **l**, and **R** the absolute pole of ρ. By 9.41, **R** does not lie on **l′**. The planes through **l** meet ρ in a flat pencil of lines, whose absolute polar lines join **R** to the points of **l′**. Hence the planes through the lines of the flat pencil (i.e. the planes through **A**) have for absolute poles the points of the plane $\alpha =$ **Rl′**. This plane is ideal, as every point in it is the absolute pole of an ordinary plane.

We call α the *absolute polar plane* of **A**.

Thus every *ordinary* point, line. or plane has a definite absolute polar plane, line, or point (pole), with incidences corresponding dually.* But we must not assume that every *ideal* element is the pole or polar of an ordinary element.

We saw, in §8.9, that a line may contain *one* ideal point or *more than one*. By the Axioms of Congruence, any line is congruent to every line; hence either every line contains just one ideal point, or every line contains infinitely many ideal points. In other words, these axioms exclude the coexistence of the two kinds of point at infinity defined on page 178. Accordingly, we make our geometry categorical by stating one further axiom, either affirming or denying the existence of a *unique* parallel to a given line through a given point.

9.5. The Euclidean case. One of the alternative statements is:†

THE EUCLIDEAN AXIOM OF PARALLELISM

9.51. *There is at least one line* **q** *and at least one point* **A**, *not on* **q**, *such that not more than one line can be drawn through* **A** *coplanar with but not meeting* **q**.

This implies that one (and so every) ordinary line contains just one ideal point, its "point at infinity." Hence the join of two such points must be an ideal line. An ordinary plane contains only one ideal line (since two such would meet an ordinary line in two ideal points); hence the plane determined by two such "lines at infinity" must be an ideal plane. Similarly, there is only one ideal plane (in three-dimensional space). This "plane at infinity" contains every ideal point, and is the absolute polar plane of every ordinary point. Thus the "absolute polarity" does not operate uniformly throughout the whole projective space, but is degenerate.

*Pasch and Dehn [1], p. 147.

†Moore [1], p. 489; Forder [1], p. 307.

Let **A**, $\mathbf{A}_1$ be two ordinary points, and ρ, ρ_1 the planes perpendicular to $\mathbf{AA}_1$ through **A**, $\mathbf{A}_1$, respectively. Then ρ and ρ_1 have the same absolute pole, namely the point at infinity on $\mathbf{AA}_1$; and every line perpendicular to ρ is also perpendicular to ρ_1. Hence the absolute polar line of any such line is (ρ, ρ_1), and every ordinary plane through the line at infinity in ρ is perpendicular to every ordinary line through the point at infinity on $\mathbf{AA}_1$. In other words, we have an improper pencil of ("horizontal") planes perpendicular to an improper bundle of ("vertical") lines.

This correspondence between pencils and bundles can be regarded as a two-dimensional polarity in the plane at infinity, two ordinary lines (or planes) being perpendicular if their points (or lines) at infinity are conjugate in this polarity. Since self-perpendicular lines are excluded, the polarity is elliptic, and we have the projective definition for Euclidean geometry.

Since every ideal point is now a point at infinity, the lines and planes perpendicular to any given plane form a bundle of parallels, and the planes perpendicular to any given line form a pencil of parallels.

A large part of the above theory can be developed without using continuity, provided we insert some extra axioms of congruence. However, we must then abandon the theory of parallelism as developed in §8.9, and the above deductions from it. In fact, Dehn has developed a "semi-Euclidean geometry" in which all ordinary points have the same absolute polar plane, although ideal points exist outside this plane.*

9.6. The hyperbolic case. The other alternative is:†

THE HYPERBOLIC AXIOM OF PARALLELISM

9.61. *There is at least one line* **q** *and at least one point* **A**, *such that two distinct lines can be drawn through* **A** *coplanar with but not meeting* **q**.

*Dehn [1], pp. 436-438; Forder [1], pp. 337-338.

†Moore [1], p. 503.

This implies that one (and so every) ordinary line contains two (and so infinitely many) ideal points. We shall deduce, in §9.7, that in this case the absolute polarity is non-degenerate, i.e. that every point, ideal as well as ordinary, has a definite polar plane. In preparation for this, we shall prove some of the classical theorems of two-dimensional hyperbolic geometry,* culminating in the famous result about the angle-sum of a triangle (1.32, 9.66).

By 9.61, the two rays that can be drawn from **A** parallel to **q** are not supplementary, and a ray from **A** meets **q** if and only if it lies within the angle formed by these particular rays. In other words, the flat pencil of lines through **A** in **Aq** contains two special lines, **p** and **p**′, which separate all other lines not meeting **q** from all lines which meet **q**. (See Fig. 1.2A.) According to our definition, the lines **p** and **p**′ are parallel to **q**. It is convenient to describe the "other lines not meeting **q**" as *ultra-parallel* to **q**; they lie in the "external" angle formed by the two parallels.

The figure consisting of two parallel lines with a transversal **AB** may be regarded as a triangle **ABM** with **M** at infinity; it is therefore called a *singly-asymptotic triangle.* Similarly, a triangle with two or three vertices at infinity is said to be doubly- or trebly-asymptotic. The analogy with ordinary triangles is exemplified in 9.62 and 9.64, which are adapted from Euclid I, 26 and 16.

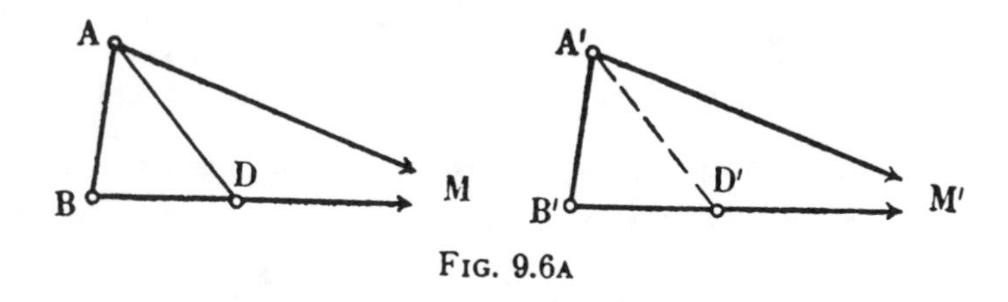

FIG. 9.6A

*Carslaw [1], pp. 48-54.

9.62. *Two singly-asymptotic triangles* **ABM**, **A′B′M′**, *such that* **AB** = **A′B′** *and* ∠**ABM** = ∠**A′B′M′**, *are congruent.*

PROOF. If the angles at **A** and **A′** are unequal, suppose the former to be the greater. Draw a ray **AD** so that ∠**BAD** = ∠**B′A′M′**, and let **D** be the point where this ray meets **BM**, as in Fig. 9.6A. On **B′M′**, take a point **D′** so that **B′D′** = **BD**. Then the triangles **ABD** and **A′B′D′** are congruent. Therefore

$$\angle \mathbf{B'A'D'} = \angle \mathbf{BAD} = \angle \mathbf{B'A'M'},$$

which is absurd. Hence in fact ∠**BAM** = ∠**B′A′M′**.

Applying this theorem to the singly-asymptotic triangles **ABM** and **ABN**, where **AM** and **AN** are the parallels from **A** to **q**, and **AB** is perpendicular to **q** (as in Fig. 1.2A), we conclude that ∠**BAM** and ∠**BAN** are equal, and therefore acute. Their common value (which, by 9.62, is a function of the distance **AB**) is called the *angle of parallelism* for **AB**, and is denoted by Π(**AB**). (This is Lobatschewsky's notation.) It follows also that the line through **A** perpendicular to **AB** is ultra-parallel to **q**. Thus

9.63. *Two lines which are perpendicular to the same line are ultra-parallel to one another.*

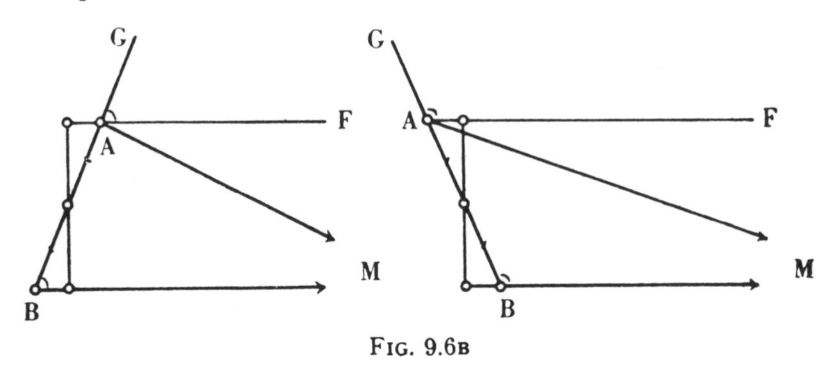

FIG. 9.6B

9.64. *In a singly-asymptotic triangle* **ABM**, *the external angle at* **A** *is greater than the internal angle at* **B**.

PROOF. Produce **BA** to **G**, and draw **AF** so that $\angle$**FAG** = $\angle$**MBA**, as in Fig. 9.6B. Then the perpendicular to **BM** from the mid-point of **AB** is also perpendicular to **AF**. By 9.63, **AF** is ultra-parallel to **BM**, and $\angle$**FAG** < $\angle$**MAG**; that is,

$$\angle \mathbf{MAG} > \angle \mathbf{MBA}.$$

Our next theorem concerns Saccheri's isosceles birectangle.

9.65. *If a simple quadrangle* **ABED** *has right angles at* **D** *and* **E**, *while* **AD** = **BE**, *then the angles at* **A** *and* **B** *are equal acute angles.*

PROOF. By constructing pairs of congruent triangles, we easily see that the angles at **A** and **B** are equal. To show that they are acute, we draw **AM** and **BM** parallel to **D/E**, as in Fig. 9.6C, and apply 9.62 to the triangles **BEM**, **ADM**. Since **BE** = **AD**, while the angles at **E** and **D** are equal, we conclude that $\angle$**EBM** = $\angle$**DAM**. By 9.64, we have $\angle$**MBA** < $\angle$**MAG**, where **G** is any point on **A/B**. Hence

$$\angle \mathbf{BAD} = \angle \mathbf{EBA} < \angle \mathbf{DAG},$$

and $\angle$**BAD** is acute.

We now come to the theorem which Gauss used in his unsuccessful attempt to determine the nature of physical space, when he set up theodolites on three mountain peaks.

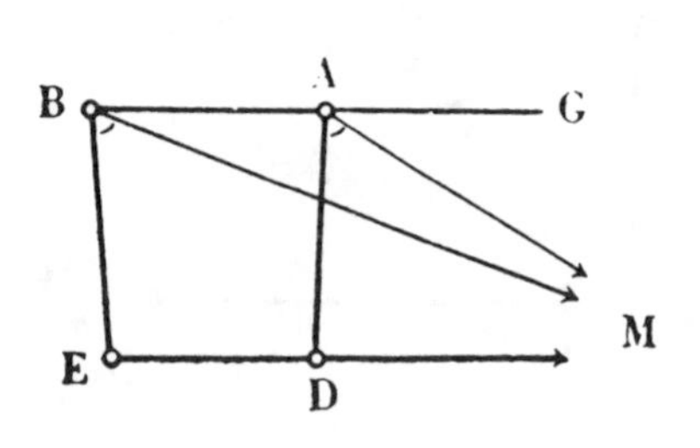

FIG. 9.6C

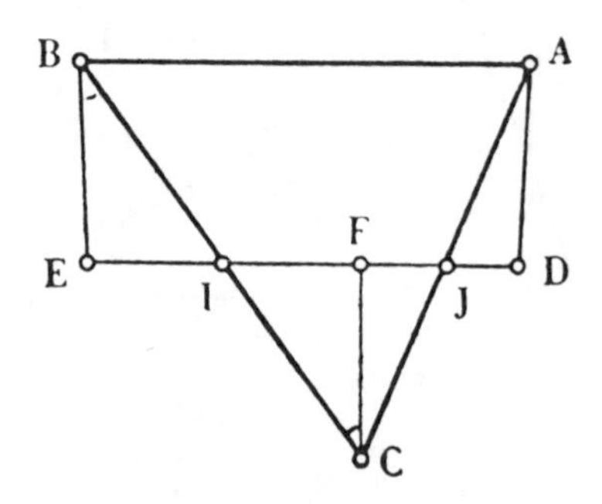

FIG. 9.6D

9.66. *The sum of the angles of any triangle is less than two right angles.*

PROOF. By Euclid I, 17, the triangle cannot have two angles that are right or obtuse; in other words, at least two of the three angles must be acute. Let **ABC** be any triangle with acute angles at **A** and **B**. Let **I** and **J** be the mid-points of **BC** and **CA**, as in Fig. 9.6D. Draw **AD**, **BE**, **CF**, perpendicular to **IJ**. Then the right-angled triangles **ADJ**, **CFJ** are congruent; so also are **BEI**, **CFI**. Hence

9.67. $$\mathbf{AD} = \mathbf{CF} = \mathbf{BE},$$

and $\angle\mathbf{ACB} = \angle\mathbf{JCF} + \angle\mathbf{FCI} = \angle\mathbf{JAD} + \angle\mathbf{EBI}$.

The angle-sum of the triangle **ABC** is

$$\angle\mathbf{BAC} + \angle\mathbf{ACB} + \angle\mathbf{CBA} = \angle\mathbf{BAJ} + \angle\mathbf{JAD} + \angle\mathbf{EBI} + \angle\mathbf{IBA}$$
$$= \angle\mathbf{BAD} + \angle\mathbf{EBA}.$$

By 9.65, these last angles are both acute. Thus the angle-sum is less than two right angles. (Cf. 1.14.)

Since any simple quadrangle can be dissected into two triangles, it follows that the angle-sum of a quadrangle is less than four right angles. In particular, there are no rectangles:

9.68. *Two coplanar lines cannot have two common perpendiculars.*

We saw, in 9.63, that two coplanar lines which have a common perpendicular are ultra-parallel. Conversely,

9.69. *Any two ultra-parallel lines have a common perpendicular.*

CONSTRUCTION.* Let **r** and **s** be two ultra-parallel lines. From any two points **A** and **C** on **s**, draw **AB** and **CB′** perpendicular to **r**. If it happens that **AB** = **CB′**, the desired common perpendicular joins the mid-points of **AC** and **BB′** (by the symmetry of the isosceles birectangle **ACB′B**). If not, suppose **AB** < **CB′**. Take **A′** on **CB′** so that **A′B′** = **AB**, as in Fig. 9.6E.

*Hilbert [1], p. 149.

Through **A′** draw a line **s′**, making the same angle with **A′B′** that **s** makes with **AB**. Then (as we shall prove) **s** meets **s′** in an ordinary point **D**. Take a point **D′** on **A/C** so that **AD′** = **A′D**. Then the perpendicular bisector of **DD′** is also perpendicular to **r**.

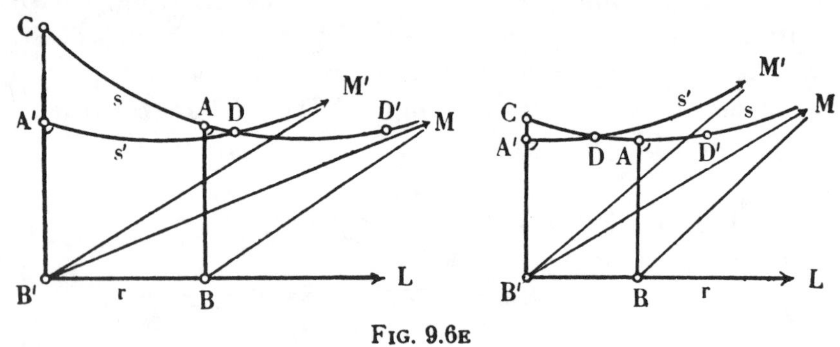

FIG. 9.6E

PROOF. Let **L**, **M**, **M′** be the points at infinity on **r**, **s**, **s′**, on the same side of **CB′** as **A**. Draw the parallels **BM**, **B′M**, **B′M′**. By 9.62, the singly-asymptotic triangles **ABM**, **A′B′M′** are congruent. By 9.64,

$$\angle \mathbf{LB'M} < \angle \mathbf{LBM} = \angle \mathbf{LB'M'}.$$

Hence **B′M′**, lying within ∠**MB′C**, must meet **CM**; so also must **A′M′**. This gives **D**, and we have constructed **D′** in such a way that **D** and **D′** are two vertices of an isosceles birectangle with its base on **r**. The desired result now follows easily.

9.7. The Absolute. We now return to three-dimensional geometry, extended by the postulation of ideal elements: a point at infinity for each bundle of parallels, a *line at infinity* for each pencil of parallels, an *ultra-infinite point* for each bundle of ultra-parallels (9.31), an *ultra-infinite line* for each pencil of ultra-parallels (9.32), and an ideal plane determined by any ideal point and line that do not determine an ordinary plane. Assuming 9.61, we obtain the following important theorem (which does not hold in Euclidean geometry):

9.71. *Any two ordinary planes have distinct absolute poles.*

PROOF. If two ordinary planes ρ and ρ_1 had the same absolute pole, every line perpendicular to ρ would be perpendicular to ρ_1. Two such lines would form, with the sections of ρ and ρ_1 by their plane, a rectangle, contradicting 9.68.

It follows that the planes perpendicular to any two planes ρ and σ, being the common planes of two distinct bundles, form a pencil, whose axis joins the absolute poles of ρ and σ. The planes perpendicular to any two planes of this pencil form a second pencil, which includes ρ and σ and so consists of all the planes through (ρ, σ). Hence each plane of the second pencil is perpendicular not merely to two planes of the first, but to all. Two such pencils are said to be *reciprocal*. They include as special cases the pencils whose axes are absolute polar lines, as defined in §9.4. In other cases (namely, when both axes are ideal), they provide a *definition* for absolute polar lines. Thus two lines (one or both ideal) are absolute polars if the ordinary planes through one are perpendicular to the ordinary planes through the other. In this sense every line, ordinary or ideal, has a definite absolute polar line. Consequently, every point has an absolute polar plane, whose lines are polar to the lines through the point; and similarly every plane has an absolute pole.

9.72. *The absolute polarity is of type* (3, 3).

PROOF. Of the four possible types of polarity (§3.8), two are ruled out immediately: this cannot be a null polarity, since an ordinary plane does not contain its pole; and it cannot be of type (2, 4), since the plane joining an ordinary point to a self-polar line would be self-perpendicular.

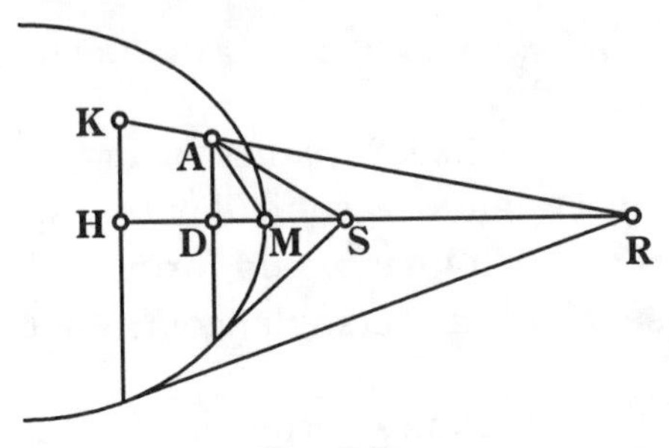

FIG. 9.7A

To exclude the remaining possibility, of a uniform or (6, 0) polarity, consider the involution of conjugate points on the line **HD** in Fig. 9.7A. Here **ADHK** is a simple quadrangle with right angles at **D**, **H**, **K**, and therefore an acute angle at **A**. (We may take **H** and **K** to be the mid-points of **DE** and **AB** in Fig. 9.6C.) The absolute polarity in space induces a two-dimensional polarity in the plane of this "trirectangle." Thus the ideal point **R** = (**KA**, **HD**) is the pole of **KH**. Let **S** be the pole of **AD**, and **M** the point at infinity on **D/H**. Then ∠**DAR**, ∠**DAS**, and ∠**DAM** are respectively obtuse, right, and acute. Hence

$$S(\mathbf{RSM}) \neq S(\mathbf{HDM})$$

and the order of the ordinary points on **HD** is opposite to that of their ideal conjugates. By 2.96, the involution of conjugate points is hyperbolic, and has two (ideal) double points, one of which will later be identified with **M**. Thus the absolute polarity admits self-conjugate points, and is not of type (6, 0).

The above proof depends essentially on 9.65, which tells us that the fourth angle of a trirectangle is acute. This, in turn, depends on the properties of parallelism, and so ultimately on the Axiom of Continuity. By denying continuity, Dehn has developed a "non-Legendrian geometry" in which pairs of coplanar lines may still have no ordinary intersection although the absolute polarity is uniform (from which it follows that the fourth angle of a trirectangle is obtuse, and the angle-sum of a triangle is greater than two right angles). Following a different procedure, Hilbert rules out this possibility (without assuming continuity) by strengthening 9.61.*

By 9.72, there is an oval quadric, the *Absolute*, whose points are self-conjugate; but we have not yet identified these points with the points at infinity determined by bundles of parallels.

9.73. *The points on the Absolute are the points at infinity.*

PROOF. By 9.69, every improper bundle is either a bundle of parallels or else the class of lines and planes perpendicular to a plane. Hence every ideal point which is not a point at infinity has an ordinary polar plane. Also, by 9.42, every ordinary point has an ideal polar plane. Thus, on an ordinary line, whose points at infinity are **M** and **N**, all the ordinary points are conjugate to ideal points, and all the ideal points except **M** and **N** are conjugate to ordinary points. Hence **M** and **N**, which separate the two classes of points, are the double points of the involution of conjugate points; i.e. they are self-conjugate. Conversely, any self-conjugate point can be joined to an ordinary point by an ordinary line, and occurs as one of the points at infinity on that line.

Hence, also, the interior and exterior points of the Absolute are the ordinary and ultra-infinite points, respectively, and its secant lines and planes are the ordinary lines and planes. Thus any two points at infinity are joined by an ordinary line; in other words, the common planes of two bundles of parallels form a proper pencil.

9.74. *The tangent lines of the Absolute are the lines at infinity.*

PROOF. A "line at infinity," being the axis of a pencil of parallels, is an ideal line which contains a point at infinity;

Dehn [1], pp. 431-436; Forder [1], p. 338; Hilbert [1], p. 147.

i.e. it is a tangent line of the Absolute. Conversely, if **m** is any tangent, its polar line **m**′ is another tangent having the same point of contact. The reciprocal pencils determined by these lines both belong to the same bundle of parallels, and each is a pencil of parallels (as defined in §9.3). Hence **m** and **m**′ are lines at infinity.

It follows that the lines and planes through a point **A**, parallel to a plane ρ (not through **A**), are the generators and tangent planes of a *cone*, which joins **A** to the section of the Absolute by ρ.

By a natural extension of the above terminology, we define a *plane at infinity* as a tangent plane of the Absolute, or as the polar plane of a point at infinity. Thus a plane at infinity is traced out by the lines at infinity in the planes of a bundle of parallels. Other ideal planes are said to be *ultra-infinite*; they are the polar planes of ordinary points.

The points at infinity on an ordinary plane form a conic, which separates ordinary (interior) points from ultra-infinite points. Secants of the conic are ordinary lines, tangents are lines at infinity, and exterior lines are ultra-infinite. A plane at infinity contains just one point at infinity, namely its pole (point of contact), and contains a flat pencil of *lines* at infinity. Its other points and lines are ultra-infinite. Finally, an ultra-infinite plane consists solely of ultra-infinite points and lines. Thus the pairs of polar elements may be summarized as follows:

Ordinary points	Ultra-infinite planes
Points at infinity	Planes at infinity
Ultra-infinite points	Ordinary planes
Ordinary lines	Ultra-infinite lines
Lines at infinity	Lines at infinity

This completes the proof that hyperbolic geometry is both consistent and categorical. It is categorical (or unique), since Axioms 8.3, 9.1, 9.61 suffice for the construction of Cayley's

Absolute (§8.1); and it is consistent, since the Cayley-Klein model satisfies these axioms (as the reader can easily verify). We have seen also that Euclidean geometry can be based on Axioms 8.3, 9.1, 9.51. Hence either axiom of parallelism is independent of the axioms of order, continuity, and congruence. This means that Euclid is vindicated: his decision to include "Postulate V" among his basic assumptions is entirely justified.

9.8. The geometry of a bundle. It is interesting to see how Theorems 8.54 and 8.81 are affected by the introduction of congruence. The absolute polarity induces a two-dimensional polarity in any bundle whose centre is not self-conjugate. This polarity will be hyperbolic or elliptic according as the bundle does or does not contain self-conjugate lines (tangents of the Absolute), i.e., according as the centre is exterior or interior to the Absolute. Hence

9.81. *The geometry of lines and planes through an ordinary point can be identified with the elliptic geometry of points and lines in a plane,* and

9.82. *The geometry of lines and planes through an ultra-infinite point can be identified with the hyperbolic geometry of points and lines in an ordinary plane.*

In 9.81, the elliptic polarity is the correspondence between perpendicular lines and planes through the given point. In both cases, perpendicular planes are represented by perpendicular lines. In 9.82, the given lines and planes are all perpendicular to one ordinary plane (the polar of the ultra-infinite point), and can be represented on that plane by their sections. Analogously,

9.83. *The geometry of lines and planes through a point at infinity can be identified with the Euclidean geometry of points and lines in a plane.*

PROOF. Let **M** be the given point at infinity, and μ its polar plane. Apart from μ, all the planes through **M** are ordinary. In μ itself, the pairs of polar lines through **M** constitute an ***absolute involution***, which establishes a Euclidean metric in the bundle of parallels.

This result is remarkable, as showing that every theorem of Euclidean plane geometry has its counterpart in hyperbolic solid geometry. For instance, the counterpart of 9.51 is:*

9.84. *If a plane* ρ *is parallel to a line* **l**, *there is only one plane through* **l** *which does not meet* ρ *in an ordinary line.*

This unique parallel plane is, of course, **lm**, where **m** is the line in which ρ meets the polar plane of the point at infinity (**l**, ρ). In other words, it is the plane through **l** perpendicular to σ, where σ is the plane through **l** perpendicular to ρ.

It is interesting to notice that 9.81, 9.83, and 9.84 hold also for Euclidean geometry, and 9.81 for elliptic geometry. To sum up, the geometry of a bundle is elliptic, or Euclidean, or hyperbolic, according as the bundle is proper, or a bundle of parallels, or a bundle of ultra-parallels.

---

*Sommerville [2], p. 50.

CHAPTER X

HYPERBOLIC GEOMETRY IN TWO DIMENSIONS

10.1. Ideal elements. As a sufficient set of axioms for plane hyperbolic geometry (based on *point*, *intermediacy*, and *congruence*) we may take 8.311, 8.313-8.317, 8.32, 9.11-9.15, and 9.61 (along with the denial of 8.318). It is, of course, possible to prove such theorems as 8.92 and 9.69 without using ideal elements.* But the advantage of *points at infinity* has already been seen, and the reader will find that many propositions can be handled very expeditiously with the aid of the powerful machinery of projective geometry.

By considering flat pencils of parallels (namely, lines parallel to a given ray) and flat pencils of ultra-parallels (namely, lines perpendicular to a given line) it is possible to introduce ideal points into the plane, and to distinguish certain classes of them as forming ideal lines.† But the three-dimensional treatment of Chapters VIII and IX is more satisfactory, as it allows all kinds of point, ordinary and ideal, to be covered by a single definition (§§8.6, 8.7). Accordingly, we have used Axioms 8.318, 8.319, and defined the absolute polarity in terms of reciprocal pencils of planes (§9.7), obtaining an oval quadric as the locus of points at infinity. When we restrict consideration to a single ordinary plane, the points at infinity that remain form a conic (the section of the quadric by the plane). From now on, we shall reserve the name *Absolute* for this conic, as we shall be concerned almost entirely with two-dimensional geometry.

Two ordinary lines are perpendicular if and only if they

*See, for instance, Carslaw [1], pp. 45, 55.

†Bonola [2], p. 231. Cf. Owens [1].

are conjugate with respect to the Absolute. The lines perpendicular to an ordinary line **r** form a flat pencil whose centre is an ultra-infinite point **R**, the pole of **r**. Conversely, any two ultra-parallel lines (through **R**, say) have a common perpendicular (namely **r**). The poles of the lines through an ordinary point **A** are the points of an ultra-infinite line **a**, the polar of **A**. The poles of the lines of a flat pencil of parallels (i.e., of the lines through a point at infinity, **M**) are the points of a line at infinity (namely the polar, **m**). Each line at infinity, being a tangent to the Absolute, contains its pole, while the rest of its points are ultra-infinite. Thus the pairs of polar elements (cf. page 196) are as follows:

Ordinary points	Ultra-infinite lines
Points at infinity	Lines at infinity
Ultra-infinite points	Ordinary lines

10.2. Angle-bisectors. As a **striking** instance of the difference between the classical and projective methods, let us take the familiar theorem which gives a triangle an in-centre, and show that this is Brianchon's Theorem in disguise.

10.21. *The internal angle-bisectors of a triangle are concurrent.*

CLASSICAL PROOF. By considering congruent triangles, we see that every point on the bisector of an angle is equidistant from the two arms of the angle. Hence the point of intersection of two (internal) angle-bisectors of a triangle is equidistant from all three sides, and so lies on the third angle-bisector too.

PROJECTIVE PROOF. Since the two parallels to a line through a point are equally inclined to the perpendicular, the internal bisector of an angle **NAM**, with **M** and **N** at infinity, is perpendicular to **MN**, and so joins **A** to the pole of **MN**, which is the point of intersection of the tangents at **M** and **N** to the Absolute. (See frontispiece.) Hence the angle-bisectors of a triangle **ABC** are the diagonals of the simple hexagon whose sides are the polars of the **six** points at infinity on the three sides of the triangle. Their concurrence now follows from 3.36.*

10.3. Congruent transformations. We have seen that the ordinary and ultra-infinite points on an ordinary line are separated by two points at infinity, say **M** and **N** (p.175), which are the double points of the involution of conjugate points on the line (see 9.73). In other words, conjugate points (such as **H**, **R**, or **D**, **S**, in Fig. 9.7A) are merely harmonic conjugates with respect to **M** and **N**. A one-dimensional congruent transformation is thus a projectivity which either preserves or interchanges **M** and **N**, as in §5.8 (only now these are *real* points). The *translation* $_{\mathbf{A}}\Psi_{\mathbf{B}}$, taking **A** to **B**, is the projectivity by which $\mathbf{A\,M\,N} \barwedge \mathbf{B\,M\,N}$, and the *reflection* $_{\mathbf{A}}\Phi_{\mathbf{B}}$ is the involution $(\mathbf{A\ B})(\mathbf{M\ N})$. This aspect of translation shows at once that two segments **AB** and **CD** are congruent if

$$\mathbf{A\,B\,M\,N} \barwedge \mathbf{C\,D\,M\,N}.$$

It follows (see §8.1) that the *distance* **AB** is proportional to $|\log\{\mathbf{AB}, \mathbf{MN}\}|$. In order that our unit of measurement may agree with Lobatschewsky's, we choose the factor of proportionality to be $\kappa = 2$, so that

10.31. $$\mathbf{AB} = \tfrac{1}{2}\,|\log\{\mathbf{AB}, \mathbf{MN}\}|.$$

We turn now to the consideration of two-dimensional transformations. In marked contrast to 6.42, which shows that there is essentially only one kind of congruent transformation in the elliptic plane, there are *four* kinds in the hyperbolic plane:

(i) Rotation,
(ii) Parallel displacement,
(iii) Translation,
(iv) Glide-reflection.

*For a similar treatment of external angle-bisectors, see Coxeter [8], p. 13.

Each of these can be expressed as a product of *reflections*, the reflection in a line **OO′** being the transformation that preserves every point on this line, but replaces every other point **A** by a different point **B**, such that **AOO′** and **BOO′** are congruent triangles. A *rotation* is, as usual, the product of reflections in two intersecting lines. A *parallel displacement* is the product of reflections in two parallel lines; it can be regarded as a rotation whose centre is at infinity, since it shifts the lines of a pencil of parallels just as an ordinary rotation shifts those of a proper pencil. A *translation* is the product of reflections in two ultra-parallel lines; it induces a one-dimensional translation in the common perpendicular of the two lines, which is called the *axis* of the translation. A *glide-reflection* is the product of a translation with the reflection in its axis. The reflection itself can be regarded as the special case when the extent of the component translation reduces to zero. Being composed of an even number of reflections, (i), (ii), (iii) are *direct* transformations; but (iv) is *opposite*.

The only essential difference in Euclidean geometry is that there the lines perpendicular to a given line form a pencil of parallels, so that the distinction between a translation and a parallel displacement is lost, and a translation has infinitely many "axes" instead of only one.

The following projective considerations suffice to show that the above list is exhaustive. Since a congruent transformation is a collineation which preserves the Absolute, it is completely determined by the projectivity it induces in the Absolute itself.* (The transform of any line is determined by the transforms of its two points at infinity.) We saw, in §3.4, that any projectivity on a conic has a centre and an axis (which are pole and polar), and that in the hyperbolic case the projectivity is opposite or direct according as two corresponding points do or

*Veblen and Young [2], II, pp. 353, 355.

do not separate the double points. Hence the various possibilities, in the same order as above, are as follows:

(i) Elliptic projectivity,
(ii) Parabolic projectivity,
(iii) Direct hyperbolic projectivity,
(iv) Opposite hyperbolic projectivity.

In the first case the centre is an ordinary point; in the second, both centre and axis are at infinity; in the last two the axis is an ordinary line. Thus Fig. 3.4A illustrates a glide-reflection.

We saw, in §3.3, that a conic is preserved by any harmonic homology whose centre is the pole of its axis. When the conic is the Absolute, such a homology is a congruent transformation of period two, namely the reflection in a point or in a line according as the centre is ordinary or ultra-infinite. (See §3.4.) Thus again, as in §6.2, a reflection is a harmonic homology.* This may be seen directly in Fig. 3.1A, if we take **B** and **B**′ to be the images of **A** and **A**′ by reflection in **OO**′ (so that the angles at **O** and **O**′ are right angles, and **O**″ is the pole of **OO**′).

The above remarks provide a one-dimensional projective model for two-dimensional hyperbolic geometry, congruent transformations of the hyperbolic plane being represented by projectivities on a conic, or on a line. Among these projectivities, any involution, elliptic or hyperbolic, represents the reflection in a point or in a line, respectively. In this sense, points and lines are themselves represented by involutions, and the condition for incidence is that the involutions be permutable. We may say alternatively that each line is represented by a pair of points, the double points of a hyperbolic involution. Essentially, we are representing a line by its two points at infinity, and then transforming the Absolute into a line by "stereographic projection."

*Ibid., p. 352.

10.4. Some famous constructions. In 9.69 we considered Hilbert's construction for the common perpendicular to two ultra-parallel lines. We may now describe this more simply by observing that **s′** is derived from **s** by applying a translation with axis **r**, while **D′** is derived from **D** by applying the inverse translation. The following three problems are concerned with drawing lines parallel to a given ray.

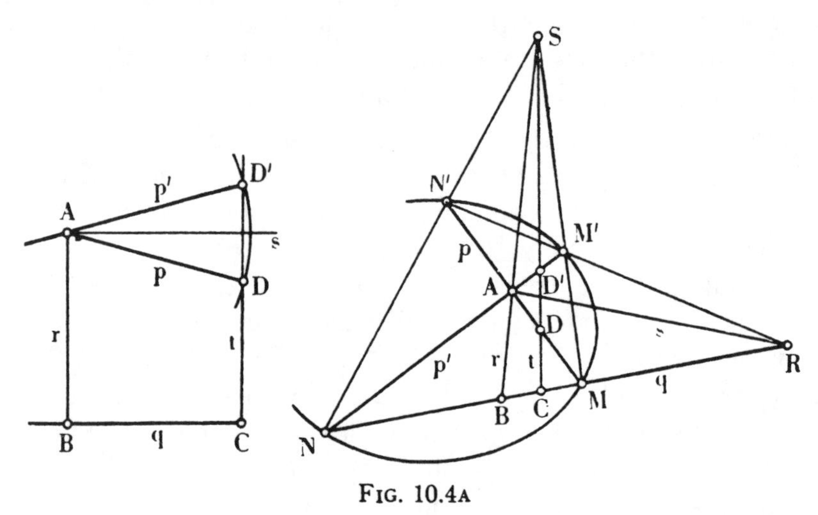

FIG. 10.4A

10.41. *Given a point* **A** *and a line* **q**, *construct the parallels to* **q** *through* **A**.

CONSTRUCTION.* Through **A** draw **r** perpendicular to **q**, and **s** perpendicular to **r**. (See Fig. 10.4A.) Through any point **C** on **q**, draw **t** perpendicular to **s**. On **t**, locate **D** and **D′** so that **AD** = **AD′** = **BC**, where **B** is (**q**, **r**). Then **AD** and **AD′** are parallel to **q**.

PROOF.† Let **M** and **N** be the points at infinity on **q**, and **M′** and **N′** the remaining points at infinity on **AN** and **AM**.

*Bolyai [2], §34.

†Baldus [1], p. 102. Mohrmann [1], p. 108.

Construct **R** = (**MN**, **M′N′**) and **S** = (**MM′**, **NN′**). Then **AS**, the polar of **R**, is **r**; and **AR**, the polar of **S**, is **s**. Being perpendicular to **s**, **t** passes through **S**; and the points where **t** meets **AM** and **AN** satisfy the relations

$$\mathbf{N'ADM} \;\overline{\wedge}\; \mathbf{NAD'M'} \;\overline{\wedge}\; \mathbf{NBCM},$$

which are the projective equivalent of **AD** = **AD′** = **BC**. These points, therefore, are the same as the **D** and **D′** previously obtained by drawing a circle arc ınd **A**.

10.42. *Given two non-parallel rays,* p *and* p′, *construct their common parallel.*

CONSTRUCTION.* Let **M** be the point at infinity on p, and **N** on p′. Through any point **A** on p, draw **AN** parallel to p′. Through any point **B** on p′, draw **BM** parallel to p. (See Fig. 10.4B.) Then the bisectors of ∠**NAM** and ∠**NBM** are ultra-parallel, and their common perpendicular is the desired line **MN**.

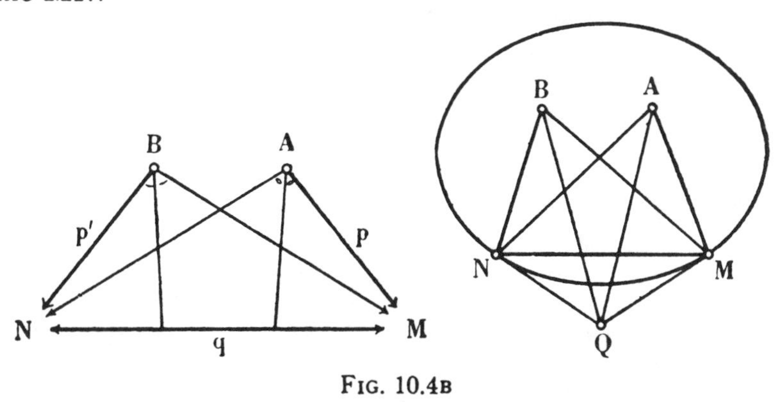

FIG. 10.4B

PROOF. We have already seen (in the frontispiece) that the bisector of an angle **NAM** is **AQ**, where **Q** is the pole of **MN**. Hence the bisectors of ∠**NAM** and ∠**NBM** meet in the ultra-infinite point **Q**, and **MN** is their common perpendicular.

*Cf. Hilbert [1], p. 151.

10.43. *Given a ray* p *and a line* **r** *which is neither parallel nor perpendicular to* p, *construct the line which is parallel to* p *and perpendicular to* **r**.

CONSTRUCTION.* Draw p′, the image of p by reflection in **r**. Then, clearly, the desired line is the common parallel to p and p′.

10.5. An alternative expression for distance. It is sometimes convenient to replace 10.31 by a formula resembling 5.76 instead of 6.91. Such a formula can most easily be obtained by using *abscissae*, with the involution of conjugate points in the form

$$x + x' = 0$$

(which is 4.29 with $s = 0$), so that the points at infinity, **M** and **N**, have abscissae 0 and ∞, while ordinary points have positive abscissae. Let x and y be the abscissae of **A** and **B**, so that $-x$ and $-y$ are those of their respective conjugates, **A**′ and **B**′. Then, in terms of Lobatschewsky's unit of measurement, we have, by 4.33,

$$e^{\pm 2\mathbf{AB}} = \{\mathbf{AB}, \mathbf{MN}\} = \frac{x}{y},$$

whence

$$\begin{aligned} \cosh \mathbf{AB} &= \tfrac{1}{2}(e^{\mathbf{AB}} + e^{-\mathbf{AB}}) \\ &= \frac{1}{2}\left(\sqrt{\frac{x}{y}} + \sqrt{\frac{y}{x}}\right) \\ &= \frac{x+y}{2\sqrt{(xy)}} \\ &= \sqrt{\frac{(x+y)\,(y+x)}{(x+x)\,(y+y)}} \\ &= \sqrt{\{\mathbf{AB}, \mathbf{B'A'}\}}\,. \end{aligned}$$

*Carslaw [1], p. 76.

This is, of course, the same as the expression for cos **AB** in elliptic geometry. In fact, every metrical formula of elliptic geometry leads to a corresponding formula for hyperbolic geometry when we multiply each distance by i. There is nothing surprising in this principle, which arises because both these geometries are special cases of the complex non-Euclidean geometry whose Absolute is the general quadric in complex projective space.

By 9.81 and 5.76, the angles between two intersecting lines **a** and **b** are given by

10.52. $$\cos \angle(\mathbf{ab}) = \pm\sqrt{\{\mathbf{ab}, \mathbf{b'a'}\}}\ ,$$

where **a′** and **b′** are the respective perpendiculars through the point (**a**, **b**) in the plane **ab**.

10.6. The angle of parallelism. Problem 10.41 enables us to construct the angle $\Pi(c)$ for a given segment of length c. Conversely, 10.43 provides the length corresponding to a given (acute) angle. The functional relation between c and $\Pi(c)$ may be found as follows.

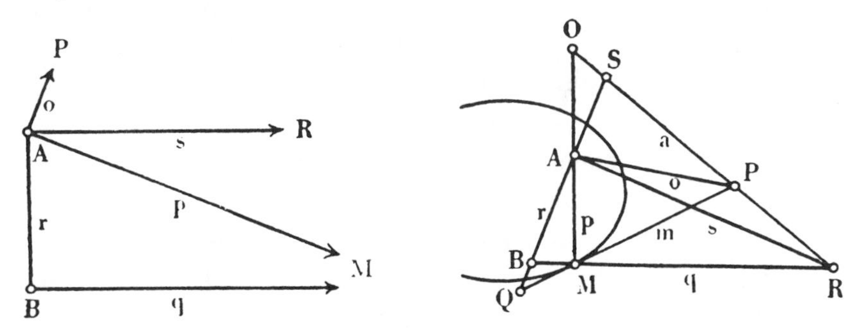

FIG. 10.6A

Let **AM** or **p** be one of the parallels from a point **A** to a line **q**. Through **A**, draw **r** perpendicular to **q**, **s** perpendicular to **r**, and **o** perpendicular to **p**, as in Fig. 10.6A. Then **a**, the polar of **A**, meets **p**, **q**, **r**, **o** in **O**, **R**, **S**, **P** (the poles of **o**, **r**, **s**, **p**).

The polar of **M** contains **P**, and meets **r** in **Q** (the pole of **q**). Hence

$$\mathbf{ABQS} \overset{\mathbf{M}}{\overline{\wedge}} \mathbf{ORPS} \,\overline{\wedge}\, \mathbf{psor}.$$

Since **S** and **Q** are conjugate to **A** and **B**, respectively, we have

$$\cosh^2 \mathbf{AB} = \{\mathbf{AB}, \mathbf{QS}\}.$$

Since **o** and **s** are perpendicular to **p** and **r**, respectively, we have

$$\cos^2 \angle(\mathbf{pr}) = \{\mathbf{pr}, \mathbf{so}\}, \quad \sin^2 \angle(\mathbf{pr}) = \{\mathbf{ps}, \mathbf{ro}\},$$

and

$$\operatorname{cosec}^2 \angle(\mathbf{pr}) = \{\mathbf{ps}, \mathbf{or}\}.$$

Hence the relation between the distance $c = \mathbf{AB}$ and its angle of parallelism $c' = \Pi(c) = \angle(\mathbf{pr})$ is

$$\cosh c = \operatorname{cosec} c',$$

or $\sinh c = \cot c'$, or $\tanh c = \cos c'$. Another form of the same relation is $\cosh c - \sinh c = \operatorname{cosec} c' - \cot c'$, or

$$e^{-c} = \tan \tfrac{1}{2} c'.$$

Thus*

$$\Pi(c) = 2 \operatorname{arc} \tan e^{-c}.$$

10.7. Distance and angle in terms of poles and polars. If **a** and **b** are the polars of **A** and **B**, we may put 10.51 into a form analogous to 6.71, namely

10.71. $$\mathbf{AB} = \operatorname{arg} \cosh \sqrt{\{\mathbf{AB}, \mathbf{ba}\}},$$

Since the shortest distance from a point **A** to a line **q** is along the perpendicular **AQ**, it is

$$\operatorname{arc} \cosh \sqrt{\{\mathbf{Aq}, \mathbf{Qa}\}} = \operatorname{arg} \sinh \sqrt{(-\{\mathbf{AQ}, \mathbf{qa}\})}.$$

Similarly, the distance between two ultra-parallel lines **q** and **s**, measured along their common perpendicular, is

10.72. $$\operatorname{arg} \cosh \sqrt{\{\mathbf{qs}, \mathbf{SQ}\}}.$$

For the angle between two intersecting lines, 6.72 can be carried over in the form

10.73. $$\angle(\mathbf{pr}) = \operatorname{arc} \cos(\pm \sqrt{\{\mathbf{pr}, \mathbf{RP}\}}).$$

---

*Lobatschewsky [1], p. 633.

10.8. Canonical coordinates. The simplest coordinates, for most purposes, are given by taking the Absolute in its canonical form 4.56, so that the polar of (x_0, x_1, x_2) is $[-x_0, x_1, x_2]$, and the condition for lines $[X]$ and $[Y]$ to be perpendicular is

$$X_0 Y_0 = X_1 Y_1 + X_2 Y_2.$$

Then the point (x) is ordinary, at infinity, or ultra-infinite, according as

$$x_0^2 - x_1^2 - x_2^2$$

is positive zero, or negative; and the nature of the line $[X]$ depends similarly on the sign of $-X_0^2 + X_1^2 + X_2^2$. Since these coordinates are homogeneous, there is no loss of generality in assuming that $x_0 \geq 0$; then x_0 is positive for every ordinary point.

By 10.71, the distance between (x) and (y) is*

10.81. $$\operatorname{arg\,cosh} \sqrt{\frac{\{xY\}\{yX\}}{\{xX\}\{yY\}}}$$
$$= \operatorname{arg\,cosh} \frac{x_0 y_0 - x_1 y_1 - x_2 y_2}{\sqrt{(x_0^2 - x_1^2 - x_2^2)}\sqrt{(y_0^2 - y_1^2 - y_2^2)}}$$

The condition for this expression to vanish is

$$(x_0 y_0 - x_1 y_1 - x_2 y_2)^2 = (x_0^2 - x_1^2 - x_2^2)(y_0^2 - y_1^2 - y_2^2),$$

or $$-(x_1 y_2 - x_2 y_1)^2 + (x_2 y_0 - x_0 y_2)^2 + (x_0 y_1 - x_1 y_0)^2 = 0.$$

This can happen either when (x) and (y) coincide, or when their join is a line at infinity. The latter possibility (which suggests the alternative name "null lines" for lines at infinity) is not so paradoxical as it looks, when we remember that it involves applying 10.81 to the ideal region, where distance has not hitherto been defined.†

Similarly, the distance from (x) to $[Y]$ is

---

*Klein [3], p. 185.

†Schilling [2]. Physicists will recognize this ideal region as a two-dimensional de Sitter's world, whose time-like and space-like lines are our ordinary and ultra-infinite lines.

10.82. $$\operatorname{arg\,sinh} \sqrt{\left(-\frac{\{xY\}\{yX\}}{\{xX\}\{yY\}}\right)} = \operatorname{arg\,sinh} \frac{|x_0 Y_0 + x_1 Y_1 + x_2 Y_2|}{\sqrt{(x_0^2 - x_1^2 - x_2^2)}\sqrt{(-Y_0^2 + Y_1^2 + Y_2^2)}}.$$

This vanishes *only* when (x) lies on $[Y]$.

The distance between non-intersecting lines $[X]$ and $[Y]$ is

10.83. $$\operatorname{arg\,cosh} \sqrt{\frac{\{xY\}\{yX\}}{\{xX\}\{yY\}}} = \operatorname{arg\,cosh} \frac{|-X_0 Y_0 + X_1 Y_1 + X_2 Y_2|}{\sqrt{(-X_0^2 + X_1^2 + X_2^2)}\sqrt{(-Y_0^2 + Y_1^2 + Y_2^2)}},$$

which vanishes if

10.84. $$(X_1 Y_2 - X_2 Y_1)^2 - (X_2 Y_0 - X_0 Y_2)^2 - (X_0 Y_1 - X_1 Y_0)^2 = 0,$$

i.e., *if the lines are parallel.* Hence, when a variable line, ultra-parallel to a fixed line and passing through a fixed point, rotates towards either of the positions of parallelism, the distance between the lines tends to zero. In other words, *parallel lines approach one another asymptotically.**

Finally, the acute angle between intersecting lines $[X]$ and $[Y]$ is†

10.85. $$\arccos \sqrt{\frac{\{xY\}\{yX\}}{\{xY\}\{yY\}}} = \arccos \frac{|-X_0 Y_0 + X_1 Y_1 + X_2 Y_2|}{\sqrt{(-X_0^2 + X_1^2 + X_2^2)}\sqrt{(-Y_0^2 + Y_1^2 + Y_2^2)}}.$$

For instance, this is zero for two parallel lines and is $[1, \pm 1, 1]$.‡ Hence, when two parallel lines arise as the limiting form of two intersecting lines whose point of intersection recedes to infinity, the angle between the lines tends to zero. Thus a singly-, doubly-, or trebly-asymptotic triangle may be regarded as having one, two, or three zero angles.

---

*This is one of Saccheri's theorems. See Carslaw [1], pp. 56-58.

†Kline [3], p. 185.

‡Bonola [2], p. 171.

10.9. Euclidean geometry as a limiting case. The formulae of §§6.7 and 10.8 can be unified by inserting a parameter K, which takes the value 1 for elliptic geometry and -1 for hyperbolic (cf. § 15.5). Then the distance between points (x) and (y) is

10.91.
$$\frac{1}{\sqrt{K}}\operatorname{arc\,cos}\frac{|x_0y_0 + Kx_1y_1 + Kx_2y_2|}{\sqrt{(x_0^2 + Kx_1^2 + Kx_2^2)}\sqrt{(y_0^2 + Ky_1^2 + Ky_2^2)}} =$$
$$\frac{1}{\sqrt{K}}\operatorname{arc\,sin}\sqrt{\frac{K\{K(x_1y_2 - x_2y_1)^2 + (x_2y_0 - x_0y_2)^2 + (x_0y_1 - x_1y_0)^2\}}{(x_0^2 + Kx_1^2 + Kx_2^2)(y_0^2 + Ky_1^2 + Ky_2^2)}},$$

the distance from (x) to $[Y]$ is

10.92.
$$\frac{1}{\sqrt{K}}\operatorname{arc\,sin}\frac{|x_0Y_0 + x_1Y_1 + x_2Y_2|\sqrt{K}}{\sqrt{(x_0^2 + Kx_1^2 + Kx_2^2)}\sqrt{(KY_0^2 + Y_1^2 + Y_2^2)}},$$

and the acute angle between lines $[X]$ and $[Y]$ is

10.93.
$$\operatorname{arc\,cos}\frac{|KX_0Y_0 + X_1Y_1 + X_2Y_2|}{\sqrt{(KX_0^2 + X_1^2 + X_2^2)}\sqrt{(KY_0^2 + Y_1^2 + Y_2^2)}}.$$

There is no harm in giving K other positive or negative values. This merely means changing the "unit point" of the coordinate system, so that the absolute polarity takes the form

$$x_0 = KX_0, \quad x_1 = X_1, \quad x_2 = X_2.$$

By making K tend to zero* we obtain Euclidean geometry with Cartesian coordinates $x_1/x_0, x_2/x_0$. For, the above expressions become, respectively,

$$\sqrt{\left(\frac{x_1}{x_0} - \frac{y_1}{y_0}\right)^2 + \left(\frac{x_2}{x_0} - \frac{y_2}{y_0}\right)^2}, \quad \frac{\left|Y_1\frac{x_1}{x_0} + Y_2\frac{x_2}{x_0} + Y_0\right|}{\sqrt{(Y_1^2 + Y_2^2)}},$$

and
$$\operatorname{arc\,cos}\frac{|X_1Y_1 + X_2Y_2|}{\sqrt{(X_1^2 + X_2^2)}\sqrt{(Y_1^2 + Y_2^2)}},$$

*Bonola [2], p. 162; Klein [3], p 190.

The factor $1/\sqrt{K}$ in 10.91 and 10.92 indicates an infinite magnification of scale as K approaches zero. Our conclusion may therefore be expressed as follows:

10.94. *The geometry of an infinitesimal region is Euclidean.*

The analogous results in three dimensions are the same, save that $[X]$ is now a plane and the formulae acquire extra terms involving the fourth coordinate. When K tends to zero, the absolute polarity

$$x_0 = KX_0, \quad x_1 = X_1, \quad x_2 = X_2, \quad x_3 = X_3$$

degenerates in the manner described in §9.5, so that every plane has its absolute pole in the plane at infinity $x_0 = 0$.

There is thus a continuous transition from hyperbolic geometry to elliptic, with Euclidean (or "parabolic") geometry occurring instantaneously on the way. The absolute quadric

$$KX_0^2 + X_1^2 + X_2^2 + X_3^2 = 0$$

disappears when K ceases to be negative. But if we enlarge our space by including complex points, we may say that the quadric degenerates into an imaginary conic in the plane at infinity when $K = 0$, and becomes an imaginary quadric when $K > 0$. (See §7.9.)

CHAPTER XI

CIRCLES AND TRIANGLES

11.1. Various definitions for a circle. We have seen that both elliptic geometry and hyperbolic geometry can be derived from real projective geometry by singling out a polarity. In the present chapter, so far as is possible, we give the definitions and theorems in such a form as to apply equally well in either of these non-Euclidean geometries.

In §8.6 we generalized the concepts "bundle" and "axial pencil" (§2.1) in such a way that any line and plane belong to a bundle, any two planes to a pencil. Those lines of a bundle which lie in a plane of the bundle are said to form a *flat pencil* (§10.1). Thus any two coplanar lines determine a flat pencil. In the proof of 8.61, we saw that any flat pencil can be constructed as a plane section of an axial pencil. We now make the analogous generalization of the concept "circle."

*A circle is the class of images of a point by reflection in the lines of a flat pencil.** The lines are called *diameters*, their common point the *centre*, and the absolute polar of this point the *axis*. In elliptic geometry this will be seen to agree with our previous definition (§6.5). But in hyperbolic geometry we have to distinguish three cases: a *proper circle* (the "circle" of §9.1) has an ordinary centre and an ultra-infinite axis; a *horocycle* has parallel diameters, so that its centre and axis are at infinity; and an *equidistant curve* has ultra-parallel diameters, all perpendicular to the axis, while the centre is ultra-infinite. (Some authors use the distinctive word *cycle* for the general circle in hyperbolic geometry, and distinguish the three kinds as a circle, a horocycle, and a hypercycle.)

*Baldus [1], p. 140.

Since the product of reflections in two diameters is a displacement which can vary continuously, an alternative definition for a circle in elliptic geometry, or for a proper circle in hyperbolic, is *the locus of a point under continuous rotation about the centre*. So also a horocycle or an equidistant curve is the locus of a point under continuous *parallel displacement* or *translation*, respectively. (See §10.3.) It follows that a proper circle is the locus of a point whose distance from the centre is constant, say R, while an equidistant curve is the locus of a point whose distance from the axis is constant, say D. According to this description, a circle in elliptic geometry is both a proper circle and an equidistant curve, with $R+D=\frac{1}{2}\pi$, and "diametrically opposite" points are images of one another by reflection in the axis.

In hyperbolic geometry, on the other hand, an equidistant curve is not obviously symmetrical by reflection in its axis. To achieve this desirable symmetry, we have to combine the reflections in two conjugate diameters, one of which is necessarily ultra-infinite. When this is done, the curve is seen to have two separate branches, like a Euclidean hyperbola (or, better, like the two parallel lines that can arise from an ellipse by making the eccentricity tend to 1). It consists of the points distant D from the axis *on both sides*. Returning to our first definition, we can describe the second branch as the class of images of the same point by reflection in the points of the axis (which is the common perpendicular of all the diameters).

Thus a circle of any kind meets each diameter twice, though in the case of the horocycle one of the points of intersection is the centre (at infinity). In this respect the horocycle resembles the Euclidean parabola.

Since a circle is symmetrical by reflection in any diameter, the diameters through two points **A** and **B** on the circle make equal angles with the "chord" **AB**. Conversely, this property is sometimes used to define a circle.* After Gauss, points **A** and **B** on lines **AA′** and **BB′** are said to *correspond* when $\angle$**A′AB** = $\angle$**ABB′**. If the lines intersect (in an ordinary point), this means that **A** and **B** are equidistant from the point of intersection. If they are ultra-parallel, it means that **A** and **B** are equidistant from the common perpendicular. But the definition remains significant in the "intermediate" case when the lines are parallel; for it is still true† that the property of correspondence is transitive for points on all lines of the consequent pencil of parallels. Thus in all three cases, the locus of corresponding points on the lines of a pencil is a circle.

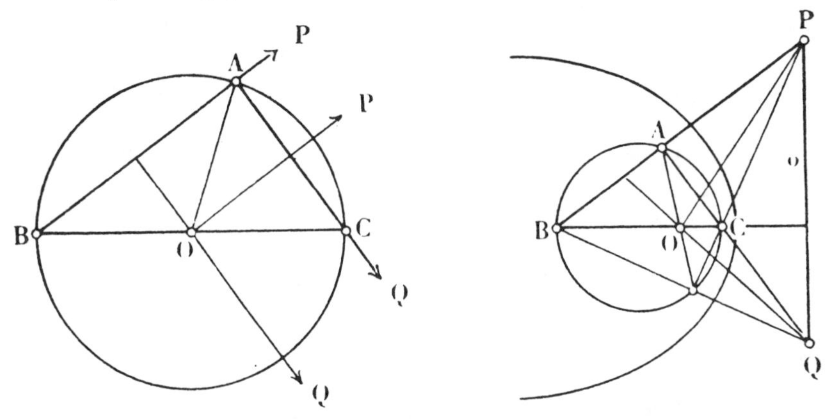

FIG. 11.2A

11.2. The circle as a special conic. Many properties of a circle are easily deduced from the fact that *every circle is a conic.*‡ This was proved for elliptic geometry in 6.51, where we obtained the circle as the locus of **A** = (**BP**, **CQ**), **P** and **Q** being a variable pair of conjugate points on the axis. The same proof can be applied to a proper circle in hyperbolic geometry, and to an equidistant curve. In the former case

*Bonola [2], p. 74.

†Gauss [1], p. 207; Bolyai [2], §§8-12; Carslaw [1], pp. 69-71, 80.

‡Klein [3], p. 177.

(Fig. 11.2A), **OP** and **OQ** are the bisectors of the supplementary angles **COA** and **AOB** (or the internal and external bisectors of ∠**COA**) and are evidently perpendicular to the respective chords **AC** and **AB**. Thus a proper circle is the locus of the point of intersection of perpendiculars from **B** and **C** to a variable pair of perpendicular lines (**OQ** and **OP**) through the mid-point of **BC**.

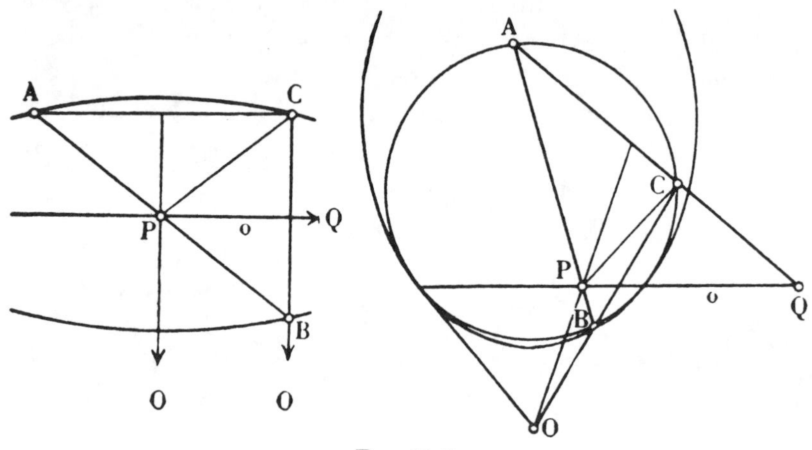

FIG. 11.2B

In the case of the equidistant curve (Fig. 11.2B), the axis **o** is the perpendicular bisector of **BC**. On **o**, we have **P** ordinary and **Q** ultra-infinite for one branch, and *vice versa* for the other. For the first branch, **PO** is perpendicular to **o**, and the curve is the locus of the point where **BP** meets the perpendicular from **C** to **PO**. When **P** is either of the points at infinity on **o**, **Q** and **A** coincide with it; thus the points at infinity on **o** lie on the curve (and connect its two branches). The proof of 6.52 remains valid, and shows that the absolute pole of any diameter is also its pole with respect to the circle (or equidistant curve); whence the axis (which contains such poles) is the polar of the centre. It follows that the tangents to an equidistant curve from its centre are also tangents to the

Absolute (i.e. lines at infinity). Hence

11.21. *An equidistant curve has double contact* with the Absolute.*

(In complex geometry, the same statement can be made for a proper circle. Cf. 6.92.)

The horocycle has to be treated differently, since in this case **B** and **P** coincide with the centre **O**, at infinity. As before, let **C** and **A** be two points on the curve, fixed and variable respectively. Let **N** and **L** be the remaining points at infinity on **OC** and **OA**, as in Fig. 11.2c. Then the reflection that carries **C** to **A** is a harmonic homology whose centre, **Q**, lies on the axis **o**, as well as on **CA** and **NL**. Since **O**, **N**, **C** are fixed points, we can apply 3.33 to the Absolute, obtaining the projectively related pencils

$$\mathbf{OL} \,\overline{\wedge}\, \mathbf{NQ} \,\overline{\wedge}\, \mathbf{CQ}.$$

Hence, by 3.34, *the horocycle is a conic.*

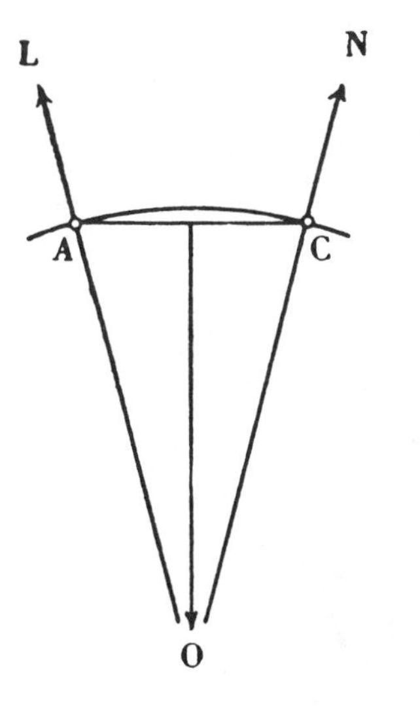

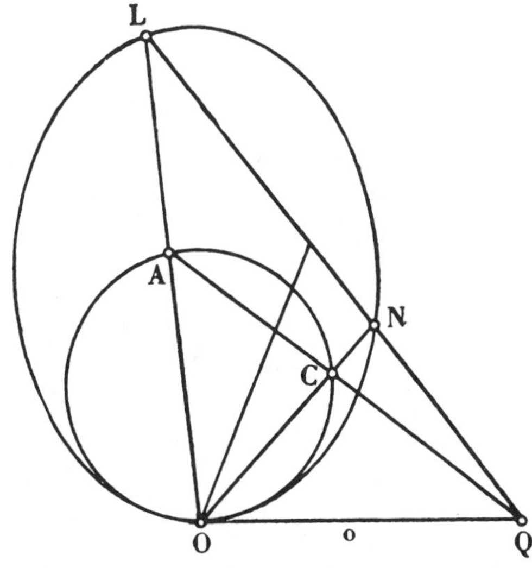

FIG. 11.2c

*Chasles [2], p. 324.

One line of the pencil **OL** is **o**, the tangent at **O** to the Absolute. The corresponding line of the pencil **CQ** is **CO**. Hence **o** is the tangent at **O** to the horocycle also. In any other position, **OL** is an ordinary line, and **A** an ordinary point. Thus the horocycle has only the point **O** in common with the Absolute.

When **OL** coincides with **ON**, **NQ** is the tangent at **N** to the Absolute, and **CQ** is the tangent at **C** to the horocycle. Thus 6.52 continues to hold, and the contact of the horocycle with the Absolute is such that each point on their common tangent has the same polar with respect to both conics. Two conics so related are said to have *contact of the third order*.* This is the limiting form of double contact when the two points of contact become coincident. The horocycle is thus exhibited as a limiting form of an equidistant curve. It is obviously also a limiting form of a proper circle, when the centre recedes to infinity. In fact Lobatschewsky sometimes called it simply the *limiting-curve*.

11.3. Spheres. The extension of the above results from two to three dimensions presents no serious difficulty. A *sphere* is defined as the class of images of a point by reflection in the planes of a bundle. Its *axial plane* is the absolute polar of its centre (which is the centre of the bundle). In hyperbolic geometry, a *proper sphere* has an ordinary centre (from which all its points are distant R, say), a *horosphere* has its centre and axial plane at infinity, and an *equidistant surface* has an ordinary axial plane (from which all its points are distant D, say) and consists of two separate sheets.

To show that a proper sphere or an equidistant surface is a quadric, we consider it as the locus of (**BP**, **Cp**), where **B** and **C** are two fixed diametrically opposite points, **P** is a variable

*Veblen and Young [2], I, pp. 133-134.

point on the axial plane ω, and **p** is the line in which ω meets the absolute polar plane of **P**.

In the case of a proper sphere, we have a variable line **OP** through the centre **O**, and a plane **Op** which varies in such a way as to remain perpendicular to **OP**. Then **BP** is the line through **B** perpendicular to **Op**, and **Cp** is the plane through **C** perpendicular to **OP**.

In the case of the equidistant surface, ω is the perpendicular bisector of **BC**. On ω, we have **P** ordinary and **p** ultra-infinite for one sheet, and *vice versa* for the other. For the first sheet, **PO** is perpendicular to ω, and the surface is the locus of the point where **BP** meets the plane through **C** perpendicular to **PO**. For the second sheet, **pO** is perpendicular to ω, and the surface is the locus of the point where **Cp** meets the line through **B** perpendicular to **pO**.

In either case, **BP** and **Cp** trace *correlated bundles*.* Therefore the locus of their point of intersection is a quadric according to Seydewitz's definition, which can be reconciled with von Staudt's by means of a chain of theorems analogous to 3.31-3.34.

As for the horosphere, let **C** and **A** be two points on it, fixed and variable respectively, as in Fig. 11.2c. Let **O** be the centre, **N** and **L** the remaining points at infinity on **OC** and **OA**, and **S** the absolute pole of the plane **OCA**. Then the reflection that carries **C** to **A** is a harmonic homology whose centre, **Q** = (**CA**, **NL**), lies in the axial plane. Since the Absolute is a quadric through the fixed points **O**, **N**, and the variable point **L**, the analogue of Steiner's Theorem shows that the plane **OSL** and line **NL** trace correlated bundles. But the latter bundle is collineated with that traced by **CQ**. Hence **OSL** and **CQ** (which meet at **A**) trace correlated bundles, and *the horosphere is a quadric.*

---

*Seydewitz [1], p. 158; Reye [1], pp. 26-38.

The diameters of a horosphere are the lines of a bundle of parallels. Any plane of the bundle meets the other planes in a flat pencil of parallels. Hence the section of the horosphere by any "diametral plane" is a horocycle. Theorem 9.83 now gives

11.31. *The geometry of points and horocycles on a horosphere can be identified with the Euclidean geometry of points and lines in a plane.*

This theorem, due to Wachter (§1.3), was rediscovered by both Lobatschewsky and Bolyai, and is fundamental in their treatment of hyperbolic trigonometry.*

11.4. The in- and ex-circles of a triangle. In discussing the general triangle we shall use the customary terminology and notation, letting a, b, c denote the *sides*, A, B, C the *angles*, h_a, h_b, h_c the *altitudes*, r the *in-radius*, and r_a, r_b, r_c, R, R_a, R_b, R_c the three *ex-radii* and four *circum-radii* (when they exist).

We have seen (2.61) that three lines **BC**, **CA**, **AB** decompose the real projective plane into four regions, any one of which can be singled out as "the triangle **ABC**" by naming an exterior line (or an interior point). The remaining three are then called the associated or *colunar* triangles.† In elliptic geometry, the colunar triangle that has the same side a has for its other two sides $\pi-b$ and $\pi-c$, while its angles are A, $\pi-B$, $\pi-C$. In hyperbolic geometry, we take **A**, **B**, **C** to be ordinary points, and consider the one triangle **ABC** whose interior points are all ordinary (i.e., the triangle **ABC/p**, where **p** is any ideal line).

The internal and external bisectors of the angles A, B, C, are the loci of points equidistant from pairs of sides. In elliptic geometry, they accordingly concur in sets of three at four points, each of which is equidistant from all three sides. One

*Bolyai [2], §21; Klein [3], p. 252.

†McClelland and Preston [1], I, p. 15.

of these, being interior to the triangle, is called the ***in-centre*, I**. The distance of this point from any side is called the ***in-radius***, r, since it is the radius of the ***inscribed circle***. The rest of the four points are the three ***ex-centres***, $\mathbf{I}_a$, $\mathbf{I}_b$, $\mathbf{I}_c$; their distances from a side are the ***ex-radii***, r_a, r_b, r_c. The ***escribed circles*** are the inscribed circles of the colunar triangles.

Most of these definitions remain valid in hyperbolic geometry. But some pairs of external angle-bisectors may be parallel or ultra-parallel. In such a case the triangle has an escribed horocycle or equidistant curve, whose tangents are the images of any side by reflection in the lines of the pencil determined by those two angle-bisectors.

It is easily proved, as in Euclidean geometry,* that the lengths of the tangents from **A** to the inscribed and escribed circles are respectively

$$s-a,\ s,\ s-c,\ s-b,$$

where $s=\frac{1}{2}(a+b+c)$.

11.5. The circum-circles and centroids. The above results can be dualized, as follows. The internal and external mid-points of the sides a, b, c, are the envelopes of lines equidistant from pairs of vertices. In elliptic geometry, they accordingly lie by threes on four lines, each of which is equidistant from all three vertices. The "trilinear poles" of these four lines, being the points of concurrency of the ***medians***, are the ***centroids*** $\mathbf{G}$, $\mathbf{G}_a$, $\mathbf{G}_b$, $\mathbf{G}_c$. (See Fig. 2.4B.) The absolute poles of the same four lines are called ***circum-centres***, since each is the centre of a circumscribed circle. One of the lines is exterior to the triangle, and the corresponding circle is called the ***principal circum-circle***. Similarly, there is a ***principal centroid***.

In hyperbolic geometry, the mid-points of **BC**, **CA**, **AB** are

*See, e.g., Johnson [2], p. 184.

joined in pairs by three lines, each of which is, by 9.67, the axis of a circumscribed equidistant curve (with two vertices on one branch and one on the other). The external mid-points of the sides are three ultra-infinite points lying on a line that may or may not be ordinary. Hence there is either a fourth circumscribed equidistant curve (with all three vertices on one branch), or a circumscribed horocycle, or a proper circum-circle.* As for the centroids, only the "principal" one is necessarily ordinary; any of the other three may be at infinity or ultra-infinite.

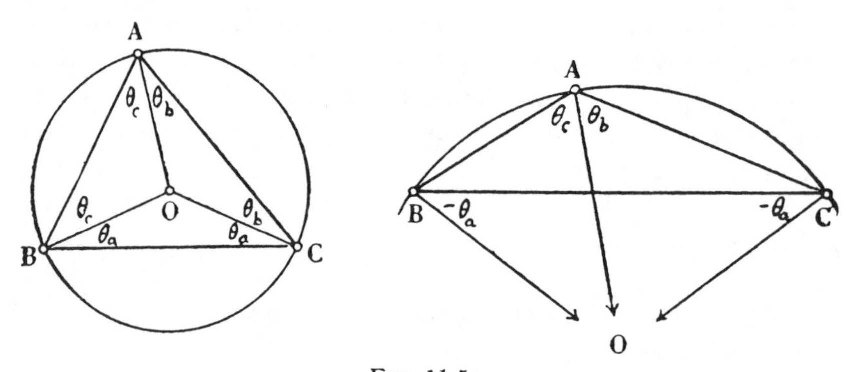

FIG. 11.5A

O being the principal circum-centre, let θ_a denote the angle **OBC** or **OCB**, considered negative if **O** is inside the first colunar triangle (i.e., if **O** is "beyond" the side **BC**), with analogous definitions for θ_b and θ_c, as in Fig. 11.5A. Then

$$\theta_b+\theta_c=A,\ \theta_c+\theta_a=B,\ \theta_a+\theta_b=C,$$

and therefore

11.51. $\theta_a=S-A,\ \theta_b=S-B,\ \theta_c=S-C,$

where $S=\theta_a+\theta_b+\theta_c=\frac{1}{2}(A+B+C).$

Thus **O** lies on **BC** if $S=A$, i.e. if $B+C=A$, and **O** is inside the first colunar triangle if $B+C<A$. Hence

*Sommerville [2], pp. 54, 143.

11.52. *The principal circum-centre is interior to the triangle* **ABC** *if and only if each of the angles A, B, C is less than the sum of the other two.*

11.6. The polar triangle and the orthocentre. In elliptic geometry, the absolute polars of the points **A**, **B**, **C** intersect in pairs at the absolute poles of **BC**, **CA**, **AB**, forming four new triangles. One of these, having sides $\pi-A$, $\pi-B$, $\pi-C$, and angles $\pi-a$, $\pi-b$, $\pi-c$, is called the *polar triangle* of **ABC**. Clearly, its in-radius and principal circum-radius are $\frac{1}{2}\pi-R$ and $\frac{1}{2}\pi-r$. The relation between the original triangle and its polar triangle is symmetrical. The in- and ex-centres of either are the circum-centres of the other.* The same definition can be applied in hyperbolic geometry; but then, of course, the polar triangle is entirely ultra-infinite.

By Hesse's Theorem (3.21), if the respective perpendiculars from **A** and **B** to **BC** and **CA** meet at **H**, then **CH** will be perpendicular to **AB**. In other words, the "altitude lines" (which join **A**, **B**, **C** to the corresponding vertices of the polar triangle, as in 3.22) concur at a point **H**, called the *orthocentre*. In hyperbolic geometry, the orthocentre of an obtuse-angled triangle may be ordinary, at infinity, or ultra-infinite.

*McClelland and Preston [1], I, p. 27; II, p. 8.

CHAPTER XII

THE USE OF A GENERAL TRIANGLE OF REFERENCE

12.1. Formulae for distance and angle. For many purposes, such as the development of trigonometry, it is desirable to take the absolute polarity in its general form 4.52, namely

12.11. $$X_\mu = c_{\mu\nu} x_\nu, \quad x_\mu = C_{\mu\nu} X_\nu \qquad (\mu = 0, 1, 2),$$

where $$c_{\mu\nu} = c_{\nu\mu} \text{ and } c_{\lambda\nu} C_{\mu\nu} = \delta_{\lambda\mu}.$$

We shall find it convenient to make the abbreviations*

$$(xy) = c_{\mu\nu} x_\mu y_\nu, \quad [X\,Y] = C_{\mu\nu} X_\mu Y_\nu,$$

so that the hyperbolic Absolute is the conic $(xx) = 0$ or $[X\,X] = 0$. Since the polarity is not changed by reversing the signs of all the $c_{\mu\nu}$ and $C_{\mu\nu}$, there is no loss of generality in assuming that $(xx) > 0$ for one ordinary point, and therefore (by continuous variation) for *all* ordinary points. In hyperbolic geometry this implies that $(xx) < 0$ for all ultra-infinite points.

Let $[X]$ and $[Y]$ be the respective polars of (x) and (y), so that, besides 12.11, we have a similar relation between the Y's and y's. Then since

$$c_{\mu\nu} x_\mu y_\nu = x_\mu Y_\mu \text{ and } C_{\mu\nu} X_\mu Y_\nu = x_\nu Y_\nu,$$

we have

12.12. $$(xy) = \{x\,Y\} = [X\,Y] = \{X y\}.$$

Factors of proportionality have been suppressed in 12.11, and so also here. This makes no confusion, provided we keep our equations homogeneous in the coordinates of each point or line.

In particular, $[X\,Y] = 0$ is the condition for lines $[X]$ and $[Y]$ to be perpendicular, and $(xy) = 0$ is the condition for points

*In the notation of Klein [3], pp. 167-169, these are Ω_{xy} and Φ_{uv}.

(x) and (y) to be perpendicular (or conjugate). More generally, the squared cosine of the angle between lines $[X]$ and $[Y]$ is

$$\frac{\{xY\}\{yX\}}{\{xX\}\{yY\}} = \frac{[XY]^2}{[XX][YY]},$$

and the cosine (or hyperbolic cosine) of the distance between points (x) and (y) is

$$\sqrt{\frac{\{xY\}\ \{yX\}}{\{xX\}\ \{yY\}}} = \frac{|(xy)|}{\sqrt{(xx)}\sqrt{(yy)}}.$$

In elliptic geometry we have $[XX]=(xx)>0$ for every line; in fact both (xx) and $[XX]$ are positive definite forms. In hyperbolic geometry we have $[XX]=(xx)>0$ for the polar of any ordinary point, i.e. for any ultra-infinite line; therefore $[XX]<0$ for every ordinary line. Hence, according to the convention of §6.7, the "internal" angle between intersecting lines $[X]$ and $[Y]$ is

12.13. $$\text{arc cos}\,\frac{-[XY]}{\sqrt{[XX]}\sqrt{[YY]}} \quad \text{or} \quad \text{arc cos}\,\frac{[XY]}{\sqrt{([XX][YY])}}$$

in the two geometries respectively. Also the distance from the point (x) to the line $[Y]$ is

12.14. $$\text{arc sin}\,\frac{|\{xY\}|}{\sqrt{(xx)}\sqrt{[YY]}} \quad \text{or} \quad \text{arg sinh}\,\frac{|\{xY\}|}{\sqrt{(xx)}\sqrt{(-[YY])}},$$

and, as we have already remarked, the distance between points (x) and (y) is

12.15. $$\text{arc cos}\,\frac{|(xy)|}{\sqrt{(xx)}\sqrt{(yy)}} \quad \text{or} \quad \text{arg cosh}\,\frac{|(xy)|}{\sqrt{(xx)}\sqrt{(yy)}}.$$

Finally, by 10.72, the distance between ultra-parallel lines $[X]$ and $[Y]$ is

12.16. $$\text{arg cosh}\,\frac{|[XY]|}{\sqrt{([XX][YY])}}.$$

It is interesting to see how 12.15 was obtained *before* Klein had defined distance in terms of cross ratio. Cayley showed that this expression has the additive property for juxtaposed segments, by observing that the relation

$$\arc \cos \frac{(xy)}{\sqrt{(xx)}\sqrt{(yy)}} + \arc \cos \frac{(yz)}{\sqrt{(yy)}\sqrt{(zz)}} = \arc \cos \frac{(xz)}{\sqrt{(xx)}\sqrt{(zz)}}$$

implies the vanishing of the determinant

$$\begin{vmatrix} (xx) & (xy) & (xz) \\ (yx) & (yy) & (yz) \\ (zx) & (zy) & (zz) \end{vmatrix} = \begin{vmatrix} \{xX\} & \{xY\} & \{xZ\} \\ \{yX\} & \{yY\} & \{yZ\} \\ \{zX\} & \{zY\} & \{zZ\} \end{vmatrix} = \begin{vmatrix} x_0 & x_1 & x_2 \\ y_0 & y_1 & y_2 \\ z_0 & z_1 & z_2 \end{vmatrix} \cdot \begin{vmatrix} X_0 & Y_0 & Z_0 \\ X_1 & Y_1 & Z_1 \\ X_2 & Y_2 & Z_2 \end{vmatrix}$$

which is the condition for the points (x), (y), (z) to be collinear* (or for the lines $[X]$, $[Y]$, $[Z]$ to be concurrent, which amounts to the same thing).

12.2. The general circle. To obtain the equation for a circle, we use the first definition of §11.1, reflecting the fixed point (y) in the variable line $[U]$ through the centre (z). If (u) is the absolute pole of $[U]$, so that $(uz) = \{Uz\} = 0$, 4.41 gives the circle as the locus of (x) where

12.21. $x_\mu = y_\mu - 2u_\mu \{yU\}/\{uU\} = y_\mu - 2u_\mu (yu)/(uu)$,

and similarly $y_\mu = x_\mu - 2u_\mu (xu)/(uu)$.

Multiplying the first of these relations by $c_{\mu\nu} x_\nu$, the second by $c_{\mu\nu} y_\nu$, and either of them by $c_{\mu\nu} z_\nu$, we deduce

$$(xx) = (xy) - 2(xu)(yu)/(uu) = (yy), \text{ and } (zx) = (zy).$$

These lead to the single homogeneous equation

$$(yy)(zx)^2 = (zy)^2(xx).$$

(A factor of proportionality may be appended to x_μ in 12.21

*Cayley [3], pp. 69, 81.

if desired. Then the final step consists in eliminating it.) Hence *the circle through* (y) *with centre* (z) *has the equation**

12.22. $$(zx)^2 = g'(xx),$$

where $$g' = (zy)^2/(yy).$$

In other words, *the circle through* (y) *with axis* $[Z]$ *has the equation*

12.23. $$\{Zx\}^2 = g(xx),$$

where $$g = \{Zy\}^2/(yy).$$

In hyperbolic geometry, the above equations represent a proper circle, horocycle, or equidistant curve, according as $(zz) = [ZZ]$ is positive, zero, or negative. Applying 12.15 to (y) and (z), and 12.14 to (y) and $[Z]$, we see that

$$g' = (zz)\cos^2 R \text{ or } (zz)\cosh^2 R \quad \text{(for a proper circle)}$$

and $g = [ZZ]\sin^2 D$ or $-[ZZ]\sinh^2 D$ (for an equidistant curve). No such expression for g' or g can be looked for in the case of a horocycle. The form of equation 12.23 is clearly in agreement with 6.92 and 11.21, the chord of contact being the axis $\{Zx\} = 0$.

In particular, when the absolute polarity is expressed in its canonical form, a proper circle with centre $(1, 0, 0)$ is

12.24. $x_0^2 = (x_0^2 + x_1^2 + x_2^2)\cos^2 R$ or $x_0^2 = (x_0^2 - x_1^2 - x_2^2)\cosh^2 R$;

the horocycle through $(1, 0, 0)$ with centre $(1, 0, 1)$ is

12.25. $$(x_0 - x_2)^2 = x_0^2 - x_1^2 - x_2^2,$$

which may alternatively be expressed as

$$x_1^2 = 2(x_0 - x_2)x_2;$$

and an equidistant curve with axis $[0, 0, 1]$ is

12.26. $$x_2^2 = (x_0^2 - x_1^2 - x_2^2)\sinh^2 D.$$

As an example of the use of general coordinates, consider the familiar theorem that *the polar of a point with respect to a*

*Sommerville [2], p. 136.

circle is perpendicular to the diameter through the point. This can be proved (for all kinds of circle at once) as follows. The polar of the point (w) with respect to the conic 12.23 is the line

$$\{Zw\}\ \{Zx\} = g(wx).$$

From the form of its equation, we see that this is concurrent with the lines $\{Zx\} = 0$ and $(wx) = 0$, namely the axis of the circle and the absolute polar of (w). Hence its absolute polar is collinear with the centre of the circle and (w) itself. The desired result is now evident.

12.3. Tangential equations. The harmonic homology with centre (u) and axis $[U]$ may be expressed as a transformation of tangential coordinates by dualizing the steps that led to 4.41. The result is

$$X'_\mu = X_\mu - 2U_\mu\{uX\}/\{uU\}.$$

This enables us to dualize the derivation of 12.22, so that *the circle touching* $[Y]$ *with axis* $[Z]$ *has the tangential equation*

12.31. $$[ZX]^2 = G'[XX],$$

where $$G' = [ZY]^2/[YY].$$

In other words, *the circle touching* $[Y]$ *with centre* (z) *has the tangential equation*

12.32. $$\{zX\}^2 = G[XX],$$

where $$G = \{zY\}^2/[YY].$$

Applying 12.13 or 12.16 to $[Z]$ and $[Y]$, and 12.14 to (z) and $[Y]$, we see that

$G' = [ZZ]\cos^2 D$ or $[ZZ]\cosh^2 D$ (for an equidistant curve)

and $G = (zz)\sin^2 R$ or $-(zz)\sinh^2 R$ (for a proper circle). Thus, in both geometries, $G + g' = (zz)$ and $g + G' = [ZZ]$. In fact, we could verify directly that the tangents to the conic 12.23 satisfy the equation $[ZX]^2 = ([ZZ] - g)[XX]$.

12.4. Circum-circles and centroids. There are two possible ways of applying coordinates to the study of the general triangle. One* is to take the absolute polarity in its canonical form, letting the vertices of the triangle be three general points (x), (y), (z). The other, preferred here, is to take the absolute polarity in its general form 12.11 and to study the triangle of reference, whose sides and angles are definite functions of the coefficients $c_{\mu\nu}$.

In elliptic geometry, (xx) and $[XX]$ being positive definite forms, the "diagonal" coefficients c_{00}, c_{11}, c_{22}, C_{00}, C_{11}, C_{22} are positive, and we shall find it convenient to denote their positive square roots by

$$c_0,\ c_1,\ c_2,\ C_0,\ C_1,\ C_2.$$

The reciprocal determinants γ and Γ are positive also.

In hyperbolic geometry, the vertices and sides of the triangle of reference being ordinary points and lines, we have c_{00}, c_{11}, c_{22} positive, and C_{00}, C_{11}, C_{22} negative; accordingly we write

$$c_0 = \sqrt{c_{00}},\quad C_0 = \sqrt{|C_{00}|}\ ,$$

and so on. The absolute polars of the vertices, namely the lines

$$[c_{00},\ c_{10},\ c_{20}],\ [c_{01},\ c_{11},\ c_{21}],\ [c_{02},\ c_{12},\ c_{22}]\ ,$$

are ultra-infinite, and so external; hence the $c_{\mu\nu}$ are all positive. The points at infinity on the side $x_0 = 0$ are given by the quadratic equation

$$c_{11}x_1^2 + 2c_{12}x_1x_2 + c_{22}x_2^2 = 0;$$

hence $\gamma C_{00} = c_{11}c_{22} - c_{12}^2 < 0$, and γ is again positive.

Since the coefficient of x_0^2 in (xx) is c_{00} or c_0^2, the general circle 12.23 passes through the point $(1, 0, 0)$ if $Z_0^2 = gc_0^2$, and

*Coolidge [1], p. 102; Sommerville [2], p. 139.

through the other two vertices if $Z_1^2 = gc_1^2$ and $Z_2^2 = gc_2^2$. Hence the four circum-circles are

12.41. $$(c_0x_0 \pm c_1x_1 \pm c_2x_2)^2 = (xx);$$

and their axes (the absolute polars of the circum-centres) are

12.42. $$[c_0, c_1, c_2], [-c_0, c_1, c_2], [c_0, -c_1, c_2], [c_0, c_1, -c_2].$$

Thus the principal circum-centre is $[cX]=0$, or

$$(C_{\mu 0}\, c_\mu, C_{\mu 1}\, c_\mu, C_{\mu 2}\, c_\mu).$$

In hyperbolic geometry, there is a proper circum-circle if the axis $[c_0, c_1, c_2]$ is ultra-infinite, i.e. if $[cc] > 0$; and there is a circumscribed horocycle if $[cc]=0$. In the former case the circum-radius R, being the distance from $(1, 0, 0)$ to the circum-centre, is given by

$$\cosh R = \frac{c_{0\nu} C_{\mu\nu} c_\mu}{\sqrt{c_{00}}\sqrt{[cc]}}.$$

Since $c_{0\nu} C_{\mu\nu} c_\mu = \delta_{0\mu} c_\mu = c_0$, this reduces to

12.43. $$R = \arg\cosh\, [cc]^{-\frac{1}{2}},$$

In elliptic geometry the principal circum-radius is found more easily. Since $\frac{1}{2}\pi - R$ is the distance from the point $(1, 0, 0)$ to the line $[c_0, c_1, c_2]$, we have

$$\sin\left(\tfrac{1}{2}\pi - R\right) = c_0/\sqrt{c_{00}}\sqrt{[cc]},$$

so that

12.44. $$R = \arccos\, [cc]^{-\frac{1}{2}}.$$

The centroids, being the trilinear poles of the lines 12.42, are

12.45.

$$\left(\frac{1}{c_0}, \frac{1}{c_1}, \frac{1}{c_2}\right), \left(-\frac{1}{c_0}, \frac{1}{c_1}, \frac{1}{c_2}\right), \left(\frac{1}{c_0}, -\frac{1}{c_1}, \frac{1}{c_2}\right), \left(\frac{1}{c_0}, \frac{1}{c_1}, -\frac{1}{c_2}\right).$$

In hyperbolic geometry, the last three are not necessarily ordinary points; e.g. the fourth is ultra-infinite if

12.46. $$\frac{c_{12}}{c_1c_2} + \frac{c_{20}}{c_2c_0} - \frac{c_{01}}{c_0c_1} > \frac{3}{2}.$$

12.5. In- and ex-circles. The circle 12.32 touches the line [1, 0, 0] if $z_0^2 = GC_{00} = |G|C_0^2$, and also the other two sides if $z_1^2 = |G|C_1^2$ and $z_2^2 = |G|C_2^2$. (In hyperbolic geometry, G is negative.) Hence the in-circle and ex-circles are

12.51. $$(C_0 X_0 \pm C_1 X_1 \pm C_2 X_2)^2 = |\,[X\,X]\,|,$$

and their centres are

12.52. $(C_0, C_1, C_2), (-C_0, C_1, C_2), (C_0, -C_1, C_2), (C_0, C_1, -C_2)$.

The in-radius, being the distance from (C_0, C_1, C_2) to [1, 0, 0], is

12.53. $$r = \text{arc sin } (CC)^{-\frac{1}{2}} \text{ or arg sinh } (CC)^{-\frac{1}{2}}.$$

12.6. The orthocentre. The vertices of the polar triangle of the triangle of reference, being the absolute poles of the sides [1, 0, 0], [0, 1, 0], [0, 0, 1], are

$$(C_{00}, C_{10}, C_{20}),\ (C_{01}, C_{11}, C_{21}),\ (C_{02}, C_{12}, C_{22}).$$

These are joined to the corresponding vertices of the original triangle by the altitude-lines

$$[0, C_{20}, -C_{10}],\ [-C_{21}, 0, C_{01}],\ [C_{12}, -C_{02}, 0],$$

which concur at the orthocentre $(C_{20}C_{01}, C_{01}C_{12}, C_{12}C_{20})$. If $C_{12}C_{20}C_{01} \neq 0$, this is simply

12.61. $$\left(\frac{1}{C_{12}}, \frac{1}{C_{20}}, \frac{1}{C_{01}}\right).$$

If one of C_{12}, C_{20}, C_{01} vanishes, the triangle is right-angled, and the orthocentre coincides with a vertex. If more than one of them vanishes, the triangle has two (or three) right angles, and the orthocentre is indeterminate. If one of C_{12}, C_{20}, C_{01} differs in sign from the other two, the orthocentre is exterior. (This is, of course, the case of an obtuse-angled triangle.)

In hyperbolic geometry, the orthocentre is ultra-infinite if

$$\frac{c_{00}}{C_{12}^2} + \frac{c_{11}}{C_{20}^2} + \frac{c_{22}}{C_{01}^2} + \frac{2c_{12}}{C_{20}\,C_{01}} + \frac{2c_{20}}{C_{01}\,C_{12}} + \frac{2c_{01}}{C_{12}\,C_{20}} < 0.$$

Since $c_{00}=\gamma(C_{11}C_{22}-C_{12}^2)$ and $c_{12}=\gamma(C_{20}C_{01}-C_{00}C_{12})$, this condition is equivalent to

$$3+\frac{C_{11}C_{22}}{C_{12}^2}+\frac{C_{22}C_{00}}{C_{20}^2}+\frac{C_{00}C_{11}}{C_{01}^2}$$
$$-\frac{2C_{00}C_{12}}{C_{20}C_{01}}-\frac{2C_{11}C_{20}}{C_{01}C_{12}}-\frac{2C_{22}C_{01}}{C_{12}C_{20}}<0$$

or
$$\left(\frac{C_{01}}{C_{20}}-\frac{C_{11}}{C_{12}}\right)\left(\frac{C_{20}}{C_{01}}-\frac{C_{22}}{C_{12}}\right)+\left(\frac{C_{12}}{C_{01}}-\frac{C_{22}}{C_{20}}\right)\left(\frac{C_{01}}{C_{12}}-\frac{C_{00}}{C_{20}}\right)$$
$$+\left(\frac{C_{20}}{C_{12}}-\frac{C_{00}}{C_{01}}\right)\left(\frac{C_{12}}{C_{20}}-\frac{C_{11}}{C_{01}}\right)<0$$

or
$$\frac{c_{20}}{C_{20}C_{12}}\,\frac{c_{01}}{C_{01}C_{12}}+\frac{c_{01}}{C_{01}C_{20}}\,\frac{c_{12}}{C_{12}C_{20}}+\frac{c_{12}}{C_{12}C_{01}}\,\frac{c_{20}}{C_{20}C_{01}}<0$$

or

12.62.
$$\frac{C_{20}C_{01}}{c_{12}}+\frac{C_{01}C_{12}}{c_{20}}+\frac{C_{12}C_{20}}{c_{01}}<0.$$

12.7. Elliptic trigonometry. By considering the model of §6.1, we see that the formulae of elliptic trigonometry must be identical with the familiar formulae of spherical trigonometry.* Nevertheless, it is interesting to derive them differently.

When the points (x) and (y) of 12.15 are $(0, 1, 0)$ and $(0, 0, 1)$, we have $(xx)=c_{11}$, $(xy)=c_{12}$, $(yy)=c_{22}$. Hence, in elliptic geometry, the sides of the triangle of reference are given by†

12.71.
$$\cos a=\frac{c_{12}}{c_1c_2},\quad \cos b=\frac{c_{20}}{c_2c_0},\quad \cos c=\frac{c_{01}}{c_0c_1}.$$

Since $c_{11}c_{22}-c_{12}^2=C_{00}\gamma$, it follows that

$$\sin a=\frac{C_0\sqrt{\gamma}}{c_1c_2},\quad \sin b=\frac{C_1\sqrt{\gamma}}{c_2c_0},\quad \sin c=\frac{C_2\sqrt{\gamma}}{c_0c_1}.$$

*See, e.g., McClelland and Preston [1], I, pp. 36-39, 50, 58.
†Baker [1], II, p. 205.

Similarly, when the lines $[X]$ and $[Y]$ of 12.13 are [0, 1 0] and [0, 0, 1], we have $[XX]=C_{11}$, $[XY]=C_{12}$, $[YY]=C_{22}$. Hence the angles are given by

12.72.

$$\cos A = -\frac{C_{12}}{C_1 C_2}, \quad \cos B = -\frac{C_{20}}{C_2 C_0}, \quad \cos C = -\frac{C_{01}}{C_0 C_1}.$$

Since $C_{11}C_{22}-C_{12}^2 = c_{00}\Gamma$, it follows that

$$\sin A = \frac{c_0\sqrt{\Gamma}}{C_1 C_2}, \quad \sin B = \frac{c_1\sqrt{\Gamma}}{C_2 C_0}, \quad \sin C = \frac{c_2\sqrt{\Gamma}}{C_0 C_1}.$$

To obtain the three colunar triangles, we merely have to reverse the signs of two of c_0, c_1, c_2, and of the corresponding two of C_0, C_1, C_2. To obtain the polar triangle, we interchange $c_{\mu\nu}$ and $C_{\mu\nu}$, γ and Γ.

When the point (x) and line $[Y]$ of 12.14 are (1, 0, 0) and [1, 0, 0], we have $(xx)=c_{00}$, $\{xY\}=1$, $[YY]=C_{00}$. Hence the altitudes are given by

$$\operatorname{cosec} h_a = c_0C_0, \quad \operatorname{cosec} h_b = c_1C_1, \quad \operatorname{cosec} h_c = c_2C_2.$$

Thus $\sin a \sin h_a = \sin b \sin h_b = \sin c \sin h_c = \sqrt{\gamma}/c_0c_1c_2$,

$$\sin A \sin h_a = \sin B \sin h_b = \sin C \sin h_c = \sqrt{\Gamma}/C_0C_1C_2,$$

and we have the famous "rule of sines":

12.73.
$$\frac{\sin a}{\sin A} = \frac{\sin b}{\sin B} = \frac{\sin c}{\sin C}.$$

To express the angle C in terms of the three sides, we have the formula

$$\sin a \sin b \cos C = -C_{01}\gamma/c_1c_{22}c_0 = (c_{22}c_{01}-c_{12}c_{20})/c_1c_{22}c_0$$

12.74.
$$= \cos c - \cos a \cos b.$$

Again, to express the side c in terms of the three angles, we have

$$\sin A \sin B \cos c = c_{01}\Gamma/C_1C_{22}C_0 = (-C_{22}C_{01}+C_{12}C_{20})/C_1C_{22}C_0$$

12.75.
$$= \cos C + \cos A \cos B.$$

From each such formula we can derive others by permuting a, b, c, and simultaneously A, B, C; e.g. 12.75 gives

$$\sin B \sin C \cos a = \cos A + \cos B \cos C.$$

In a similar manner we may prove that

$$\cot c \sin a - \cot C \sin B - \cos a \cos B = c_{\mu 1} C_{\mu 0} / c_1 c_2 C_2 C_0 = 0,$$

$$\cot a \sin b - \cot A \sin C - \cos b \cos C = 0,$$

and so on.

By putting $C = \frac{1}{2}\pi$, we derive the following formulae for a right-angled triangle:

$$\sin a / \sin A = \sin b / \sin B = \sin c,$$
$$0 = \cos c - \cos a \cos b,$$
$$\sin A \sin B \cos c = \cos A \cos B,$$
$$\sin B \cos a = \cos A,$$
$$\cot c \sin a - \cos a \cos B = 0,$$
$$\cot a \sin b - \cot A = 0.$$

Collecting these into a more convenient form, we have

12.76. $\sin a = \sin c \sin A, \quad \sin b = \sin c \sin B,$

12.77. $\cos c = \cos a \cos b = \cot A \cot B,$

12.78. $\cos A = \cos a \sin B, \quad \cos B = \cos b \sin A,$

12.79. $\begin{cases} \tan a = \tan c \cos B = \sin b \tan A, \\ \tan b = \tan c \cos A = \sin a \tan B. \end{cases}$

It was noticed by Napier that the formulae for a right-angled triangle are permuted among themselves by cyclic permutation of the five "parts" A, $\frac{1}{2}\pi - a$, c, $\frac{1}{2}\pi - b$, B. If we remember this rule, we can deduce all the ten formulae from 12.77. By drawing the absolute polars of the vertices **A** and **B**, Gauss showed that the five "parts" in the above order are the supplements of the angles of a pentagon whose five pairs of alternate sides are perpendicular.* They are also the halves

*For a nicely illustrated description of this *pentagramma mirificum*, see Mohrmann [1], p. 72.

of the angles of a *hyperbolic* pentagon whose five pairs of alternate sides are parallel.*

The result of using a general unit of length, as in 5.76, would be to replace 12.74 by

$$\cos\frac{\pi c}{2\lambda}=\cos\frac{\pi a}{2\lambda}\cos\frac{\pi b}{2\lambda}+\sin\frac{\pi a}{2\lambda}\sin\frac{\pi b}{2\lambda}\cos C,$$

where λ is the length of a "right segment." Making λ tend to infinity, we obtain the familiar formula

$$c^2=a^2+b^2-2ab\cos C.$$

This provides an alternative proof for 10.94 (in the elliptic case).

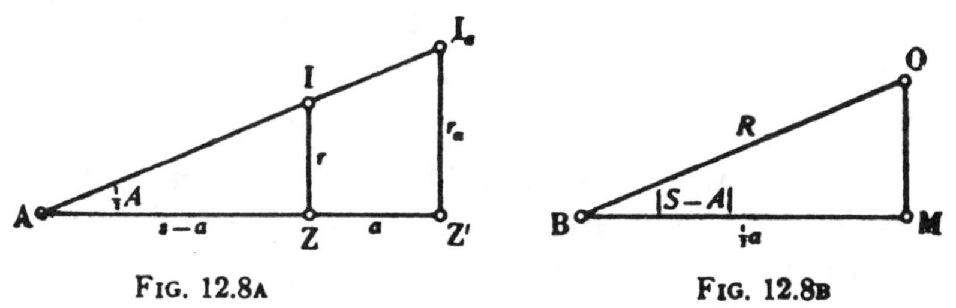

FIG. 12.8A FIG. 12.8B

12.8. The radii. Continuing our investigation of the elliptic geometry of the general triangle **ABC**, let **Z** and **Z′** be the points of contact of the side **AB** with the in-circle and the first ex-circle, so that the triangles **AIZ** and **AI$_a$Z′** are right-angled at **Z** and **Z′**, as in Fig. 12.8A. Also let **M** be the internal mid-point of **BC**, so that the triangle **OBM** is right-angled at **M**, as in Fig. 12.8B. Applying 12.79 to these triangles, we have, in the notation of §§11.4 and 11.5,†

12.81. $\tan r=\sin(s-a)\tan\frac{1}{2}A,$

12.82. $\tan r_a=\sin s\tan\frac{1}{2}A,$

and $\tan\frac{1}{2}a=\tan R\cos(S-A),$ whence

*See §15.2.

†McClelland and Preston [1], II, pp. 4-12.

12.83. $\cot R = \cos(S-A)\cot \tfrac{1}{2}a.$

Formula 12.82 could have been obtained by applying 12.81 to the colunar triangle whose sides are a, $\pi - b$, $\pi - c$; we merely have to replace $s-a$ by $\pi - s$. Similarly, replacing $S-A$ by $\pi - S$, we deduce from 12.83,

12.84. $\cot R_a = -\cos S \cot \tfrac{1}{2}a.$

(This expression is positive, since S is obtuse.)

Alternatively, 12.83 and 12.84 could have been obtained by applying 12.81 and 12.82 to the polar triangle, whose sides, angles, and circum-radii are

$$\pi - A,\ \pi - B,\ \pi - C,\ \pi - a,\ \pi - b,\ \pi - c,$$
$$\tfrac{1}{2}\pi - r,\ \tfrac{1}{2}\pi - r_a,\ \tfrac{1}{2}\pi - r_b,\ \tfrac{1}{2}\pi - r_c.$$

By permuting a, b, c, and A, B, C, we derive alternative formulae for r and R, and analogous formulae for r_b, r_c, R_b, R_c.

Since a triangle is determined by its three sides, we lose no generality by putting either $C_0 = C_1 = C_2 = 1$ or $c_0 = c_1 = c_2 = 1$. In the former case we deduce

$$C_{12} = -\cos A,\ C_{20} = -\cos B,\ C_{01} = -\cos C,$$

12.85. $\Gamma = 1 - \cos^2 A - \cos^2 B - \cos^2 C - 2\cos A \cos B \cos C,$

and $\sin B \sin C \sin a = \sqrt{\Gamma}.$

By 12.44, we have

$$\begin{aligned}\Gamma \sec^2 R &= \Gamma\,(c_0^2 + c_1^2 + c_2^2 + 2C_{12}c_1c_2 + 2C_{20}c_2c_0 + 2C_{01}c_0c_1)\\ &= \sin^2 A + \sin^2 B + \sin^2 C - 2\cos A \sin B \sin C\\ &\qquad -2\cos B \sin C \sin A - 2\cos C \sin A \sin B,\end{aligned}$$

whence $\Gamma \tan^2 R = 2 + 2\cos(A+B+C) = 4\cos^2 S$. But $\cos S < 0$; hence

12.86. $\tan R = -2\Gamma^{-\frac{1}{2}} \cos S.$

Moreover, the in-centre, orthocentre, and (principal) centroid are now

$$(1, 1, 1),\ (\sec A, \sec B, \sec C),\ (\operatorname{cosec} A, \operatorname{cosec} B, \operatorname{cosec} C),$$

precisely as in Euclidean *trilinear* coordinates.* In fact, we can derive Euclidean geometry by making Γ tend to zero. (Cf. §10.9.)

The analogue of *areal* coordinates* is obtained by making the alternative specialization $c_0=c_1=c_2=1$, whence

$$c_{12}=\cos a,\ c_{20}=\cos b,\ c_{01}=\cos c,$$

12.87. $\gamma=1-\cos^2 a-\cos^2 b-\cos^2 c+2\cos a\cos b\cos c,$

and† $\sin b\sin c\sin A=\sqrt{\gamma}.$

From 12.53 we readily deduce

12.88. $\cot r=2\gamma^{-\frac{1}{2}}\sin s.$

The coordinates of any point are the same as in the previous case, only multiplied respectively by $\sin A$, $\sin B$, $\sin C$ (or by $\sin a$, $\sin b$, $\sin c$). The principal circum-circle, in the form

$$(c_1c_2-c_{12})x_1x_2+(c_2c_0-c_{20})x_2x_0+(c_0c_1-c_{01})x_0x_1=0,$$

reduces to

$$(\sin^2 \tfrac{1}{2}a)x_1x_2+(\sin^2 \tfrac{1}{2}b)x_2x_0+(\sin^2 \tfrac{1}{2}c)x_0x_1=0.$$

This time we derive Euclidean geometry‡ by making γ tend to zero. (Since we have made two different specializations, we no longer have $\gamma\Gamma=1$.)

12.9. Hyperbolic trigonometry. By applying the methods of §12.7 to hyperbolic geometry, we obtain the sides, angles, and altitudes of the triangle of reference in the form

12.91. $\cosh a=\dfrac{c_{12}}{c_1c_2}$, *etc.*, $\sinh a=\dfrac{C_0\sqrt{\gamma}}{c_1c_2}$, *etc.*,

12.92. $\cos A=\dfrac{C_{12}}{C_1C_2}$, *etc.*, $\sin A=\dfrac{c_0\sqrt{\Gamma}}{C_1C_2}$, *etc.*,

---

*Sommerville [4], p. 19.

†Thus γ and Γ are the $4n^2$ and $4N^2$ of McClelland and Preston [1], I, pp. 42, 55.

‡Milne [1], p. 106.

$$\text{cosech } h_a = c_0 C_0 \text{, } etc.,$$

and deduce*

12.93. $$\frac{\sinh a}{\sin A} = \frac{\sinh b}{\sin B} = \frac{\sinh c}{\sin C},$$

12.94. $\sinh b \sinh c \cos A = \cosh b \cosh c - \cosh a,$ *etc.*,

12.95. $\sin B \sin C \cosh a = \cos B \cos C + \cos A,$ *etc.*

For a right-angled triangle (with $C = \frac{1}{2}\pi$), we have

12.96. $\sinh a = \sinh c \sin A, \quad \sinh b = \sinh c \sin B,$

12.97. $\cosh c = \cosh a \cosh b = \cot A \cot B,$

12.98. $\cos A = \cosh a \sin B, \quad \cos B = \cosh b \sin A,$

12.99. $$\begin{cases} \tanh a = \tanh c \cos B = \sinh b \tan A, \\ \tanh b = \tanh c \cos A = \sinh a \tan B. \end{cases}$$

All these formulae were discovered by Lobatschewsky. But he preferred to express them in terms of angles of parallelism; e.g., he wrote 12.98 in the form

$$\sin B = \sin \Pi(a) \cos A, \quad \sin A = \sin \Pi(b) \cos B.$$

The result of using a general unit of length, as in §8.1, would be to replace 12.94 by

$$\cosh \tfrac{1}{2}\kappa a = \cosh \tfrac{1}{2}\kappa b \cosh \tfrac{1}{2}\kappa c - \sinh \tfrac{1}{2}\kappa b \sinh \tfrac{1}{2}\kappa c \cos A.$$

Making κ tend to zero, we see again that the geometry of an infinitesimal region is Euclidean. On the other hand, putting $\kappa = 2i$ instead of 2, we obtain the elliptic formula

$$\cos a = \cos b \cos c + \sin b \sin c \cos A.$$

Conversely,† hyperbolic formulae can be derived from elliptic formulae by multiplying each symbol for distance (such as a or h_a) by i.

We find, as in §12.8, that the radii of the general triangle are given by

*Lobatschewsky [1], pp. 633-638.

†This is the device used by Taurinus; see §1.3.

$$\tanh r = \sinh(s-a) \tan \tfrac{1}{2}A,$$

$$\tanh r_a = \sinh s \tan \tfrac{1}{2}A, \text{ etc.},$$

$$\coth R = \cos(S-A) \coth \tfrac{1}{2}a.$$

The condition for a proper ex-circle beyond the side **BC** (or for a real r_a)* is found to be

$$\cos \tfrac{1}{2}A < \sin \tfrac{1}{2}B + \sin \tfrac{1}{2}C.$$

Thus the conditions for three escribed horocycles are

$$A = B = C = 2 \text{ arc tan } \tfrac{1}{2} = \text{arc tan } \tfrac{4}{3},$$

implying†

$$a = b = c = \text{arg cosh } \tfrac{3}{2}\,.$$

By putting $C_{00} = C_{11} = C_{22} = -1$, so that $C_{12} = \cos A$, *etc.*, we find the hyperbolic analogue of 12.86 to be

$$\tanh R = 2\Gamma^{-\frac{1}{2}} \cos S,$$

where $\Gamma = -1 + \cos^2 A + \cos^2 B + \cos^2 C + 2 \cos A \cos B \cos C$. Similarly, the analogue of 12.88 is

$$\coth r = 2\gamma^{-\frac{1}{2}} \sinh s,$$

where $\gamma = 1 - \cosh^2 a - \cosh^2 b - \cosh^2 c + 2 \cosh a \cosh b \cosh c$.

The condition 12.46 for the fourth centroid $\mathbf{G}_c$ to be ultra-infinite is clearly equivalent to

$$\cosh a + \cosh b - \cosh c > \tfrac{3}{2}.$$

Thus all three centroids are at infinity if $a = b = c = \text{arg cosh } \tfrac{3}{2}$. In this case there are three escribed horocycles; for, it follows from 12.91 that any triangle whose principal centroid and in-centre coincide is equilateral, and that then the other centroids coincide with the ex-centres.

By 12.92, an obtuse-angled triangle has $C_{12}C_{20}C_{01} < 0$. Hence, by 12.62, the condition for the orthocentre of an obtuse-angled triangle to be ultra-infinite is

*Coxeter [3].

†Sommerville [2], p. 85 (Example 13).

$$\frac{1}{c_{12}C_{12}}+\frac{1}{c_{20}C_{20}}+\frac{1}{c_{01}C_{01}}>0,$$

or $\quad \operatorname{sech} a \tan A+\operatorname{sech} b \tan B+\operatorname{sech} c \tan C > 0,$

or $\quad \tanh a \sec A+\tanh b \sec B+\tanh c \sec C > 0.$

We can study a triangle that is singly-, doubly-, or trebly-asymptotic by making one or more of c_0, c_1, c_2 tend to zero. Thus a trebly-asymptotic triangle has

$$c_0=c_1=c_2=0,\ C_{12}=C_1C_2, \ldots,\ c_{12}=C_0\sqrt{\gamma}, \ldots,$$

and $\gamma=2c_{12}c_{20}c_{01}=2C_0C_1C_2\gamma^{3/2}$. In fact, the Absolute takes the form

$$C_0x_1x_2+C_1x_2x_0+C_2x_0x_1=0 \quad \text{or}$$
$$\sqrt{(C_0X_0)}\pm\sqrt{(C_1X_1)}\pm\sqrt{(C_2X_2)}=0.$$

For such a triangle, 12.53 gives

$$\operatorname{cosech}^2 r=(CC)=2c_{12}C_1C_2+2c_{20}C_2C_0+2c_{01}C_0C_1$$
$$=6C_0C_1C_2\sqrt{\gamma}=3.$$

Thus* the in-radius of a trebly-asymptotic triangle is $\frac{1}{2}\log 3$. More simply, it is geometrically evident that $\Pi(r)=\frac{1}{3}\pi$, whence $r=\log\cot\frac{1}{6}\pi$.

*Sommerville [2], p. 85 (Example 10).

CHAPTER XIII

AREA

13.1. Equivalent regions. Two polygonal regions in a plane are said to be equivalent if they can be dissected into parts which are respectively congruent.* For instance, in Fig. 9.6D, the triangle **ABC** is equivalent to the isosceles birectangle **ABED**, since the parts **CFJ** and **CFI** of the former are congruent to the parts **ADJ** and **BEI** of the latter. That the relation of equivalence is transitive may be seen by superposing two dissections to make a finer dissection. Regions bounded by curves can be treated similarly, by regarding them as limiting cases of polygonal regions.

This notion enables us to define the *area* of any region in terms of a standard *unit* region, as follows. A region is said to be of area $1/n$ (where n is a positive integer) if the unit region can be dissected into n parts each equivalent to the given region; and a region is said to be of area m/n if it is equivalent to m juxtaposed replicas of a region of area $1/n$. By a natural limiting process, we obtain a real number as the area of *any* given region.

13.2. The choice of a unit. In Euclidean geometry the unit of area is the square of unit side. In non-Euclidean geometry there is no square, but only a "regular quadrangle" whose angles are not right. By drawing the diagonals of this

*For the further refinements required when continuity is not assumed, see Hilbert [1], pp. 58-60; Carslaw [1], pp. 84-90. It is remarkable that the above simple definition for equivalence cannot be extended to three-dimensional space. Two polyhedral regions may have equal volume without being equivalent by dissection into a finite number of parts.

figure, we dissect it into four isosceles right-angled triangles (whose base-angles differ from $\frac{1}{4}\pi$). For such a triangle **ABC**, with $A = B$, formulae 12.77 and 12.97 give

$$\cos c = \cos^2 a = \cot^2 A \qquad \text{(elliptic)},$$

$$\cosh c = \cosh^2 a = \cot^2 A \qquad \text{(hyperbolic)}.$$

If we make a tend towards zero, this more and more closely resembles a Euclidean triangle with $c^2 = 2a^2$ and $A = \frac{1}{4}\pi$. (See 10.94.) Accordingly, we adjust our unit so as to make the area Δ of the non-Euclidean triangle satisfy

13.21. $$\lim_{a \to 0} \frac{\Delta}{\frac{1}{2}a^2} = 1.$$

We shall see later that, in elliptic geometry, our unit is in fact the area of a birectangular triangle whose third angle is one radian (or whose base is of unit length), while in hyperbolic geometry it is the area of a horocyclic sector of unit arc. (The latter figure has the disadvantage of extending to infinity, but can easily be replaced by other figures that are entirely accessible.)

13.3. The area of a triangle in elliptic geometry. Two lines decompose the real projective plane into two angular regions called *lunes*. (Such a lune has two sides but only one vertex, unlike a lune on a sphere.) In elliptic geometry, the two supplementary lunes are congruent if the lines are perpendicular. By repeated bisection of such a right-angled lune, we see that *the area of a lune is proportional to its angle.*

If $\mu\theta$ is the area of a lune of angle θ, the area of the whole plane is

$$\mu\theta + \mu(\pi - \theta) = \mu\pi.$$

Let Δ, Δ_a, Δ_b, Δ_c denote the areas of any triangle **ABC** and its colunar triangles. (See Fig. 2.6A on page 35.) Then

$$\Delta + \Delta_a = \mu A, \quad \Delta + \Delta_b = \mu B, \quad \Delta + \Delta_c = \mu C,$$

and $$\Delta+\Delta_a+\Delta_b+\Delta_c=\mu\pi.$$
Hence* $$2\Delta=\mu(A+B+C-\pi).$$

This argument provides the simplest proof of the famous inequality

13.31. $$A+B+C>\pi,$$

which can alternatively be deduced from 12.75 (just as 1.32 was deduced from 1.31).

To evaluate μ, we consider a right-angled triangle with $A=B=\frac{1}{4}\pi+\epsilon$, so that

$$\cos a=\cot A=(1-\tan\epsilon)/(1+\tan\epsilon)$$

and $$\sin^2 a=4\tan\epsilon\,/(1+\tan\epsilon)^2.$$

The ratio of the area of this triangle to that of a Euclidean right-angled triangle with equal sides a is

$$\frac{\mu\epsilon}{\frac{1}{2}a^2}=\frac{\mu}{2}\left(\frac{\sin a}{a}\right)^2\frac{\epsilon}{\tan\epsilon}(1+\tan\epsilon)^2.$$

The limit, as a and ϵ tend to zero, is $\mu/2$. Hence $\mu=2$, and

13.32. *In elliptic geometry, the area of any triangle* **ABC** *is*

$$A+B+C-\pi.$$

13.4. Area in hyperbolic geometry. The corresponding result in hyperbolic geometry does not come so easily. Since the plane is now infinite, our only hope of an analogous method will be to work within some region whose area is known to be finite. The main steps in the argument are given by the following chain of theorems.

13.41. *A horocyclic sector has a finite area.*

PROOF.† Let $\mathbf{H}$, $\mathbf{H}_1$, $\mathbf{H}_2$, $\mathbf{H}_3$, ... be points evenly spaced along a line **HM**. Let these be joined to corresponding points on a parallel line **AM** by horocyclic arcs **HA**, $\mathbf{H}_1\mathbf{A}_1$, $\mathbf{H}_2\mathbf{A}_2$, $\mathbf{H}_3\mathbf{A}_3$,

---

*Klein [3], pp. 200-201.

†Suggested by Carslaw [1], p. 120. See also Coxeter [9], p. 295.

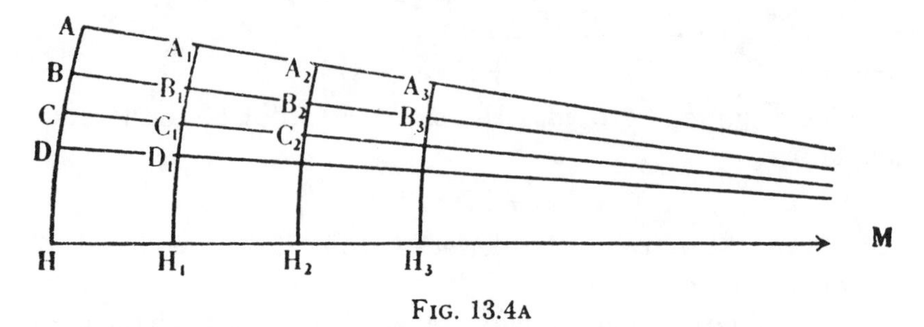

FIG. 13.4A

. . . , with centre **M** (at infinity). On the arc **HA**, take points **B**, **C**, **D**, . . . , so that

$$\mathbf{HB} = \mathbf{H_1A_1},\ \mathbf{HC} = \mathbf{H_2A_2},\ \mathbf{HD} = \mathbf{H_3A_3},\ \ldots,$$

and let the parallel lines **BM**, **CM**, **DM**, . . . meet the other arcs at points $\mathbf{B}_r$, $\mathbf{C}_r$, $\mathbf{D}_r$, . . . , as in Fig. 13.4A. Then the translation along **HM** that takes **H** to $\mathbf{H}_1$ (and $\mathbf{H}_r$ to $\mathbf{H}_{r+1}$) takes $\mathbf{BB_1B_2}\ldots$ to $\mathbf{A_1A_2A_3}\ldots$, $\mathbf{CC_1C_2}\ldots$ to $\mathbf{B_1B_2B_3}\ldots$, and so on.* Thus the "curvilinear rectangles" $\mathbf{AB_1}$, $\mathbf{BC_1}$, $\mathbf{CD_1}$, . . . are respectively congruent to the curvilinear rectangles $\mathbf{AB_1}$, $\mathbf{A_1B_2}$, $\mathbf{A_2B_3}$, . . . , which together make up the horocyclic sector **ABM**. By a natural limiting process, we deduce that this horocyclic sector is equivalent to the curvilinear rectangle $\mathbf{AH_1}$. The area of the given sector **AHM**, being greater than that of **ABM** in the ratio of the arcs **AH**:**AB**, is still finite.

13.42. *A trebly-asymptotic triangle has a finite area.*

PROOF. Let **LMN** be a trebly-asymptotic triangle. Draw **LH** perpendicular to **MN**, **HK** perpendicular to **LM**, and let the horocycle through **H** with centre **M** meet **LM** at **A**, as in Fig. 13.4B. By 13.41, the sector **AHM** has a finite area; *a fortiori*, so has the singly-asymptotic triangle **KHM**. By reflection in **HK**, the area of the doubly-asymptotic triangle **LHM** is twice that of **KHM**. Finally, doubling again by reflection in **LH**, the area of **LMN** is still finite.

*We see in this manner that *all horocycles are congruent.*

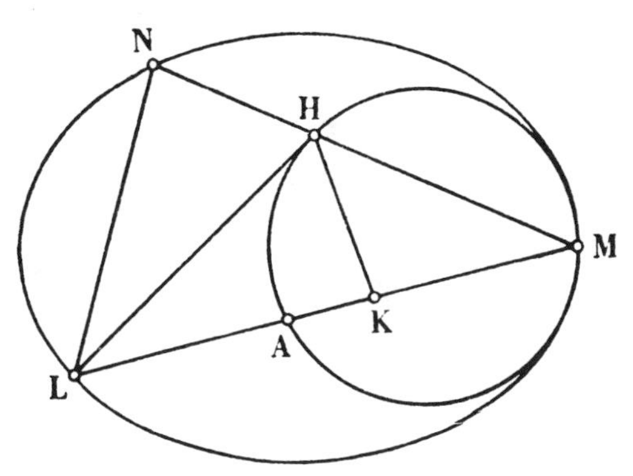

FIG. 13.4B

Since any three points at infinity can be transformed into any other three points at infinity by a projectivity in the Absolute, i.e. by a congruent transformation, we see that *all trebly-asymptotic triangles are congruent.* Accordingly, we let $\mu\pi$ denote the area of such a triangle, μ being a constant which remains to be calculated. (The π is inserted for convenience.) It follows that the area of a right-angled doubly-asymptotic triangle, such as **LHM**, is $\frac{1}{2}\mu\pi$.

13.43. *The area of a doubly-asymptotic triangle of angle* θ *is* $\mu(\pi-\theta)$.

PROOF.* Let **H** be any point on the "doubly infinite" side **MN** of the doubly-asymptotic triangle **AMN**, and let **L** be the point at infinity on the ray **H/A**, as in Fig. 13.4C. Then the areas of the various triangles combine as follows:

$$\mathbf{ALM}+\mathbf{ALN}=\mathbf{AMN}+\mathbf{LMN}.$$

Hence, letting $f(\theta)$ denote the area of a doubly-asymptotic triangle of angle θ, we have

$$f(\theta)+f(\phi)=f(\theta+\phi)+\mu\pi.$$

Differentiating with respect to θ, while keeping ϕ constant, we obtain

*Cf. Liebmann [1], p. 44; Schilling [1], II, p. 198.

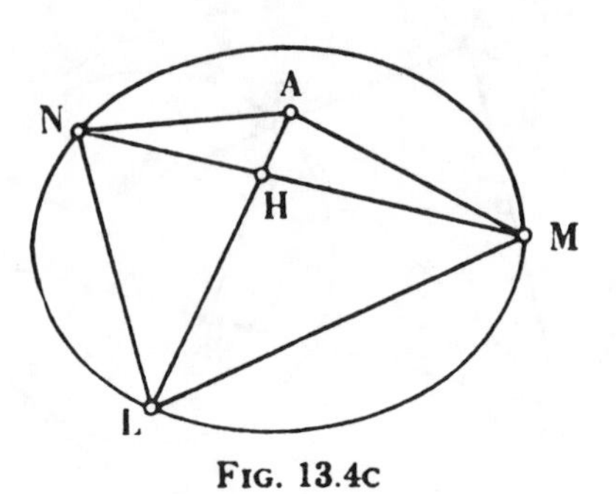

FIG. 13.4C

$$f'(\theta) = f'(\theta + \phi).$$

(Here we are letting **M** vary while the triangle **ALN** remains fixed.) Since ϕ is arbitrary, we conclude that $f'(\theta)$ is constant, so that $f(\theta)$ is linear. The two special values $f(0) = \mu\pi$ and $f(\pi) = 0$ now determine

$$f(\theta) = \mu(\pi - \theta).$$

13.44. *In hyperbolic geometry, the area of any triangle* **ABC** *is*

$$\pi - A - B - C.$$

PROOF.* Let **L**, **M**, **N** be the points at infinity on **B/C**, **C/A**, **A/B**, as in Fig. 13.4D. Then the trebly-asymptotic triangle **LMN** is dissected into four parts: the original triangle **ABC**, and the doubly-asymptotic triangles **AMN**, **BNL**, **CLM** whose angles are respectively $\pi - A$, $\pi - B$, $\pi - C$. Hence the area

$$\mathbf{ABC} = \mu\pi - \mu A - \mu B - \mu C = \mu(\pi - A - B - C).$$

To evaluate μ, we consider a right-angled triangle with $A = B = \frac{1}{4}\pi - \epsilon$, so that

$$\cosh a = \cot A = (1 + \tan \epsilon)/(1 - \tan \epsilon)$$

and $$\sinh^2 a = 4 \tan \epsilon \,/(1 - \tan \epsilon)^2.$$

*Gauss [1], p. 223.

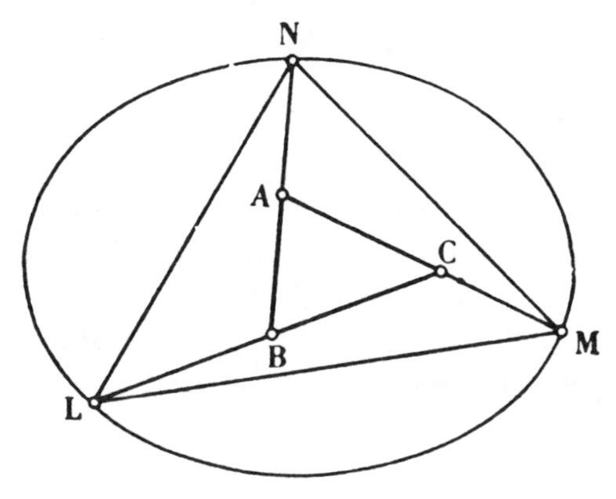

FIG. 13.4D

In this case 13.21 gives

$$\lim \mu \left(\frac{\sinh a}{a}\right)^2 \frac{\epsilon}{\tan \epsilon} (1-\tan \epsilon)^2 = 1,$$

whence $\qquad \mu = 1.$

13.5. The extension to three dimensions. The analogous problem, of finding the volume of a tetrahedron as a function of its angles, is vastly harder, since this is not one of the "elementary functions" at all. In terms of suitably defined new functions, however, the problem was solved by Lobatschewsky (for hyperbolic geometry) and Schläfli (for spherical, and hence elliptic).*

One of Schläfli's results may well be mentioned here, as it is analogous to 13.32 and 13.44 in their differentiated form

$$d\Delta = \pm(dA + dB + dC).$$

13.51. *If AB, . . . , CD are the dihedral angles at the edges* **AB**, . . . , **CD** *of a tetrahedron* **ABCD** *of volume V, then*

$$dV = \pm\tfrac{1}{2}(\mathbf{AB}\ dAB + \mathbf{AC}\ dAC + \mathbf{AD}\ dAD + \mathbf{BC}\ dBC + \mathbf{BD}\ dBD + \mathbf{CD}\ dCD).$$

*Lobatschewsky [1], pp. 608-610; Schläfli [1], p. 289 (for $n-1$ dimensions, p. 287).

A proof of this remarkable theorem will be given in §15.7.

13.6. The differential of distance. A *continuous curve* may be defined as the locus of a point whose coordinates are continuous functions of a parameter t. We assume further that the functions have continuous derivatives. If Δs is the distance between points (x) and $(x + \Delta x)$, whose parameters are t and $t+\Delta t$, we consider the curve as the limit of a sequence of broken lines, or polygons, formed by smaller and smaller chords, and define its *length** to be

$$s = \lim_{\Delta t \to 0} \sum \Delta s = \lim_{\Delta t \to 0} \sum \frac{\Delta s}{\Delta t} \Delta t = \int_{t_1}^{t_2} \dot{s}\, dt,$$

where $\quad \dot{s} = \lim(\Delta s / \Delta t)$.

For purposes of differential geometry, it is convenient to *normalize* the coordinates by making $(xx) = 1$, so that the distance between (x) and (y) is simply

$$\operatorname{arc}\cos(xy) \quad \text{or} \quad \operatorname{arg}\cosh(xy).$$

(Cf. page 122.)

13.61. *In terms of normalized coordinates, the differential of distance along any continuous curve is given by*

$$ds^2 = \pm(dx\, dx) = \pm c_{\mu\nu}\, dx_\mu\, dx_\nu,$$

with the upper sign for elliptic geometry, the lower for hyperbolic.

PROOF. In elliptic geometry, the distance from (x) to $(x + \Delta x)$ is given by†

$$\cos \Delta s = (x\, x + \Delta x) = (x\, x) + (x\, \Delta x) = 1 + (x\, \Delta x).$$

Since $(x + \Delta x\, x + \Delta x) = 1 = (x\, x)$, we have $2(x\, \Delta x) + (\Delta x\, \Delta x) = 0$, and therefore

$$4 \sin^2 \tfrac{1}{2} \Delta s = 2(1 - \cos \Delta s) = -2(x\, \Delta x) = (\Delta x\, \Delta x).$$

---

*Klein [3], p. 245.

†Coolidge [1], p. 187.

Dividing by Δt^2 and passing to the limit, we find $\dot{s}^2 = (\dot{x}\dot{x})$, or in the notation of differentials

$$ds^2 = (dx\,dx).$$

In hyperbolic geometry, we have similarly

$$\cosh \Delta s = 1 + (x\,\Delta x),$$

$$4\sinh^2 \tfrac{1}{2}\Delta s = 2(\cosh \Delta s - 1) = 2(x\,\Delta x) = -(\Delta x\,\Delta x),$$

and

$$ds^2 = -(dx\,dx).$$

Thus, in terms of normalized coordinates with the absolute polarity in canonical form, we have

13.62. $$ds^2 = \pm dx_0^2 + dx_1^2 + dx_2^2$$

in the plane, and

$$ds_2 = \pm dx_0^2 + dx_1^2 + dx_2^2 + dx_3^2$$

in space.

13.7. Arcs and areas of circles. The circle 12.24, in the form

$$x_1^2 + x_2^2 = \sin^2 R \quad \text{or} \quad \sinh^2 R$$

(cf. 6.79), has the parametric representation

$$x_0 = \cos R, \quad x_1 = \sin R \cos t, \quad x_2 = \sin R \sin t$$

or

$$x_0 = \cosh R, \; x_1 = \sinh R \cos t, \; x_2 = \sinh R \sin t$$

(in the two geometries, respectively). Hence

$$\dot{x}_0 = 0, \quad \dot{x}_1 = -x_2, \quad \dot{x}_2 = x_1,$$

and

$$\dot{s} = \sqrt{(x_1^2 + x_2^2)}$$
$$= \sin R \quad \text{or} \quad \sinh R.$$

Integrating from $t = 0$ to $t = \theta$, we deduce that *an arc of angle θ of a circle of radius R is of length**

13.71. $$s(R) = \theta \sin R \quad \text{or} \quad \theta \sinh R.$$

Theorem 10.94 enables us to calculate areas by integration, as in Euclidean geometry. Thus *a sector of angle θ of a circle of radius R has area*

---

*Carslaw [1], p. 118. See also Coxeter [5], p. 268, [13], p. 24.

13.72. $$\int_0^R s(r)dr = \theta(1-\cos R) \quad \text{or} \quad \theta(\cosh R - 1).$$

In particular,* *a circle of radius R has circumference*

$$2\pi \sin R \quad \text{or} \quad 2\pi \sinh R$$

and area

13.73. $$2\pi(1-\cos R) \quad \text{or} \quad 2\pi(\cosh R-1).$$

Putting $R=\frac{1}{2}\pi$, we deduce that *the area of the elliptic plane is* 2π, in agreement with §13.3.

By 13.71 and 13.72, the area of a sector of arc s (in hyperbolic geometry) is

$$s(\cosh R-1)/\sinh R = s \tanh \tfrac{1}{2}R.$$

Making R tend to infinity, we deduce that

13.74. *The area of a horocyclic sector is equal to its arc.*

(Cf. 13.41.) In other words, our chosen unit is the area of a horocyclic sector of unit arc.

In terms of normalized coordinates, the horocycle 12.25 may be expressed as

$$x_0 - x_2 = 1,$$

whence $dx_0 = dx_2$ and

$$ds = dx_1.$$

It follows that a horocyclic arc of length s goes from $(1, 0, 0)$ to

$$(1+\tfrac{1}{2}s^2,\ s,\ \tfrac{1}{2}s^2).$$

The diameter through the latter point meets the line $x_2=0$ at $(1, s, 0)$. But this line is the tangent at $(1, 0, 0)$. Putting $s=1$, we deduce the following characterization for a horocyclic arc of unit length: *the tangent at one end is parallel to the diameter through the other.*† Thus, in Fig. 13.4B, **AH** is a horocyclic arc of unit length, and **AHM** is a horocyclic sector of unit area.

---

*Bolyai [2], §§30, 32 (IV).

†Carslaw [1], p. 119.

13.8. Two surfaces which can be developed on the Euclidean plane. Analogously, the horosphere through (1, 0, 0, 0) with centre (1, 0, 0, 1) has the equation

$$(x_0-x_3)^2=x_0^2-x_1^2-x_2^2-x_3^2$$

or
$$x_1^2+x_2^2=2(x_0-x_3)x_3.$$

In terms of normalized coordinates, this becomes

$$x_0-x_3=1,$$

whence $dx_0=dx_3$ and

13.81. $$ds^2=dx_1^2+dx_2^2.$$

Since $x_3=\frac{1}{2}(x_1^2+x_2^2)$ and $x_0=1+x_3$, all four coordinates are expressible in terms of x_1 and x_2. Thus 13.81 provides an alternative proof that the intrinsic geometry of a horosphere is Euclidean. (See 11.31.)

The corresponding proof of the same result for a Clifford surface (7.86) is as follows. The obvious representation in terms of parameters ξ and η is

$$x_0=\cos\tfrac{1}{2}\psi\cos\xi,\quad x_1=\cos\tfrac{1}{2}\psi\sin\xi,$$
$$x_2=\sin\tfrac{1}{2}\psi\cos\eta,\quad x_3=\sin\tfrac{1}{2}\psi\sin\eta,$$

whence

$$dx_0=-x_1d\xi,\quad dx_1=x_0d\xi,\quad dx_2=-x_3d\eta,\quad dx_3=x_2d\eta,$$

and

$$ds^2=\{dx\,dx\}=(x_0^2+x_1^2)d\xi^2+(x_2^2+x_3^2)d\eta^2$$
$$=\cos^2\tfrac{1}{2}\psi\ \ d\xi^2+\sin^2\tfrac{1}{2}\psi\ \ d\eta^2.$$

This can be put into the "Euclidean" form $ds^2=dx^2+dy^2$ by taking new parameters

$$x=\xi\cos\tfrac{1}{2}\psi,\quad y=\eta\sin\tfrac{1}{2}\psi.$$

In comparing these two "developables," the one hyperbolic and the other elliptic, it is important to notice that, whereas the horosphere can be mapped on the *whole* Euclidean plane, the corresponding map of the Clifford surface covers only a finite portion of the plane, namely a rhombus of side π and angle ψ. (See §7.5.)

CHAPTER XIV

EUCLIDEAN MODELS

14.1. The meaning of "elliptic" and "hyperbolic." In ordinary Euclidean geometry, a central conic may be either an *ellipse* or a *hyperbola*. For any central conic, the pairs of conjugate diameters belong to an involution (of lines through the centre); but it is only the hyperbola that has *self*-conjugate diameters (viz. its two asymptotes). Accordingly, any involution (and so, conveniently, any one-dimensional projectivity) is said to be *hyperbolic* if it has two self-corresponding elements, and *elliptic* if it has none. Analogously, a *polarity* is said to be hyperbolic or elliptic according as it does or does not contain self-conjugate elements. Finally, a non-Euclidean geometry is said to be hyperbolic or elliptic according to the nature of its absolute polarity.

A more direct connection with ellipses and hyperbolas will be seen in Fig. 14.2A.

14.2. Beltrami's model. In the case of two-dimensional hyperbolic geometry, we are at liberty to draw the Absolute as a circle in the Euclidean plane, provided we understand that we are then using two metrics simultaneously: the Euclidean metric by which the circle is drawn, and the hyperbolic metric defined by 10.71 and 10.73. The poles and polars are taken with respect to the circle in the ordinary sense, as the constructions involved are essentially projective. This model for hyperbolic geometry is due to Beltrami (1835-1900).*

Any ordinary point of the hyperbolic plane may be identified with the centre of the circle. If the two metrics are

*Beltrami [1], [2]; Bonola [2], pp. 164-175; Baldus [1], pp. 56-148.

taken to agree at that point, they will deviate more and more as we recede from it, until at the circumference the distortion is infinite.

The hyperbolic metric can be expressed in terms of Cartesian coordinates x_1, x_2, by putting $x_0=1$ (and $y_0=1$) in 10.81. Alternatively, we can use that formula as it stands, by regarding x_0, x_1, x_2 as "homogeneous Cartesian coordinates," with $x_0=0$ as the line at infinity. The corresponding re-interpretation of 6.75 provides a similar model for elliptic geometry. The point $(x, 0)$ or $(1, x, 0)$, at Euclidean distance x from the origin, is at elliptic distance arc tan x, and at hyperbolic distance arg tanh x. This comparison of metrics establishes the following model for one-dimensional non-Euclidean geometry.*

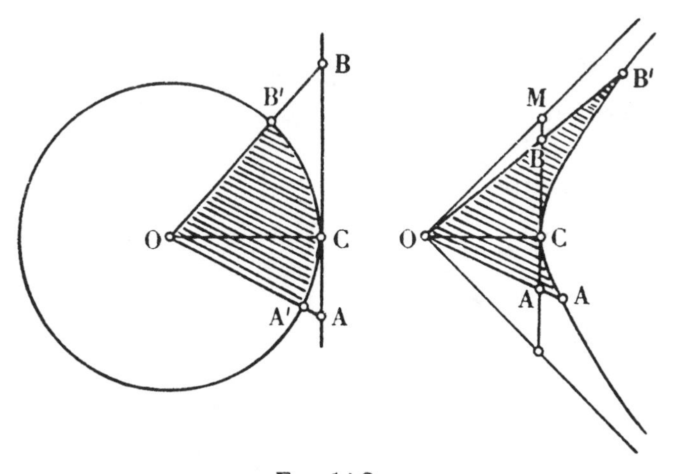

FIG. 14.2A

We introduce the elliptic or hyperbolic metric into a line **AB** of the Euclidean plane, by regarding **AB** as a tangent to an ellipse or hyperbola, onto which its points are projected from the centre **O**, as in Fig. 14.2A. The non-Euclidean distance **AB** is defined to be twice the area of the corresponding sector

*Klein [3], p. 173.

OA′B′. Thus the elliptic length of the whole line is finite, being just the area of the ellipse; but the hyperbolic distance **AM**, to a point on either asymptote, is infinite. For simplicity we may take the ellipse to be a circle, and the hyperbola to be rectangular* with **AB** as the tangent at the vertex, **C**. Then, if x, x_e, x_h are the Euclidean, elliptic, and hyperbolic distances **CB**, we have

$$x_e = \text{arc tan } x, \qquad x_h = \text{arg tanh } x,$$

as above.

Note that the two kinds of trigonometrical functions have to be called *circular* and hyperbolic, the name "elliptic functions" being already used in another connection. In fact, the latter functions arise when we consider the arc-length of an ellipse (or hyperbola), and this is not proportional to the area of the corresponding sector unless the ellipse reduces to a circle.

14.3. The differential of distance. In Beltrami's model we represented the point with canonical coordinates (x_0, x_1, x_2) by the point with Cartesian coordinates $(x_1/x_0, x_2/x_0)$. Putting $x_0 = 1$, we have the expressions

$$\text{arc cos} \frac{1 + x_1y_1 + x_2y_2}{\sqrt{(1 + x_1^2 + x_2^2)}\sqrt{(1 + y_1^2 + y_2^2)}} =$$

$$\text{arc sin} \sqrt{\frac{(x_1 - y_1)^2 + (x_2 - y_2)^2 + (x_1y_2 - x_2y_1)^2}{(1 + x_1^2 + x_2^2)(1 + y_1^2 + y_2^2)}}$$

and

$$\text{arg cosh} \frac{1 - x_1y_1 - x_2y_2}{\sqrt{(1 - x_1^2 - x_2^2)}\sqrt{(1 - y_1^2 - y_2^2)}} =$$

$$\text{arg sinh} \sqrt{\frac{(x_1 - y_1)^2 + (x_2 - y_2)^2 - (x_1y_2 - x_2y_1)^2}{(1 - x_1^2 - x_2^2)(1 - y_1^2 - y_2^2)}}$$

for the elliptic and hyperbolic distances between points (x_1, x_2), (y_1, y_2). Thus the elliptic distance between (x_1, x_2) and $(x_1 + \Delta x_1, x_2 + \Delta x_2)$ is given by

*Hobson [1], p. 329.

$$\sin^2 \Delta s = \frac{\Delta x_1^2 + \Delta x_2^2 + (x_2 \Delta x_1 - x_1 \Delta x_2)^2}{(1+x_1^2+x_2^2)\{1+(x_1+\Delta x_1)^2+(x_2+\Delta x_2)^2\}},$$

and the differential of elliptic distance is given by

14.31.

$$ds^2 = \frac{dx_1^2 + dx_2^2 + (x_2 dx_1 - x_1 dx_2)^2}{(1+x_1^2+x_2^2)^2} = \frac{(1+x_2^2)dx_1^2 - 2x_1x_2dx_1dx_2 + (1+x_1^2)dx_2^2}{(1+x_1^2+x_2^2)^2}$$

Similarly* for hyperbolic distance we have

14.32.

$$ds^2 = \frac{dx_1^2 + dx_2^2 - (x_2 dx_1 - x_1 dx_2)^2}{(1-x_1^2-x_2^2)^2} = \frac{(1-x_2^2)dx_1^2 + 2x_1x_2dx_1dx_2 + (1-x_1^2)dx_2^2}{(1-x_1^2-x_2^2)^2}$$

Analogous formulae for three-dimensional space are readily obtained. The hyperbolic absolute is then represented by a sphere, and planes by secant planes. This model† provides the earliest satisfactory proof that hyperbolic geometry is consistent (so that Euclid's Postulate V cannot possibly be deduced from his other assumptions).

14.4. Gnomonic projection. In the elliptic case, the above results have a very simple interpretation in the geometry of a bundle (§1.7). In fact, we may consider (x_0, x_1, x_2) as Cartesian coordinates in three dimensions, lines through the origin representing points of the elliptic plane. Then the result of putting $x_0 = 1$ is to take the section of the bundle by a plane.

In other words, if we represent the points of the elliptic plane by pairs of antipodal points of the sphere

$$x^2 + y^2 + z^2 = 1,$$

*Beltrami [2], pp. 287, 307.

†Beltrami [3].

we can make a ***gnomonic projection**** onto the plane $x=1$, with the result that the antipodal points (x, y, z) and $(-x, -y, -z)$ project into the single point $(1, x_1, x_2)$, where

$$x = \pm \frac{1}{\sqrt{(1+x_1^2+x_2^2)}}, \quad y = \pm \frac{x_1}{\sqrt{(1+x_1^2+x_2^2)}},$$

$$z = \pm \frac{x_2}{\sqrt{(1+x_1^2+x_2^2)}}.$$

Then 14.31 may be obtained by substitution in the familiar formula

$$ds^2 = dx^2 + dy^2 + dz^2.$$

The same analysis can be made to include a derivation of 14.32 as well. For, we can change the normalized canonical coordinates of 13.62 into homogeneous canonical coordinates by writing $x_\mu/\sqrt{(x_0^2 \pm x_1^2 \pm x_2^2)}$ for x_μ, and then we can derive Beltrami's model by putting $x_0=1$. But in order to interpret 14.32 in terms of a bundle or gnomonic projection, we would have to use a three-dimensional space that is Minkowskian instead of Euclidean. (See page 178.)

14.5. Development on surfaces of constant curvature. Since the elliptic plane can be developed (i.e. represented without distortion) on a sphere, it is natural to seek a corresponding development of the hyperbolic plane on some kind of "pseudo-sphere." This should be a surface in Euclidean space, such that all distances measured on the surface itself are related like distances in the hyperbolic plane.† In particular, the lines of the hyperbolic plane should be represented by geodesics (i.e. shortest paths, like great circles on a sphere). The point-to-point homogeneity of the hyperbolic plane re-

---

*Klein [3], p. 294.

†Minding [2], p. 324. For a very enjoyable account of this aspect of non-Euclidean geometry, see Lieber [1].

quires constant *specific curvature* K for the surface. (Cf. §1.6.)

Gauss proved that, for a triangle **ABC** formed by geodesic arcs on a surface of area S, the *total curvature* is

$$\iint K dS = A + B + C - \pi.$$

In the case where K is constant, this means that the area of the geodesic triangle is $(A + B + C - \pi)/K$, which agrees with 13.32 and 13.44 when the unit of length is generalized as in §10.9. Thus a pseudo-sphere is a surface of constant *negative* curvature.

Such surfaces were studied by Minding (1806-1885).* The simplest example is the horn-shaped *tractroid*, a surface of revolution for which cylindrical coordinates r and z are expressible in terms of a parameter t, thus:

$$r = \operatorname{sech} t, \quad z = t - \tanh t.$$

This model for hyperbolic geometry is inferior to the spherical model for elliptic geometry in two respects. First, although the maximum and minimum normal curvatures at a point on the surface have the constant product $K = -1$, they vary from one point to another, whereas every normal section of the unit sphere has curvature 1. Second, the tractroid does not represent the *whole* hyperbolic plane, but only a restricted region, namely a horocyclic sector (Fig. 13.4A). For, the meridians of the tractroid, being geodesics which approach one another asymptotically, represent parallel lines; and the circles orthogonal to them represent arcs of concentric horocycles, one of which, representing the "rim" ($t = 0$) of the tractroid, is of length 2π. The horocyclic sector is "wrapped around" the tractroid so that its two bounding diameters coincide with

*Minding [1], pp. 379-380. For good pictures of these surfaces, see Klein [3], p. 286. For an easy proof that the tractroid has constant curvature, see Sommerville [2], pp. 168-170.

one meridian. We easily verify that the total area of the surface is 2π, in agreement with 13.74.

Beltrami suspected, and it is now known,* that there is no smooth surface in ordinary space which provides an undistorted representation for the whole hyperbolic plane.

14.6. Klein's conformal model of the elliptic plane. The various ways of representing a sphere on the Euclidean plane have a practical application in mapping the surface of the Earth, and it is convenient to use the terminology suggested by that application. Thus *stereographic projection*† may be described geographically as the projection from the "north pole" onto the "equatorial plane" (or, in some treatments, onto the tangent plane at the "south pole," which gives a similar result). Using Cartesian coordinates, we project the general point (x_0, x_1, x_2) of the unit sphere $x_0^2+x_1^2+x_2^2=1$ from the point $(1, 0, 0)$ onto the plane $x_0=0$, obtaining the point $(0, u_1, u_2)$ for which

$$\frac{u_1}{x_1}=\frac{u_2}{x_2}=\frac{1}{1-x_0},$$

so that

$$\text{14.61.}\quad x_0=\frac{u_1^2+u_2^2-1}{u_1^2+u_2^2+1},\quad x_1=\frac{2u_1}{u_1^2+u_2^2+1},\quad x_2=\frac{2u_2}{u_1^2+u_2^2+1}.$$

These equations take the slightly simpler form

$$\text{14.62.}\qquad x_0=\frac{u\bar{u}-1}{u\bar{u}+1},\quad x_1=\frac{u+\bar{u}}{u\bar{u}+1},\quad ix_2=\frac{u-\bar{u}}{u\bar{u}+1}$$

in terms of the complex number $u=u_1+iu_2$. Conversely, we have here a representation of complex numbers by points of a sphere, known as the Neumann sphere. The geodesic dis-

*Lütkemeyer [1].

†Neumann [1], p. 52; Klein [3], p. 295.

tance δ between two such points (x) and (y), corresponding to complex numbers u and v, is given by

$$\cos\delta = x_0y_0 + x_1y_1 + x_2y_2$$
$$= \frac{(u\bar{u}-1)(v\bar{v}-1)+(u+\bar{u})(v+\bar{v})-(u-\bar{u})(v-\bar{v})}{(u\bar{u}+1)(v\bar{v}+1)},$$

so that

$$\cos^2 \tfrac{1}{2}\delta = \frac{(u\bar{v}+1)(v\bar{u}+1)}{(u\bar{u}+1)(v\bar{v}+1)} = \frac{(u+\bar{v}^{-1})(v+\bar{u}^{-1})}{(u+\bar{u}^{-1})(v+\bar{v}^{-1})}.$$

This is the cross ratio* of the four complex numbers u, v, $-\bar{v}^{-1}$, $-\bar{u}^{-1}$. Hence, if **A**, **B**, **B**$'$, **A**$'$ are the points in the u-plane which represent these complex numbers, we have

14.63. $$\cos^2 \tfrac{1}{2}\delta = \left|\cos^2 \tfrac{1}{2}\delta\right| = \frac{|u+\bar{v}^{-1}| \,.\, |v+\bar{u}^{-1}|}{|u+\bar{u}^{-1}| \,.\, |v+\bar{v}^{-1}|} = \frac{\mathbf{AB}'\cdot\mathbf{BA}'}{\mathbf{AA}'\cdot\mathbf{BB}'}.$$

We thus have a simple expression for the "spherical distance **AB**" in terms of the Euclidean distances between **A**, **B**, and their "negative inverse" points **A**$'$, **B**$'$. (If $u=re^{i\theta}$, then $-\bar{u}^{-1}=r^{-1}e^{i(\theta+\pi)}$. Thus **A** and **A**$'$ are at reciprocal distances from the origin in opposite directions.)

By 14.61, every linear relation among x_0, x_1, x_2 is equivalent to a linear relation among $u_1^2+u_2^2$, u_1, u_2. In other words, every circle on the sphere projects into a circle (or line) in the plane. In particular, a "meridian" (lying in the plane $X_1x_1+X_2x_2=0$, say) projects into a line through the origin, but any other great circle (such as that in the plane $x_0+X_1x_1+X_2x_2=0$) projects into a circle of the form

$$u_1^2+u_2^2+2X_1u_1+2X_2u_2-1=0,$$

for which the power of the origin is -1. Such a circle meets the "equator" $u_1^2+u_2^2-1=0$ in two diametrically opposite points (of the latter).†

This representation of spherical geometry on the Euclidean

*Cf. Sommerville [2], pp. 182-185.
†Carslaw [1], pp. 171-174.

plane is due to Klein. "Lines" of the spherical geometry, being great circles of the sphere, are represented by circles and lines which meet a fixed circle at the ends of a diameter. To derive the corresponding representation for *elliptic* geometry, we merely have to identify each pair of "negative inverse" points, or equally well to restrict consideration to the inside of the fixed circle, identifying each pair of diametrically opposite points of that circle.

Since circles are represented by circles, the above transformation (from the spherical or elliptic plane to the Euclidean plane) is *conformal*,* i.e., angles are represented without any distortion. For distances, on the other hand, we have from 14.62 the formula

14.64. $$ds^2 = dx_0^2 + dx_1^2 + dx_2^2 = \frac{4du\, d\bar{u}}{(u\bar{u}+1)^2} = \frac{4(du_1^2+du_2^2)}{(u_1^2+u_2^2+1)^2}\,.$$

14.7. Klein's conformal model of the hyperbolic plane. To obtain a representation of two-dimensional hyperbolic geometry on a sphere in Euclidean space, we take Beltrami's model in the equatorial plane, with the equator for Absolute, and project orthogonally.† Then every ordinary point of the hyperbolic plane is represented by two points on the sphere, one in the northern hemisphere and one in the southern, while the equator represents points at infinity. Lines of the hyperbolic plane are represented by circles orthogonal to the equator. Any circle on the sphere which intersects the equator projects orthogonally into an ellipse which has double contact with the equator, and so represents an equidistant

*The preservation of circles is a *sufficient* condition for conformality. See Sommerville [2], p. 237.

†For beautiful drawings of this construction, see Klein [3], p. 296. Note that we are using the equatorial plane instead of the tangent plane at the south pole. The advantage is that points at infinity now have the same representatives in both models.

curve. We shall not be surprised, therefore, to find that this representation on the sphere is conformal. Finally, we go back to the equatorial plane by stereographic projection from the north (or south) pole. The result is a representation of the hyperbolic plane on the Euclidean plane, lines being represented by circles which cut a fixed circle orthogonally, i.e. by circles for which the centre of the fixed circle has power 1.

Since two such circles meet once inside and once outside the fixed circle, each point of the hyperbolic plane is represented by two points which are inverses with respect to this circle, or equally well by one of these alone, say that which lies inside the circle.

A flat pencil is represented by a system of coaxal circles (orthogonal to the fixed circle). In the case of a proper pencil, these will of course be intersecting circles. For a pencil of parallels, they will touch one another at a point on the fixed circle, their common tangent being a diameter. For a pencil of ultra-parallels, they will be non-intersecting circles, namely the system of circles orthogonal to two given intersecting circles. In each case, the "complementary" system of coaxal circles, which are the orthogonal trajectories of the system just described, will represent the system of concentric circles (or horocycles, or equidistant curves) whose diameters are the lines of the given flat pencil. In particular, a horocycle is represented by a circle touching the fixed circle.

We must remember that the points outside the fixed circle merely represent the ordinary points over again. Ultra-infinite points are not represented by points at all (unless we extend the Euclidean plane so as to include complex points). If we restrict consideration to the inside of the fixed circle, we must say that an equidistant curve is represented by a figure consisting of two arcs which cut the fixed circle at supplementary angles. This shows clearly that an equidistant curve may meet a proper circle (or a horocycle, or another equidistant curve) in as many as four points.

We have seen that lines are represented by circular arcs joining pairs of points on the fixed circle. To return to Beltrami's model, we merely have to replace these arcs by their chords.

Analytically, if $x_0=0$ is the equatorial plane of the sphere

14.71. $$x_0^2+x_1^2+x_2^2=1,$$

the point $(0, x_1, x_2)$ of Beltrami's model projects orthogonally into the two points $(\pm\sqrt{1-x_1^2-x_2^2}, x_1, x_2)$ on the sphere. If the point $(-\sqrt{1-x_1^2-x_2^2}, x_1, x_2)$ in the southern hemisphere projects stereographically into $(0, u_1, u_2)$, we have

$$x_1=\frac{2u_1}{1+u_1^2+u_2^2},\quad x_2=\frac{2u_2}{1+u_1^2+u_2^2},$$

with $u_1^2+u_2^2<1$. In terms of complex numbers $x=x_1+ix_2$ and $u=u_1+iu_2$, these relations become

$$x=2u/(1+u\bar{u}).$$

By 10.81, the hyperbolic distance δ between points x and y (in Beltrami's model) or u and v (in the conformal model) is given by

$$\cosh\delta=\frac{1-\frac{1}{2}(x\bar{y}+y\bar{x})}{\sqrt{(1-x\bar{x})}\sqrt{(1-y\bar{y})}}=\frac{(1+u\bar{u})(1+v\bar{v})-2(u\bar{v}+v\bar{u})}{(1-u\bar{u})(1-v\bar{v})},$$

whence

$$\cosh^2\frac{\delta}{2}=\frac{(1-u\bar{v})(1-v\bar{u})}{(1-u\bar{u})(1-v\bar{v})}=\frac{(u-\bar{v}^{-1})(v-\bar{u}^{-1})}{(u-\bar{u}^{-1})(v-\bar{v}^{-1})}.$$

This cross ratio can again be expressed as

$$\frac{\mathbf{AB'}\cdot\mathbf{BA'}}{\mathbf{AA'}\cdot\mathbf{BB'}}$$

(cf. 14.63), where **A** and **B** are the points in the u-plane which represent the complex numbers u and v; only now **A'** and **B'** (representing $\bar{u}^{-1}$ and $\bar{v}^{-1}$) are their ordinary inverses with respect to the circle $u\bar{u}=1$.

14.8. Poincaré's model of the hyperbolic plane. There is clearly a very close analogy between the models described in §14.6 and §14.7. It is instructive to place corresponding results side by side; e.g., Fig. 14.8A illustrates the fact that the angle-sum of a triangle is greater than π in elliptic geometry, less than π in hyperbolic.

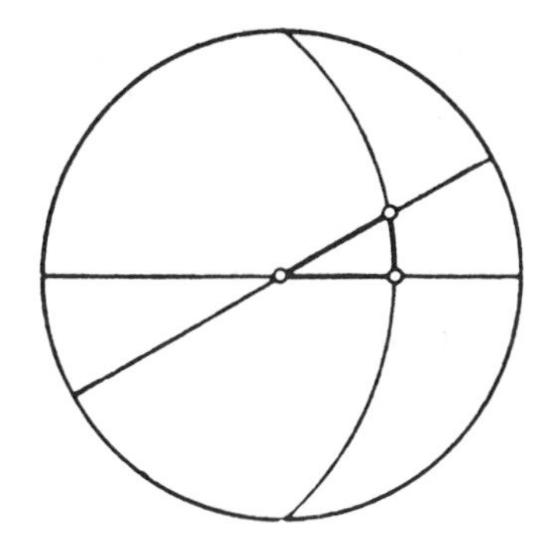

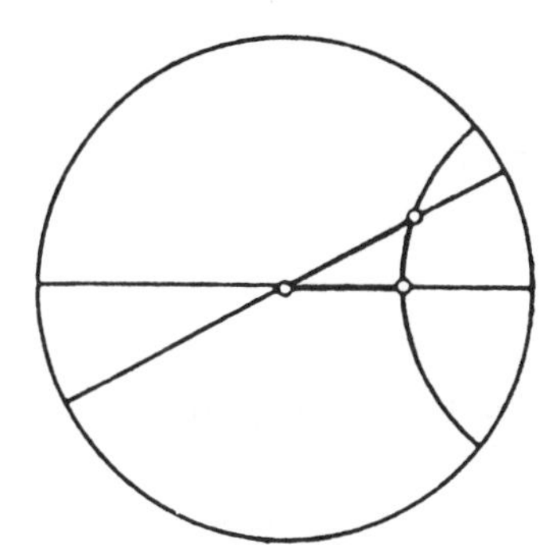

FIG. 14.8A

Since circles are represented by circles, and (consequently) angles are preserved, it is clear that *reflections* are represented by *inversions*. Hence any inversion, applied once for all to the whole system, will produce another conformal representation.

A distinction between the elliptic and hyperbolic cases arises here, since the property of cutting a circle orthogonally is maintained by inversion, whereas the property of cutting a circle diametrally is not. Thus, in the hyperbolic case, a suitable inversion will replace the "fixed circle" by a straight line, and we have Poincaré's representation of the hyperbolic plane on a Euclidean half-plane, lines being represented by semicircles based on the bounding line.* When this line is taken

*Poincaré [1], p. 8.

to be $u_2=0$, or $u=\bar{u}$, the inversion in a circle orthogonal to it is expressible in the form

$$u'=(\bar{u}+b)/(c\bar{u}-1),$$

where b and c are real.

Since a semicircle is determined by its diameter, we have incidentally represented the lines in the hyperbolic plane by pairs of points on a line, as at the end of §10.3.

Klein made the interesting observation that Poincaré's model can be derived from Beltrami's by projection without inversion. His procedure* is equivalent to the following. Given Beltrami's model in the plane $x_0=0$, we project orthogonally onto the sphere 14.71, and then stereographically from the point (0, 0, 1) onto the plane $x_2=0$ (perpendicular to the equatorial plane). Thus the equator projects into the x_1-axis, and the lines of the hyperbolic plane are represented by circles orthogonal to this line.

14.9. Conformal models of non-Euclidean space. Apart from the use of complex numbers, which was convenient but not essential, all the ideas of §§14.6-14.8 can readily be extended to space of three or more dimensions. Instead of a sphere we must use a hyper-sphere, but that hardly demands an apology nowadays.

The planes of three-dimensional "spherical space" are represented by the planes and spheres that meet a fixed sphere in great circles (of the latter). The spherical distance **AB** is still expressible in terms of "negative inverse" points **A**′ and **B**′, as in 14.63. Elliptic space can then be derived by identifying pairs of negative inverse points, or by restricting consideration to the inside of the fixed sphere and identifying pairs of antipodal points of that sphere.

*Klein [3], p. 300.

In a somewhat analogous manner, the planes of hyperbolic space are represented by the planes and spheres orthogonal to a fixed sphere or plane ω (Poincaré [2]) or by the circles that are the sections of these planes and spheres by ω (Liebmann [1]). Since ω represents the Absolute (p. 195), we have here a representation of the planes of hyperbolic space by their absolute conics, which are sections of the absolute quadric. By regarding ω as a sphere or (inversive) plane, instead of a general oval quadric, we have the advantage of a *conformal* representation: the two supplementary dihedral angles between two intersecting planes are equal to the two supplementary angles between the corresponding intersecting circles.

Stereographic projection makes it natural to regard the inversive plane (Coxeter [9], pp. 83, 91, 146) as the Euclidean plane completed by the postulation of a single *point at infinity*, so that a line is a circle through this point and two parallel lines are two circles that touch each other there. Three non-coaxal circles belong to a *bundle* of circles which contains, with every two of its members, the coaxal pencil to which they belong. The bundle is said to be *hyperbolic*, *parabolic* or *elliptic* according as its circles have a common orthogonal circle, a common point, or neither. Two non-intersecting circles have an *inversive distance* which can be defined as $\log(R/r)$, where R and r (with $R > r$) are the radii of two concentric circles into which the given circles can be inverted. Any two circles are interchanged by inversion in a *mid-circle* (or "circle of antisimilitude"). In the case of intersecting circles, there are two mid-circles bisecting the supplementary angles of intersection; in the case of non-intersecting circles, there is just one mid-circle, bisecting the inversive distance. A product of inversions is called a *homography* or an *antihomography* according as the number of inversions is even or odd. In particular, the product of inversions in two orthogonal circles is a *Möbius involution*, and the product of inversions in three mutually orthogonal circles is an *anti-inversion* (or "elliptic anti-involution"). The corresponding product of reflections in three mutually orthogonal planes is a *central inversion* (or "reflection in a point").

These remarks enable us to exhibit Liebmann's representation in the form of a "dictionary," as follows:

The inversive plane	Hyperbolic space
Circles and point pairs	Planes and lines
Tangent circles	Parallel planes
The angles between two intersecting circles, and the mid-circles that bisect these angles	Dihedral angles, and the planes that bisect these angles
The inversive distance between two non-intersecting circles, and the mid-circle that bisects it	The distance between two ultraparallel planes, and the mid-plane that bisects it
Two orthogonal pencils of tangent circles	Two reciprocal pencils of parallel planes
A pencil of intersecting circles and the orthogonal pencil of non-intersecting circles	An axial pencil of planes and the reciprocal pencil of ultraparallel planes
The limiting points of the pencil containing two non-intersecting circles	The common perpendicular of two ultraparallel planes
A hyperbolic bundle of circles	The planes perpendicular to one plane
A parabolic bundle of circles	The planes parallel to one ray
An elliptic bundle of circles	A point (or the planes through it)
Three mutually orthogonal circles	Three mutually perpendicular planes
Inversion in a circle	Reflection in a plane
Homographies and anti-homographies	Congruent transformations, direct and opposite
Möbius involutions and other elliptic homographies	Half-turns and other rotations
Parabolic and hyperbolic homographies	Parallel displacements and translations
Anti-inversions and other elliptic anti-homographies	Central inversions and other rotatory-reflections
Parabolic and hyperbolic anti-homographies	Parallel-reflections and glide-reflections
Loxodromic homographies	Screw displacements or twists

CHAPTER XV

CONCLUDING REMARKS

15.1. Hjelmslev's mid-line. If **AB** and **A′B′** are congruent point-pairs in a plane, we can find a congruent transformation that takes **AB** to **A′B′** (see pp. 113, 201-203). In fact, since **A′** and **B′** are invariant by reflection in the line **A′B′**, there are two such transformations.

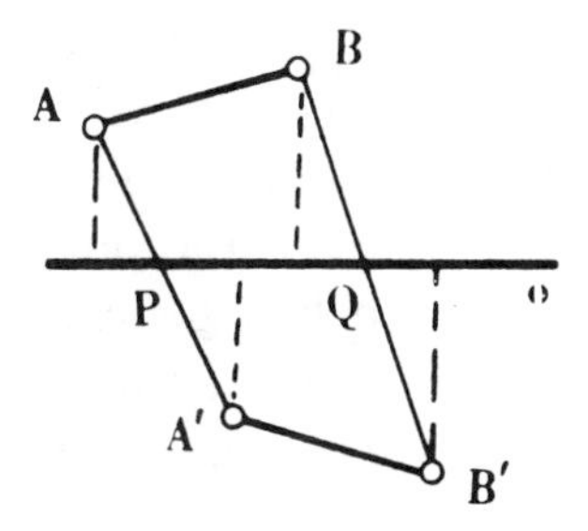

Fig. 15.1a

If the geometry is Euclidean or hyperbolic, one of these two transformations is direct and the other opposite. The latter, being the product of three reflections, cannot be anything but a *glide-reflection* (including, as a possible special case, a simple reflection). Let **o** denote the axis of the glide-reflection. Since **A**, **A′** are equidistant from **o** on opposite sides, **o** passes through **P**, the mid-point of **AA′**; similarly it passes through **Q**, the mid-point of **BB′** (see Fig. 15.1a). Moreover, **P** and **Q** are distinct unless **AB** is perpendicular to **o** (in which case the direct transformation relating **AB** to **A′B′** is simply a half-turn, i.e., a rotation through π). Since the same glide-reflection will transform any point on **AB** into the congruently related point on **A′B′**, we have proved

HJELMSLEV'S THEOREM.* *If* $\mathbf{X} \overline{\wedge} \mathbf{X}'$ *is any congruent mapping of one line on another, then the mid-points of the segments* $\mathbf{XX}'$ *are distinct and collinear or else they all coincide.*

The line **o**, which is the axis of the glide-reflection, is called the *mid-line* (or Hjelmslev line) of the congruent mapping.

If the geometry is elliptic, the two transformations relating **AB** to **A′B′** are both rotations, but it is possible to regard them as glide-reflections whose axes are the polars of their centres. Thus in the elliptic plane we have two mid-lines, the loci of the two mid-points of **XX′** (see p. 100).

By varying the segment **A′B′** on its line, and allowing its sense to be reversed, we obtain two continuously-infinite families of mid-lines belonging to the two given lines **AB** and **A′B′**.

In the elliptic plane, the centre of a rotation relating two lines, being equidistant from the lines, lies on one of their angle-bisectors. Since the mid-lines are the polars of these centres of rotation, the two families are two pencils, namely all the lines through the pole of either angle-bisector. In other words, they are the lines perpendicular to either angle-bisector.

In the Euclidean or hyperbolic plane, the axis of a glide-reflection relating two lines is equidistant from the lines or else makes equal angles with them. Hence, for two intersecting lines, the two families of mid-lines are (as before) the lines perpendicular to either angle-bisector. In the Euclidean case they are pencils of parallels; in the hyperbolic, pencils of ultra-parallels (see p. 199).

The above families of mid-lines may be described as the joins **AA′** of *corresponding* points (see p. 215) on the two given lines. Such a description remains valid (for one family of mid-lines) when the two given lines (in the Euclidean or hyperbolic

*Hjelmslev [1], p. 458. He denoted this glide-reflection by W, and expressed it as the product of the half-turn about **P** and the reflection in a line perpendicular to **o**. What particularly interested him was the fact that this is a substantial theorem not requiring any assumption about parallelism.

plane) do not intersect. In the case of parallel lines in the hyperbolic plane, every point between **A** and **A**′, on such a mid-line, is the mid-point of some segment **XX**′ (such that **A**′**X**′ is congruent to **AX**). In the case of ultra-parallel lines, the locus of mid-points of segments **XX**′ is restricted to a segment in the middle of **AA**′, cut off by the parallels to the given lines through their *symmetry point,** which is the mid-point of their common perpendicular. In particular, when **AA**′ is this common perpendicular, the symmetry point is the mid-point of every segment **XX**′. In the case of parallel lines in the Euclidean plane, this reduction of the locus to a single point occurs for every pair **AA**′ of corresponding points.

In all these cases, the congruent segments **AX** and **A**′**X**′, on the given parallel or ultra-parallel lines, are oppositely directed. A second family of mid-lines (having only one member in the Euclidean case) arises when the senses of **AX** and **A**′**X**′ agree. Now the locus of mid-points of segments **XX**′ covers the whole mid-line; the mid-line lies between the two given lines and is parallel or ultra-parallel to both. In the case of parallel lines in the hyperbolic plane, this second family of mid-lines consists of *all* the (parallel) lines between them. In the case of ultra-parallel lines, it consists of part of the pencil of lines through their symmetry point, namely, those lines through the symmetry point which are ultra-parallel to the given lines.

This last remark provides a neat construction for the symmetry point of two given ultra-parallel lines. Take two adjacent segments **AB**, **BC** on one line, both congruent to **A**′**B**′ on the other (with senses agreeing, so that **B** and **B**′ are on the same side of **AA**′). Then the symmetry point lies on the join of mid-points of **AA**′, **BB**′ and also on the join of the mid-points of **BA**′, **CB**′.

Since the symmetry point lies on the common perpendicular, this construction supersedes Hilbert's (pp. 191-192).

---

*Busemann and Kelly [1], p. 200.

If **AB** and **A′B′** are congruent point-pairs in three-dimensional space (elliptic, Euclidean, or hyperbolic), one congruent transformation taking **AB** to **A′B′** is a *half-twist* about the line **PQ** joining the mid-points of **AA′** and **BB′**. To see this, let $\mathbf{B}_1$, $\mathbf{B}_1'$ be the orthogonal projections of **B**, **B′** on the plane **AA′Q**, as in Fig. 15.1B.

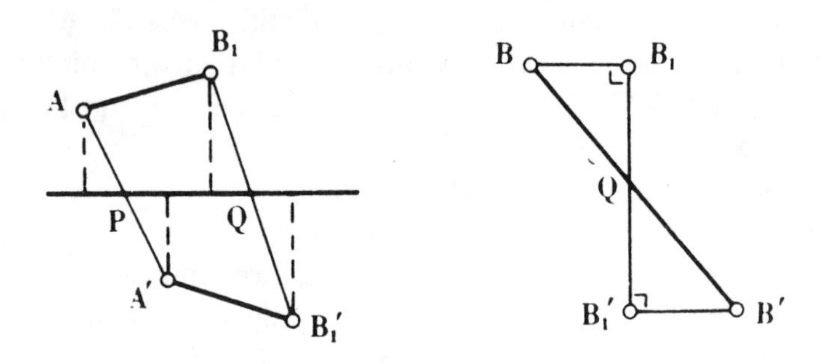

FIG. 15.1B

The congruent right-angled triangles $\mathbf{QBB}_1$, $\mathbf{QB'B}_1'$ show that **B** and **B′** are equidistant from the plane **AA′Q** on opposite sides, and that **Q** is the mid-point of $\mathbf{B}_1\mathbf{B}_1'$. The congruent right-angled triangles $\mathbf{ABB}_1$, $\mathbf{A'B'B}_1'$ show that $\mathbf{AB}_1 \equiv \mathbf{A'B}_1'$. Hence **PQ** is the mid-line of $\mathbf{AB}_1$ and $\mathbf{A'B}_1'$, i.e., it is the axis of the glide-reflection that relates $\mathbf{AB}_1$ to $\mathbf{A'B}_1'$. Regarding the reflection in the line **PQ** as a half-turn about that line, we have, instead of the glide-reflection, a half-twist (the product of the half-turn with the same translation that was used in the glide-reflection). Since this half-twist interchanges the two sides of the plane **AA′Q**, it not only takes $\mathbf{B}_1$ to $\mathbf{B}_1'$ but also **B** to **B′**, as desired.

Since the same half-twist will transform any point on **AB** into the congruently related point on **A′B′**, we see that Hjelmslev's Theorem continues to hold for point-pairs on skew lines.*

*Liebmann [1], p. 18.

As before, the different congruent mappings between two given lines yield two continuously-infinite families of mid-lines. In Euclidean space, each half-twist interchanges two parallel planes, namely the plane through each line parallel to the other; therefore the mid-lines of the given lines are the same as those of the respectively parallel lines through their symmetry point (the mid-point of their common perpendicular), namely two perpendicular pencils of parallels. But

In elliptic or hyperbolic 3-*space, the two families of mid-lines of two skew lines are generators of a ruled quadric.*

We shall prove this by showing that two mid-lines intersect or are skew according as they belong to opposite families or to the same family (see pp. 62, 69).

Consider the mid-lines of two congruent mappings $\mathbf{X} \barwedge \mathbf{X}'$ and $\mathbf{X} \barwedge \mathbf{X}''$, where $\mathbf{X}$ is a variable point on the first line $\mathbf{l}_1$ while $\mathbf{X}'$ or $\mathbf{X}''$ is the congruently related point on the second line $\mathbf{l}_2$. Suppose first that the two mid-lines belong to opposite families. Then, on $\mathbf{l}_2$, the congruent transformation $\mathbf{X}' \barwedge \mathbf{X}''$, being opposite, is simply the reflection in a point $\mathbf{A}' = \mathbf{A}''$ (see p. 98). Thus the two given mappings have a common pair of corresponding points $\mathbf{AA}'$, and the mid-point of $\mathbf{AA}'$ lies on both mid-lines. (In the elliptic case there is the slight complication that each congruent mapping yields a pair of mid-lines. The two mid-points of $\mathbf{AA}'$ lie on one pair of mid-lines from each family; therefore both mid-lines of one family meet both mid-lines of the other.)

Two mid-lines of the same family cannot intersect. For, their point of intersection would be a mid-point of some pair $\mathbf{AA}'$ related by the first mapping and also a mid-point of another pair $\mathbf{BB}''$ related by the second. (These two pairs must be distinct, since the congruent transformation $\mathbf{X}' \barwedge \mathbf{X}''$, being direct, is a translation, which has no double point.) Thus the lines $\mathbf{l}_1 = \mathbf{AB}$ and $\mathbf{l}_2 = \mathbf{A}'\mathbf{B}''$ would be coplanar, whereas we are assuming them to be skew.

In hyperbolic space, the ruled quadric may be described very simply in terms of the common parallels of the given skew lines $\mathbf{l}_1$, $\mathbf{l}_2$. Naming their points at infinity, suppose $\mathbf{l}_1 = \mathbf{M}_1\mathbf{N}_1$, $\mathbf{l}_2 = \mathbf{M}_2\mathbf{N}_2$, and let $\mathbf{p}$, $\mathbf{q}$ denote the common perpendiculars of $\mathbf{M}_1\mathbf{M}_2$ and $\mathbf{N}_1\mathbf{N}_2$, $\mathbf{M}_1\mathbf{N}_2$ and $\mathbf{M}_2\mathbf{N}_1$, as in Fig. 15.1c. Then one regulus consists of lines meeting $\mathbf{M}_1\mathbf{M}_2$, $\mathbf{N}_1\mathbf{N}_2$, $\mathbf{q}$, while the other consists of lines meeting $\mathbf{M}_1\mathbf{N}_2$, $\mathbf{M}_2\mathbf{N}_1$, $\mathbf{p}$.

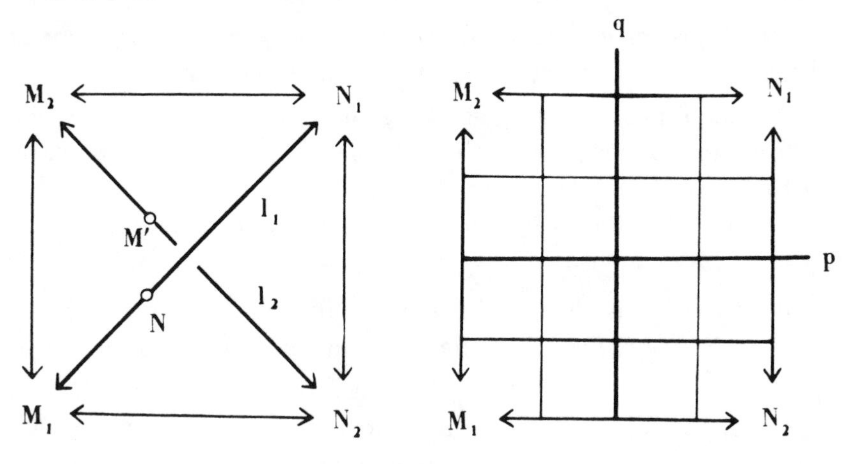

FIG. 15.1c

In fact, a limiting case of the mapping $\mathbf{X} \overline{\wedge} \mathbf{X}'$ will relate $\mathbf{M}_1$ on $\mathbf{l}_1$ to an ordinary point $\mathbf{M}'$ on $\mathbf{l}_2$, and an ordinary point $\mathbf{N}$ on $\mathbf{l}_1$ to $\mathbf{N}_2$ on $\mathbf{l}_2$. Since the mid-points of $\mathbf{M}_1\mathbf{M}'$ and $\mathbf{N}\mathbf{N}_2$ are $\mathbf{M}_1$ and $\mathbf{N}_2$, one of the mid-lines is the common parallel $\mathbf{M}_1\mathbf{N}_2$. Similarly, others are $\mathbf{M}_1\mathbf{M}_2$, $\mathbf{M}_2\mathbf{N}_1$, $\mathbf{N}_1\mathbf{N}_2$. Moreover, $\mathbf{p}$ is the mid-line for the mapping $\mathbf{M}_1\mathbf{O}_1\mathbf{N}_1 \overline{\wedge} \mathbf{M}_2\mathbf{O}_2\mathbf{N}_2$, where $\mathbf{O}_1\mathbf{O}_2$ is the common perpendicular of $\mathbf{l}_1$ and $\mathbf{l}_2$. For, in this case the half-twist reduces to a simple half-turn. Similarly, $\mathbf{q}$ is the mid-line for $\mathbf{M}_1\mathbf{O}_1\mathbf{N}_1 \overline{\wedge} \mathbf{N}_2\mathbf{O}_2\mathbf{M}_2$.

H. G. Forder has pointed out that Hjelmslev's Theorem may be extended as follows:

If $\mathbf{X} \barwedge \mathbf{X}'$ *is any congruent mapping of one plane on another, then the mid-points of the segments* $\mathbf{XX}'$ *are distinct and coplanar or else they all coincide.*

In fact, one of the two congruent transformations relating two congruent triangles $\mathbf{ABC}$, $\mathbf{A'B'C'}$ is a rotatory reflection (p. 131) whose plane, joining the mid-points of $\mathbf{AA}'$, $\mathbf{BB}'$, $\mathbf{CC}'$, is the "mid-plane" of the congruent mapping.

15.2. The Napier chain. The formulae connecting the sides and angles of a triangle in the elliptic plane (p. 234) are, of course, the same as the classical formulae for a spherical triangle on a sphere of unit radius in Euclidean 3-space. The sides of such a spherical triangle $\mathbf{ABC}$ are arcs of three great circles whose planes form a trihedral angle at the centre $\mathbf{P}_0$ of the sphere. This trihedral angle may be regarded as one corner of a special kind of tetrahedron $\mathbf{P}_0\mathbf{P}_1\mathbf{P}_2\mathbf{P}_3$ whose opposite face is perpendicular to $\mathbf{P}_0\mathbf{A}$, as in Fig. 15.2A.

This is called a *quadri-rectangular* tetrahedron or *orthoscheme*, because all its faces $\mathbf{P}_1\mathbf{P}_3\mathbf{P}_2$, $\mathbf{P}_0\mathbf{P}_3\mathbf{P}_2$, $\mathbf{P}_0\mathbf{P}_2\mathbf{P}_1$, $\mathbf{P}_0\mathbf{P}_3\mathbf{P}_1$ are right-angled triangles (with each right-angle at the point named last). Since these Euclidean triangles have respectively an angle A at $\mathbf{P}_1$, and angles a, b, c at $\mathbf{P}_0$, those formulae which do not involve B may be deduced immediately; e.g.,

$$\sin a = \frac{\mathbf{P}_2\mathbf{P}_3}{\mathbf{P}_0\mathbf{P}_3} = \frac{\mathbf{P}_1\mathbf{P}_3}{\mathbf{P}_0\mathbf{P}_3}\frac{\mathbf{P}_2\mathbf{P}_3}{\mathbf{P}_1\mathbf{P}_3} = \sin c \sin A.$$

The rest follow by Napier's rule, to the effect that any formula connecting

15.21. $$A,\ \tfrac{1}{2}\pi - a,\ c,\ \tfrac{1}{2}\pi - b,\ B$$

remains valid when these five quantities are cyclically permuted. (To be historically accurate, Napier used the "alternate" cycle

15.22. $\frac{1}{2}\pi - a,\ B,\ c,\ A,\ \frac{1}{2}\pi - b;$

but either rule implies the other.)

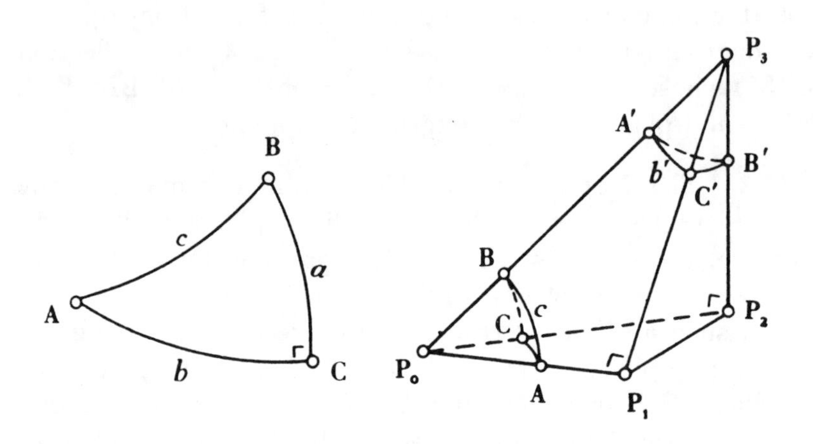

FIG. 15.2A

One way to establish Napier's rule is to draw another unit sphere, with its centre at $\mathbf{P}_3$ instead of $\mathbf{P}_0$, so as to obtain a triangle $\mathbf{A'B'C'}$ whose vertices lie on $\mathbf{P}_3\mathbf{P}_0$, $\mathbf{P}_3\mathbf{P}_2$, $\mathbf{P}_3\mathbf{P}_1$, as in Fig. 15.2A. Since the dihedral angle at the edge $\mathbf{P}_0\mathbf{P}_3$ appears as the angle B of $\mathbf{ABC}$ and as the angle A' of $\mathbf{A'B'C'}$, we have

$$A' = B.$$

Since the two acute angles of a Euclidean right-angled triangle are complementary,

$$\tfrac{1}{2}\pi - a' = A, \quad c' = \tfrac{1}{2}\pi - a, \quad \tfrac{1}{2}\pi - b' = c, \quad B' = \tfrac{1}{2}\pi - b.$$

Thus we pass from $\mathbf{ABC}$ to $\mathbf{A'B'C'}$ by a cyclic permutation. In other words, these are the first two of a cycle of five related right-angled spherical triangles, known as the *Napier chain*.

An alternative procedure is to consider a certain set of five

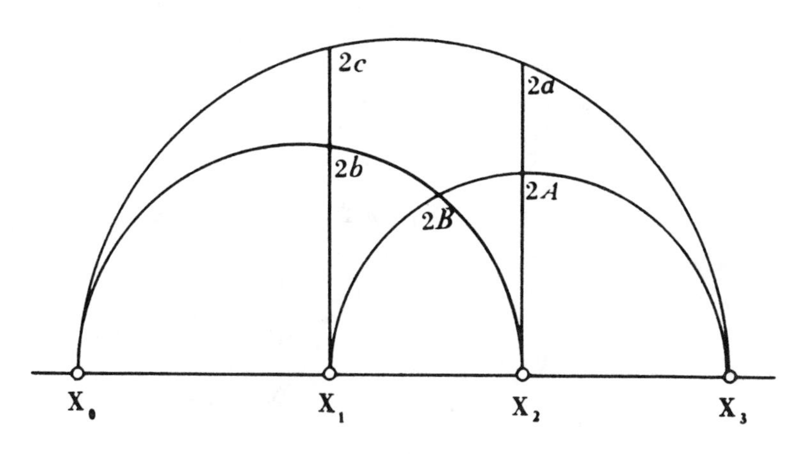

FIG. 15.2B

planes through $\mathbf{P}_0$, namely three faces of the orthoscheme, one plane parallel to the fourth face, and one perpendicular to the longest edge $\mathbf{P}_0\mathbf{P}_3$. These five planes cut out from the sphere round $\mathbf{P}_0$ a *rectangular pentagram** (Gauss's *pentagramma mirificum*) whose sides

$$\tfrac{1}{2}\pi + a. \quad \pi - B, \quad \pi - c, \quad \pi - A, \quad \tfrac{1}{2}\pi + b$$

are the supplements of Napier's cycle 15.22.

A third way is to use G. T. Bennett's diagram† (Fig. 15.2B) based on four collinear points $\mathbf{X}_0$, $\mathbf{X}_1$, $\mathbf{X}_2$, $\mathbf{X}_3$ in the Euclidean plane, so arranged that

$$\mathbf{X}_0\mathbf{X}_1 = \mathbf{P}_0\mathbf{P}_1{}^2, \quad \mathbf{X}_1\mathbf{X}_2 = \mathbf{P}_1\mathbf{P}_2{}^2, \quad \mathbf{X}_2\mathbf{X}_3 = \mathbf{P}_2\mathbf{P}_3{}^2$$

and consequently (by Pythagoras)

$$\mathbf{X}_0\mathbf{X}_2 = \mathbf{P}_0\mathbf{P}_2{}^2, \quad \mathbf{X}_0\mathbf{X}_3 = \mathbf{P}_0\mathbf{P}_3{}^2, \quad \mathbf{X}_1\mathbf{X}_3 = \mathbf{P}_1\mathbf{P}_3{}^2.$$

Semicircles and lines are drawn orthogonal to the line of the

*Sommerville [2], p. 118.

†It was stated (Coxeter [2], p. 123) that Dr. Bennett would publish this in greater detail elsewhere. But he did not live to do so.

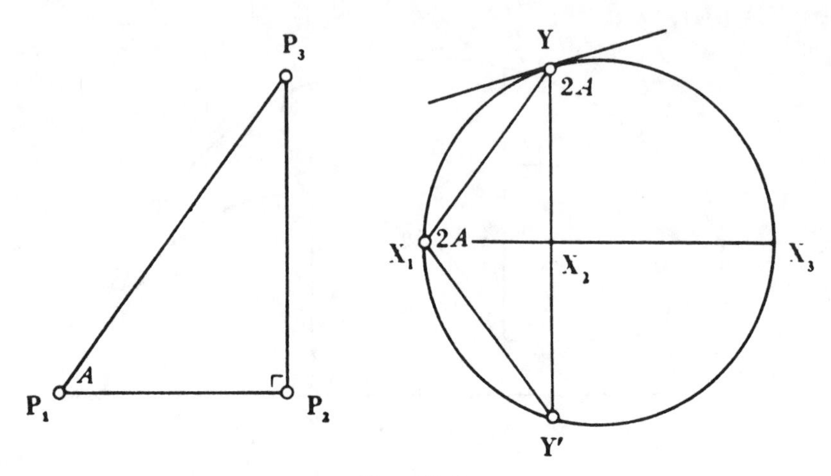

FIG. 15.2C

X's. Comparing the two parts of Fig. 15.2C, in which $\mathbf{YX_2Y'}$ is a chord perpendicular to the diameter $\mathbf{X_1X_2X_3}$, we have

$$\left(\frac{\mathbf{P_1P_2}}{\mathbf{P_2P_3}}\right)^2 = \frac{\mathbf{X_1X_2}}{\mathbf{X_2X_3}} = \frac{\mathbf{X_1X_2}}{\mathbf{X_2Y}}\,\frac{\mathbf{X_2Y}}{\mathbf{X_2X_3}} = \left(\frac{\mathbf{X_1X_2}}{\mathbf{X_2Y}}\right)^2.$$

Since $\mathbf{P_1P_3P_2}$ and $\mathbf{X_1YX_2}$ are similar triangles, the angle between the line $\mathbf{X_2Y}$ and the arc $\mathbf{YX_3}$ is $2A$. Similarly, the triangles $\mathbf{P_0P_3P_2}$, $\mathbf{P_0P_2P_1}$, $\mathbf{P_0P_3P_1}$ yield angles $2a$, $2b$, $2c$, as indicated in Fig. 15.2B. Performing an arbitrary inversion, and calling the centre of inversion $\mathbf{X_4}$, we obtain Fig. 15.2D, which shows that, since $2a$ and $2b$ are interchanged by naming $\mathbf{X_0X_1X_2X_3X_4}$ in the reverse order, the angle indicated between $\mathbf{X_0X_2}$ and $\mathbf{X_1X_3}$ is $2B$.

Thus the arcs $\mathbf{X_0X_2}$, $\mathbf{X_1X_3}$, $\mathbf{X_2X_4}$, $\mathbf{X_3X_0}$, $\mathbf{X_4X_1}$, orthogonal to the circle $\mathbf{X_0X_1X_2X_3X_4}$, form a curvilinear pentagon whose angles

$$2A,\ \pi - 2a,\ 2c,\ \pi - 2b,\ 2B$$

are the doubles of those in the cycle 15.21. Regarding the interior of the circle as a conformal model of the hyper-

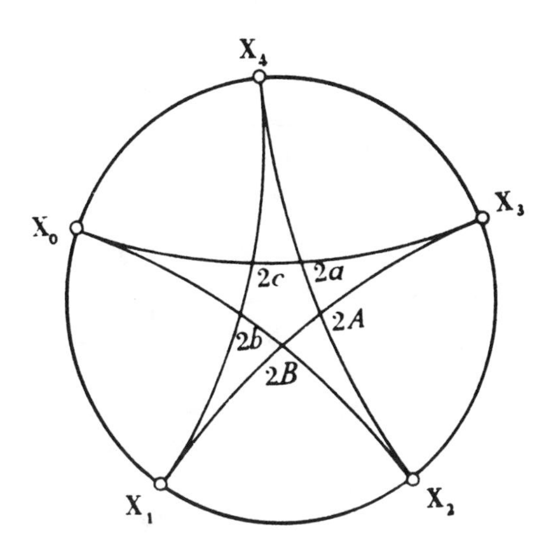

FIG. 15.2D

bolic plane (p. 261), we may say that these are the angles of a pentagon whose alternate sides are parallel.

15.3. The Engel chain. Let the angles of parallelism for lengths a, b, c, l, m be denoted by the corresponding Greek letters α, β, γ, λ, μ, and let a', b', c', l', m' denote the lengths whose angles of parallelism are the complements $\alpha' = \frac{1}{2}\pi - \alpha$, $\ldots$, $\mu' = \frac{1}{2}\pi - \mu$. By §10.6,

$$\sinh a = \cot \alpha, \qquad \tanh a = \cos \alpha,$$
$$\sinh a' = \cot \alpha' = \tan \alpha, \qquad \tanh a' = \cos \alpha' = \sin \alpha,$$

and so on. This notation is useful for discussing a remarkable connection that Lobatschewsky* discovered between the

*Lobatschewsky's treatment was trigonometrical. Engel [1] established the correspondence from purely geometrical considerations, using three-dimensional space. Liebmann ([1], pp. 37–42) gave a synthetic proof in the plane. The simpler treatment given here was inspired by Liebmann's proof of Bolyai's parallel construction (Liebmann [1], p. 36).

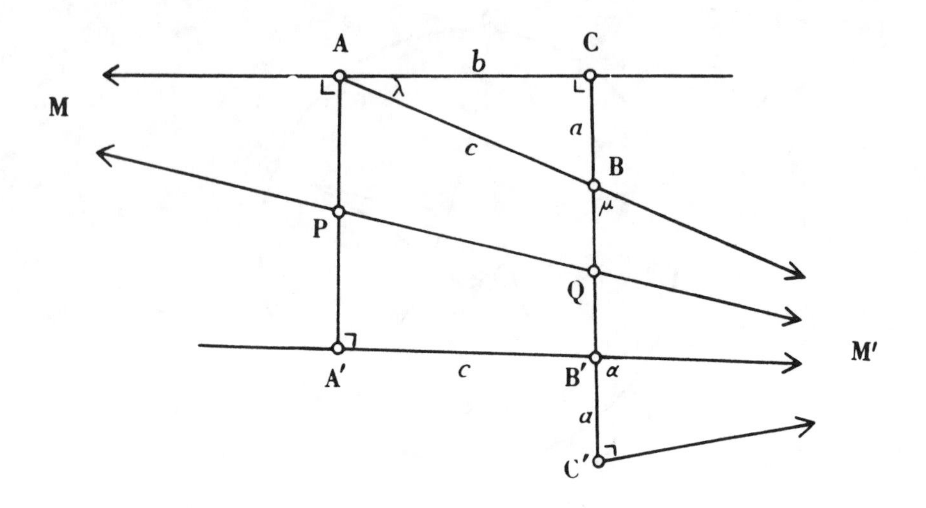

FIG. 15.3A

general trirectangle (p. 194) and the general right-angled triangle. Writing λ and μ for **A** and **B**, we may express his result by saying that the right-angled triangle **ABC** corresponds to a trirectangle with sides l', c, m, b, the second and third sides enclosing the single acute angle, which is α.

Given **ABC** (right-angled at **C**, as in Fig. 15.3A), draw **AA′** of length l' perpendicular to **MAC** on the same side as **B**, so that **A′M′**, parallel to **AB**, is perpendicular to **AA′**. By Bolyai's parallel construction (p. 204), **A′M′** meets **CB** produced in a point **B′** such that **A′B′** ≡ **AB**. Since these congruent segments are on parallel lines, their mid-line **PQM′** (§15.1) belongs to the same pencil of parallels. The half-turn about **P** interchanges **AM** and **A′M′**. The half-turn about **Q** transforms **C** into a point **C′** such that **C′M′**, parallel to **AB**, is perpendicular to **CB**. Since the acute angle at **B** is μ, we have

$$\mathbf{B'C} = \mathbf{BC'} = m;$$

and since **B′C′** = **BC** = a, the acute angle at **B′** is α. We have thus derived from the triangle **ABC** a trirectangle **AA′B′C** with

$$\mathbf{AA'} = l', \quad \mathbf{A'B'} = c, \quad \mathbf{B'C} = m, \quad \mathbf{CA} = b,$$

and angle α at **B′**.

Conversely, given the trirectangle **AA′B′C**, we can reconstruct the triangle by taking **B** on **B′C** so that **AB** ≡ **A′B′**.

Naming the five "parts" in a convenient order, we may say that we have established a correspondence between the right-angled triangle $b\lambda c\mu a$ and the trirectangle $l'c\alpha mb$. Since the five parts can equally well be named in the reverse order, the same triangle (or trirectangle) yields another trirectangle (or triangle); e.g., the trirectangle **AA′B′C** yields not only the triangle **ABC** at its corner **C** but also another triangle $l'\beta'm\gamma a$ at its opposite corner.

Carrying out the correspondence systematically, we thus obtain the *Engel chain* of alternate triangles and trirectangles:

$$\begin{array}{llllll} & & b\lambda c\mu a, & l'c\alpha mb & \text{or} & bm\alpha cl', \\ l'\beta'm\gamma a & \text{or} & a\gamma m\beta' l', & c'm\lambda' b'a & \text{or} & ab'\lambda' mc', \\ c'\alpha' b'\mu l' & \text{or} & l'\mu b'\alpha' c', & m'b'\gamma' a'l' & \text{or} & l'a'\gamma' b'm', \\ m'\lambda a'\beta' c' & \text{or} & c'\beta' a'\lambda m', & ba'\mu' lc' & \text{or} & c'l\mu' a'b, \\ b\gamma l\alpha' m' & \text{or} & m'\alpha' l\gamma b, & al\beta cm' & \text{or} & m'c\beta la, \\ a\mu c\lambda b & \text{or} & b\lambda c\mu a. & & & \end{array}$$

Since the sixth triangle is the same as the first, the chain closes; there are just five triangles and five trirectangles.

Mukhopadhyaya [1] has managed to put all the five trirectangles into a single figure by establishing the existence of a rectangular pentagon with sides l, b', c, a', m. The members of the chain are derived from one another by cyclically permuting these five sides, taken alternately, i.e., by the permutations

15.31. $$(lcmb'a') \quad (l'c'm'ba) \quad (\lambda\gamma\mu\beta'\alpha') \quad (\lambda'\gamma'\mu'\beta\alpha)\,.$$

R. C. Bose [1] has proved directly that, in a rectangular pentagon, the hyperbolic cosine of any side is equal to the product of the hyperbolic cotangents of the adjacent sides,* and equal to the product of the hyperbolic sines of the remaining sides. Thus

$$\cosh c = \coth a' \coth b' = \sinh l \sinh m,$$

i.e.,

$$\cosh c = \cosh a \cosh b = \cot \lambda \cot \mu,$$

in agreement with 12.97 (p. 238). The rest of the ten formulae for the right-angled hyperbolic triangle can be derived from these two by means of the above permutations (i.e., by applying Bose's rule to other sides of the rectangular pentagon).

Lobatschewsky described a correspondence between spherical and hyperbolic triangles† whereby the Engel chain may be derived from the Napier chain (§15.2). Regarding the orthoscheme $\mathbf{P}_0\mathbf{P}_1\mathbf{P}_2\mathbf{P}_3$ of Fig. 15.2A as being in hyperbolic space instead of Euclidean, we take the plane $\mathbf{P}_1\mathbf{P}_2\mathbf{P}_3$ (perpendicular to $\mathbf{P}_0\mathbf{A}$) to be parallel to $\mathbf{P}_0\mathbf{P}_3$, so that $\mathbf{P}_3$ is a point at infinity and the tetrahedron is "singly-asymptotic." To give the sides of the spherical triangle **ABC** their proper values, we may take the radius of the sphere round $\mathbf{P}_0$ to be arg sinh 1 (see 13.71). Thus

$$a = \Pi(\mathbf{P}_0\mathbf{P}_2), \quad c = \Pi(\mathbf{P}_0\mathbf{P}_1), \quad A = \Pi(\mathbf{P}_1\mathbf{P}_2)\ .$$

On the other hand, instead of another sphere round $\mathbf{P}_3$ we now have a horosphere, so that **A′B′C′** is a horospherical triangle, which is indistinguishable from a Euclidean triangle (see 11.31). Thus

$$B = A' = \tfrac{1}{2}\pi - B' = \tfrac{1}{2}\pi - \angle\mathbf{P}_0\mathbf{P}_2\mathbf{P}_1.$$

---

*This was anticipated by Whitehead in his *Universal Algebra* (Cambridge, 1898), p. 434.

†Lobatschewsky [2], p. 18; [3], p. 36.

Since also $b = \angle \mathbf{P}_1\mathbf{P}_0\mathbf{P}_2$, we have expressed the five parts of the spherical triangle **ABC** in terms of the five parts of the hyperbolic triangle $\mathbf{P}_2\mathbf{P}_0\mathbf{P}_1$. Changing the notation so as to identify $\mathbf{P}_2\mathbf{P}_0\mathbf{P}_1$ with the hyperbolic triangle **ABC** of our earlier discussion, we may say that the corresponding spherical triangle has hypotenuse $\Pi(a) = \alpha$ and catheti γ, μ opposite to its angles β, λ'. In other words, the hyperbolic triangle $b\lambda c\mu a$ corresponds to a spherical triangle $\mu\beta\alpha\lambda'\gamma$. The ten formulae 12.96–12.99 now follow from the ten formulae 12.76–12.79, and Napier's cycle 15.22 yields the cycle $(\gamma'\,\lambda'\,\alpha\,\beta\,\mu')$, which is the inverse of Engel's $(\lambda'\,\gamma'\,\mu'\,\beta\,\alpha)$ (see 15.31).

15.4. Normalized canonical coordinates. Another approach to hyperbolic trigonometry is the use of canonical coordinates (§10.8), normalized so that

$$x_0 > 0,\ x_0^2 - x_1^2 - x_2^2 = 1,\ X_0^2 - X_1^2 - X_2^2 = -1.$$

This normalization simplifies the formulae in a striking manner. The distance between (x) and (y) is

$$\arg\cosh\,(x_0y_0 - x_1y_1 - x_2y_2)\ ,$$

the distance from (x) to $[Y]$ is

$$\arg\sinh\,|\, x_0Y_0 + x_1Y_1 + x_2Y_2 \,|\ ,$$

and the angle between intersecting lines $[X]$ and $[Y]$ is

$$\arccos\,|\, -X_0Y_0 + X_1Y_1 + X_2Y_2 \,|\ .$$

For a right-angled triangle **ABC** with **A** at (1,0,0) and **AC** along [0,0,1], as in Fig. 15.4A, we see at once that

$$\begin{aligned} &\mathbf{C} &&\text{is} && (\cosh b,\ \sinh b,\ 0)\ ,\\ &\mathbf{BC} &&\text{is} && [\sinh b,\ -\cosh b,\ 0]\ ,\\ &\mathbf{AB} &&\text{is} && [0,\ -\sin\lambda,\ \cos\lambda]\ ,\end{aligned}$$

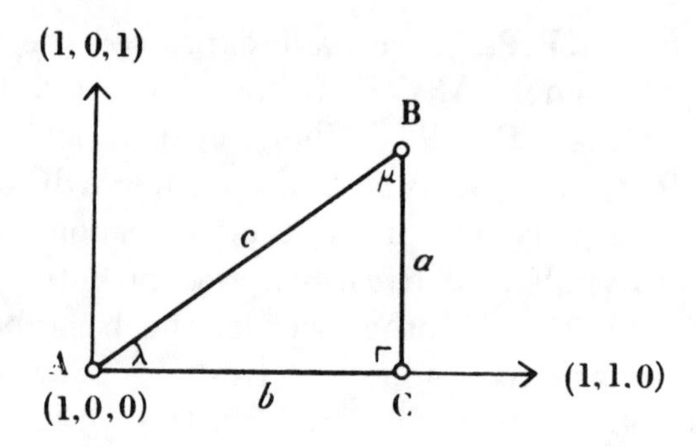

FIG. 15.4A

and **B** is ($\cosh c$, $\sinh c \cos \lambda$, $\sinh c \sin \lambda$). The distance from **B** to **AC** yields

15.41. $$\sinh a = \sinh c \sin \lambda.$$

The angle between **AB** and **BC** yields

15.42. $$\cos \mu = \cosh b \sin \lambda.$$

Making μ tend to zero, we deduce

$$\cosh b \sin \Pi(b) = 1,$$

in agreement with §10.6. Writing 15.41 and 15.42 in the form

$$\sin \lambda = \tan \gamma \tan \alpha' = \cos \mu \cos \beta',$$

and making repeated applications of the permutation $(\lambda\gamma\mu\beta'\alpha')$ (see 15.31), we deduce the rest of the ten relations 12.96–12.99 in the form*

$$\begin{aligned}
\sin \gamma &= \tan \mu \ \tan \lambda = \cos \beta' \cos \alpha',\\
\sin \mu &= \tan \beta' \tan \gamma = \cos \alpha' \cos \lambda,\\
\sin \beta' &= \tan \alpha' \tan \mu = \cos \lambda \ \cos \gamma,\\
\sin \alpha' &= \tan \lambda \ \tan \beta' = \cos \gamma \ \cos \mu.
\end{aligned}$$

*Lobatschewsky [2], p. 223.

15.5. Curvature. Instead of the normalized canonical coordinates x_0, x_1, x_2 used in 6.77, 13.62 and §§14.3, 15.4, it is sometimes convenient to use *isometric* coordinates u_1, u_2, given by

$$x_0 = \rho - 1, \quad x_1 = \rho u_1, \quad x_2 = \rho u_2,$$

where, since $x_0^2 \pm (x_1^2 + x_2^2) = 1$ (with the upper or lower sign according as the geometry is elliptic or hyperbolic),

$$\rho = \frac{2}{1 \pm (u_1^2 + u_2^2)}$$

(cf. pp. 258, 262). Since $d\rho = \mp \rho^2(u_1 du_1 + \rho_2 du_2)$, it follows that

$$ds^2 = \pm\, dx_0^2 + dx_1^2 + dx_2^2 = \rho^2(du_1^2 + du_2^2)$$
$$= \frac{4(du_1^2 + du_2^2)}{\{1 \pm (u_1^2 + u_2^2)\}^2},$$

as in 14.64 and §14.9. Comparing this with the formula

$$ds^2 = g_{\mu\nu}\, du_\mu\, du_\nu = E\, du_1^2 + 2F du_1 du_2 + G du_2^2$$

of classical differential geometry, we have $E = G = \rho^2$, $F = 0$, so that the curvature K (of the elliptic or hyperbolic plane, with the classical unit of measurement) is given by*

$$-\sqrt{EG}\, K = \frac{\partial}{\partial u_1}\left(\frac{1}{\sqrt{E}}\, \frac{\partial \sqrt{G}}{\partial u_1}\right) + \frac{\partial}{\partial u_2}\left(\frac{1}{\sqrt{G}}\, \frac{\partial \sqrt{E}}{\partial u_2}\right)$$
$$= \frac{\partial}{\partial u_1}\left(\frac{1}{\rho}\, \frac{\partial \rho}{\partial u_1}\right) + \frac{\partial}{\partial u_2}\left(\frac{1}{\rho}\, \frac{\partial \rho}{\partial u_2}\right) = \nabla^2 \log \rho$$
$$= -\, \nabla^2 \log\{1 \pm (u_1^2 + u_2^2)\} = \mp \rho^2 = \mp \sqrt{EG}.$$

We thus verify that $K = \pm 1$, in agreement with p. 211.

*Struik [1], p. 113.

Still more simply, we may use *polar* coordinates r, θ, so that

$$ds^2 = dr^2 + g\,d\theta^2,$$

where $g = \sin^2 r$ or $\sinh^2 r$ (see 13.71). Then*

$$K = -\frac{1}{\sqrt{g}}\frac{\partial^2\sqrt{g}}{\partial r^2} = \pm 1$$

(since $\sqrt{g} = \sin r$ or $\sinh r$).

15.6. Quadratic forms. From the remarks on p. 225, we see that the expression

$$\frac{(xy)^2}{(xx)\,(yy)}$$

is $\leqslant 1$ or $\geqslant 1$ according as the geometry is elliptic or hyperbolic, i.e., according as the signature of the quadratic form

$$(xx) = c_{\mu\nu}x_\mu x_\nu$$

is $\{+++\ldots\}$ or $\{+--\ldots\}$. L. J. Mordell has kindly supplied a proof for the corresponding algebraic theorem:

(i) *If* (xx) *is positive definite,* $(xy)^2 \leqslant (xx)\,(yy)$.

(ii) *If* (xx) *is indefinite, of signature* $\{+--\ldots\}$, *then for values of* (y) *satisfying* $(yy) > 0$,

$$(xy)^2 \geqslant (xx)\,(yy)\,.$$

Working in projective $(n-1)$-space, so that (xx) is an n-ary form, we may assume (x) and (y) to be two distinct points and then choose $n-2$ further points to form a simplex $(x)\,(y)\ldots(z)$, so that

$$\begin{vmatrix} x_1 & y_1 & \ldots & z_1 \\ \ldots & & \ldots & \\ x_n & y_n & \ldots & z_n \end{vmatrix} \neq 0.$$

*Struik [1], p. 138.

These n independent points determine a coordinate transformation

$$\xi_\mu = x_\mu \xi + y_\mu \eta + \ldots + z_\mu \zeta \qquad (n \text{ terms}),$$

from $(\xi_1, \xi_2, \ldots, \xi_n)$ to $(\xi, \eta, \ldots, \zeta)$ (cf. p. 83). Applying this transformation to the form $(\xi\xi)$, we obtain

$$(\xi\xi) = c_{\mu\nu}\xi_\mu\xi_\nu = c_{\mu\nu}(x_\mu\xi + y_\mu\eta + \ldots)(x_\nu\xi + y_\nu\eta + \ldots)$$
$$= (xx)\xi^2 + 2(xy)\xi\eta + (yy)\eta^2 + \ldots .$$

(i) If (xx) is positive definite, so is $(\xi\xi)$, and so is

15.61. $$(xx)\xi^2 + 2(xy)\xi\eta + (yy)\eta^2.$$

Hence $(xx)(yy) - (xy)^2 > 0$.

(ii) If (xx) is of signature $\{+--\ldots\}$, so is $(\xi\xi)$. Therefore the binary form 15.61 is either indefinite or negative definite; if $(yy) > 0$, it can only be indefinite, i.e.,

$$(xx)(yy) - (xy)^2 < 0.$$

15.7. The volume of a tetrahedron. A proof of 13.51, simpler than Schläfli's, was given in 1903 by H. W. Richmond [1]. But his presentation is so highly condensed that a full account seems desirable. Following him, we work in elliptic space; but the corresponding proof for hyperbolic space may be derived by making the appropriate changes from circular to hyperbolic functions.

We begin with a lemma concerning a spherical (or elliptic) triangle **ABC**: *For a small displacement of* **A** *along* **CA** *produced,*

15.71. $$dA = -\cos c \, dB \qquad (a \text{ and } C \text{ constant}).$$

To prove this, draw **AH** perpendicular to **BA'** as in Fig. 15.7A, so as to form an infinitesimal right-angled triangle **A'AH** in which, by 13.71, **AH** $= dB \sin c$, and therefore

$$dc = \mathbf{A'H} = dB \sin c \cot A.$$

By 12.73, $\sin c \sin A = \sin a \sin C = \text{const.}$, whence

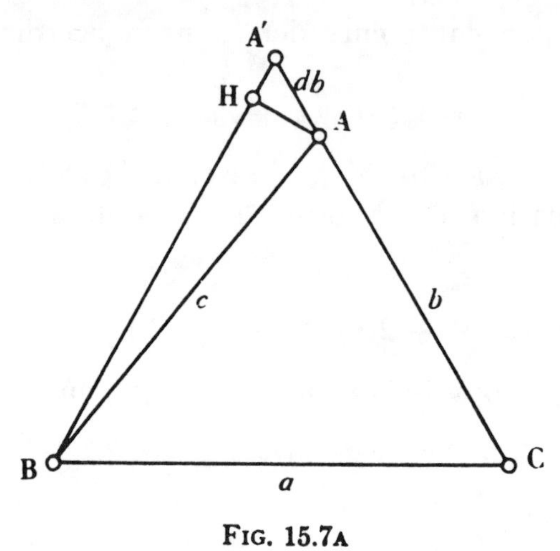

FIG. 15.7A

$$\begin{aligned}\sin c \cos A \, dA &= -\cos c \sin A \, dc \\ &= -\cos c \sin c \cos A \, dB,\end{aligned}$$

as desired.

We proceed to prove that the differential of the volume V of an elliptic tetrahedron **ABCD** is

15.72. $$dV = \tfrac{1}{2} \sum l \, d\lambda,$$

where l is the length of an edge at which the dihedral angle is λ, and the summation is taken over the six edges.

It is clearly sufficient to consider the effect of varying just one vertex. Since any small displacement of **A** may be decomposed into small displacements along the three edges at that corner, we may simply suppose **A** to be displaced to **A′** on **CA** produced, as in Fig. 15.7B. Let ω denote the dihedral angle at the edge **BD**, so that $d\omega$ is the angle between the planes **ABD** and **A′BD**.

Draw **XM** and **AN** perpendicular to **BD**, taking **Y** anywhere on **AN**, and **X** on **BY**. Set θ = **BX**, ϕ = ∠**DBX**. An

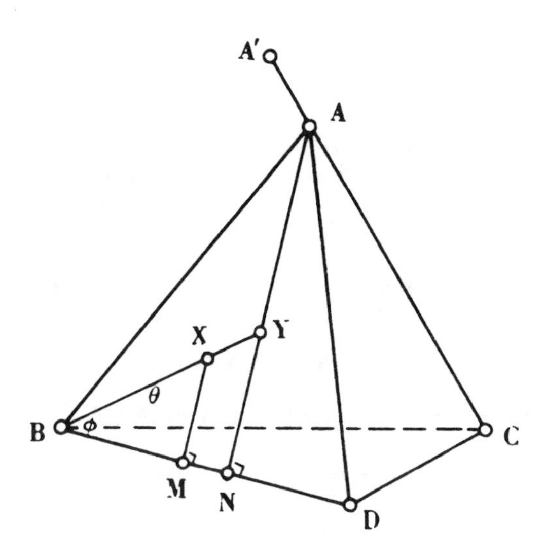

FIG. 15.7B

element of area at **X** may be expressed as $\sin \theta \, d\theta \, d\phi$. This traces out an element of volume

$$\sin \theta \, d\theta \, d\phi \, . \, d\omega \sin \mathbf{XM},$$

where, by 12.76, $\sin \mathbf{XM} = \sin \theta \sin \phi$. The increment dV is the sum of two terms, corresponding to the two parts **ABN**, **ADN** of the triangle **ABD**. (The possibility of an obtuse angle at **B** or **D** is covered by an appropriate change of sign.)

The contribution of the part **ABN** is $p \, d\omega$, where

$$p = \int_{\theta=0}^{\mathbf{BY}} \int_{\varphi=0}^{B} \sin^2 \theta \sin \phi \, d\theta \, d\phi \qquad (B = \angle \mathbf{DBA}).$$

Performing first the integration with respect to θ, and then changing the notation so that $\theta = \mathbf{BY}$, we have

$$p = \tfrac{1}{2} \int_{\phi=0}^{B} (\theta - \sin \theta \cos \theta) \sin \phi \, d\phi.$$

Applying 12.79 to the triangle **YBN**, we obtain

$$\tan \theta \cos \phi = \tan \mathbf{BN},$$

whence

$$\sin \theta \cos \theta \sin \phi \, d\phi = \cos \phi \, d\theta$$

and

$$p = \tfrac{1}{2}\int_{\phi=0}^{B} (\theta \sin \phi \, d\phi - \cos \phi \, d\theta) = -\tfrac{1}{2}\Big[\theta \cos \phi\Big]_{\phi=0}^{B}$$

$$= \tfrac{1}{2}(\mathbf{BN} - \mathbf{BA} \cos B)\,.$$

Similarly, the contribution of the part **ADN** is $q\, d\omega$, where

$$q = \tfrac{1}{2}(\mathbf{ND} - \mathbf{DA} \cos D) \quad (D = \angle\mathbf{BDA}).$$

By addition,

$$dV = p\, d\omega + q\, d\omega = \tfrac{1}{2}(\mathbf{BD} - \mathbf{BA} \cos B - \mathbf{DA} \cos D)d\omega$$

$$= \tfrac{1}{2}\{\mathbf{BD}\, d\omega + \mathbf{BA}(-\cos B\, d\omega) + \mathbf{DA}(-\cos D\, d\omega)\}.$$

Applying 15.71 to the trihedral angle at the corner **D** of the tetrahedron, we see that $d\lambda$ for the edge **DA** is $-\cos D\, d\omega$. Similarly, $d\lambda$ for the edge **BA** is $-\cos B\, d\omega$. Also $d\lambda$ for the edge **BD** is $d\omega$ itself. Hence

$$dV = \tfrac{1}{2} \sum l d\lambda,$$

summed over the three edges **BD**, **BA**, **DA** whose dihedral angles vary. For the reasons given above, this suffices to establish 15.72 in its full generality.

The computation of V itself is most readily performed by dissecting the tetrahedron into orthoschemes (Fig. 15.2A) and then employing the formulae of Coxeter [1] (where 15.72 is used on p. 25). In the hyperbolic case the explicit formula, though complicated, is elegant: If V is the volume of an orthoscheme $\mathbf{P}_0\mathbf{P}_1\mathbf{P}_2\mathbf{P}_3$ in which the dihedral angles at the edges $\mathbf{P}_0\mathbf{P}_1$, $\mathbf{P}_0\mathbf{P}_3$, $\mathbf{P}_2\mathbf{P}_3$ are $\tfrac{1}{2}\pi-\alpha$, β, $\tfrac{1}{2}\pi-\gamma$ (while the remaining three are right angles), then, as Lobatschewsky proved by an entirely different method,

$$4V = L(\delta + \alpha) + L(\delta - \alpha) - L(\delta + \beta) - L(\delta - \beta) + L(\delta + \gamma) + L(\delta - \gamma) - 2L(\delta),$$

where

$$\delta = \text{arc cos} \frac{\sin \alpha \sin \gamma}{\sqrt{(\sin^2\alpha - \sin^2\beta + \sin^2\gamma)}}$$

and

$$L(x) = \int_0^x \log \sec \theta \, d\theta .$$

15.8. A brief historical survey of construction problems. In §10.4, we saw some famous constructions in hyperbolic geometry performed by means of ruler and compasses. The construction of a line parallel to a given ray is shown in 10.41, and the construction of a line perpendicular to one line and parallel to another in 10.43. The latter can also be obtained as the altitude of infinite length passing through the intersection of the two altitudes of finite length in a singly-asymptotic triangle having the two given lines as intersecting sides.*

In 1927, Mordoukhay-Boltovskoy [1] proved a theorem which may be stated as follows:†

15.81. *By means of ruler and compasses we are able to construct, from a given point on a given line, all segments of length r, where* sinh r *belongs to the rational field extended by extraction of square roots.*

Since the proper circle is only one of the three "cycles" (§11.1), it is reasonable to consider, along with the compasses, two more instruments, namely the *hypercompass* for drawing

*Gérard [1]; Liebmann [1], p. 34.

†Cf. Jessen [1].

a hypercycle (or equidistant curve) with a given axis and a given distance (p. 214) and a *horocompass* for drawing a horocycle with a given diameter and passing through a given point. Nestorowitsch [1; 2] has proved that no further constructions are obtained by adding either or both of these new instruments. Later, Nestorowitsch [3; 4] proved that the three kinds of compasses are of equal efficacy in conjunction with a ruler. Hence

15.82. *Any construction that can be performed by means of ruler, compasses, horocompass and hypercompass can be performed by means of ruler and compasses or ruler and horocompass or ruler and hypercompass.*

A further restriction on the use of the instruments can be made without reducing their efficacy. Hjelmslev (see Forder [2], [3]) and Handest [1] proved:

15.83. *Any construction that can be performed by means of ruler and compasses can still be performed if the compasses have a fixed adjustment, or if the compasses are replaced by a hypercompass with a fixed adjustment.*

Steiner [2] proved in Euclidean geometry that a fixed circle with its centre and a ruler have the same efficacy as the ruler and compasses. This is not quite the case in hyperbolic geometry, but:

15.84. *Any construction that can be performed by means of ruler and compasses can be performed with a ruler alone if there is drawn somewhere in the plane: either a circle with its centre and two parallel lines, or a horocycle with one diameter and two parallel lines with their common end not at the centre of the horocycle, or a hypercycle with its axis and two parallel lines with their common end not on the axis.**

---

*Handest [1].

Actually, the circle with its centre and two parallel lines may be replaced by a circle of radius arg sinh 1 with its centre. For, since the angle of parallelism for this radius is $\frac{1}{4}\pi$, the ruler yields a pair of parallel lines.

Quite another instrument has the same effect, namely a *parallel-ruler*, which is an instrument for drawing a line through a given point parallel to a given ray.*

In elliptic geometry† 15.81 is valid if we write sin r instead of sinh r. But the theorems concerning new instruments are, of course, meaningless.

Georg Mohr [1] proved in 1672 (and independently of him, Mascheroni in 1797) that all the constructions in Euclidean geometry that can be performed by means of ruler and compasses can be performed with the compasses alone. (A line is then said to be constructed if two points on it are constructed.) In 1935, Fog [1] proved that in elliptic geometry the compasses alone are as effective as the ruler and compasses if two perpendicular points are given. In 1951, Smogorschevsky [1] proved that in hyperbolic geometry the three kinds of compasses together have the same effect as the ruler and ordinary compasses. Finally, in 1953, Mokriščev [1] proved:

15.85. *Any hyperbolic construction that can be performed by means of ruler and compasses can be performed by means of compasses and horocompass or by means of the (adjustable) hypercompass alone.*

---

*Handest [1].

†Bonnesen [1]; cf. Jessen [1].

15.9. Inversive distance and the angle of parallelism. Consider, in the Euclidean plane, two circles with radii a, b and centres **A**, **B**. Suppose **AB** $= c$. If each of a, b, c is less than the sum of the other two, the circles have two common points; let **C** be one of them. Then **ABC** is a triangle whose angle at **C**, is given by

15.91. $$\cos C = (a^2 + b^2 - c^2)/2ab,$$

is one of the two supplementary angles of intersection of the circles. Since angles are preserved by inversion, the expression on the right of 15.91 is an inversive invariant for all values of a, b, c.

If $a + b = c$ or $|a - b| = c$, the circles are 'tangent', that is, in contact, internally or externally, and $C = 0$ or π.

If $a + b < c$ or $|a - b| > c$, the circles are disjoint, so there is no real point **C**. Instead, there is a real number δ, invariant for inversion, such that

15.92. $$\cos \delta = \left|a^2 + b^2 - c^2\right|/2ab.$$

This δ, like C, is zero when the two circles are tangent; it becomes infinite when one of the circles shrinks to a point. Accordingly we call δ the ***inversive distance*** between the two disjoint circles.

In particular, if the circles are concentric, so that $c = 0$, we have

$$e^{\delta} + e^{-\delta} = 2\cosh \delta = \frac{a^2 + b^2}{ab} = \frac{a}{b} + \frac{b}{a}.$$

Thus, if $a > b$,

$$\delta = \log \frac{a}{b}.$$

Since any two disjoint circles can be inverted into concentric circles (Coxeter and Greitzer [1], p. 121), this expression for δ

justifies the definition for inversive distance proposed on page 265 (see also Coxeter [10]).

Consider, in Euclidean space, a uniform triangular prism. A sphere which touches all the nine edges at their midpoints contains the incircles of all the 3 + 2 faces. By stereographic projection, these incircles yield three mutually tangent circles and two disjoint circles tangent to each of those three. The inversive distance between the two disjoint circles is easily seen to be

15.93. $$\delta = 2\log(2+\sqrt{3}) = \operatorname{arg\,cosh} 7$$

(Coxeter and Greitzer [1], p. 125 with $n = 3$).

More interestingly, consider a 'loxodromic' sequence of tangent circles

$$\dots, \sigma_{-1}, \sigma_0, \sigma_1, \sigma_2, \dots$$

(Coxeter [11], pp. 112, 176–177) such that every four consecutive members are mutually tangent, as in Fig. 15.9A. Let δ_ν denote the inversive distance between σ_0 and σ_ν (or between σ_μ and $\sigma_{\mu+\nu}$ for any μ). Then it can be proved that $\cosh \delta_\nu$ is always an integer. In terms of the Fibonacci numbers

$$f_{-1} = 1, f_0 = 0, f_\mu = f_{\mu-1} + f_{\mu-2},$$

we have

15.94. $$\cosh \delta_\nu = \sum_{\mu=0}^{[\nu/2]} \binom{\nu}{2\mu} f_{\nu-\mu-2}.$$

Thus

$$\cosh \delta_1 = f_{-1} = 1,$$

$$\cosh \delta_2 = f_0 + f_{-1} = 0 + 1 = 1,$$

$$\cosh \delta_3 = f_1 + 3f_0 = 1 + 0 = 1,$$

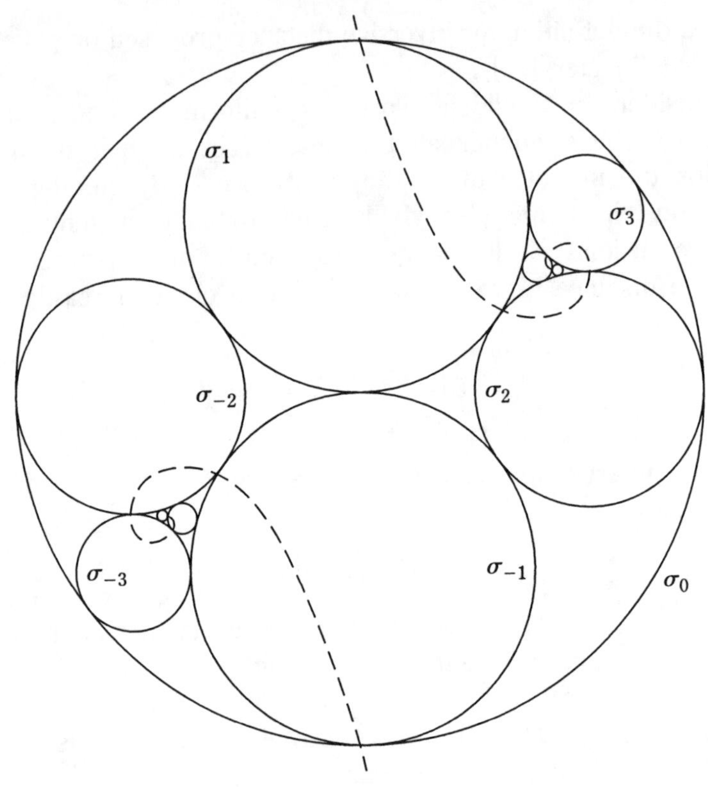

FIG. 15.9A

$$\cosh \delta_4 = f_2 + 6f_1 + f_0 = 1 + 6 + 0 = 7,$$

$$\cosh \delta_5 = f_3 + 10f_2 + 5f_1 = 2 + 10 + 5 = 17,$$

$$\cosh \delta_6 = f_4 + 15f_3 + 15f_2 + f_1 = 3 + 30 + 15 + 1 = 49,$$

and so on (Coxeter [11], pp. 117–118).

We see from page 191 that, for any two ultra-parallel lines in the hyperbolic plane, the shortest distance from a point on one to a point on the other is along the unique common perpendicular. We naturally call this the *distance* between the two lines, as in 10.72 on page 208. In other words, the length of any line segment **AB** can be measured as the distance between two ultra-parallel lines, one through **A** and one through **B**, perpendicular to the line **AB**.

In the conformal models (see page 262), instead of using cross ratios, Klein and Poincaré could more easily have defined the length of a segment **AB** as the inversive distance between the circles representing these two ultra-parallel lines.

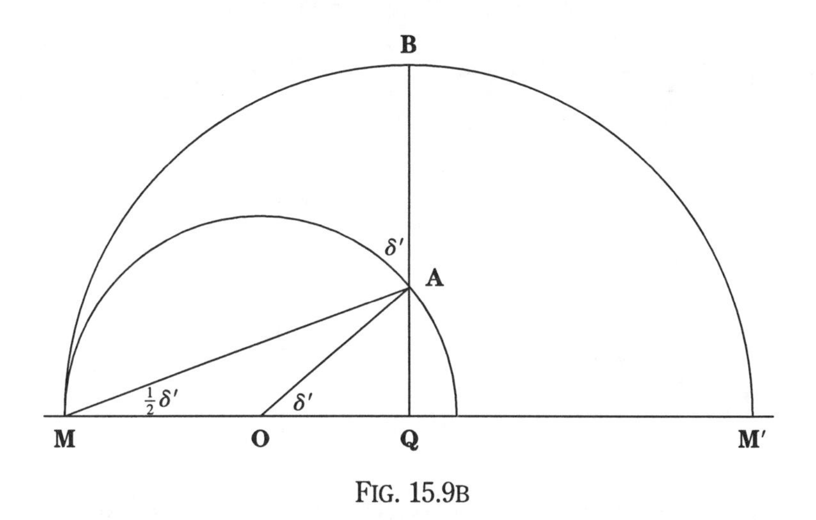

FIG. 15.9B

One spectacular application of this definition of length is P. Szász's method (Coxeter [1], p. 82) for establishing Lobatschewsky's formula

15.95. $$\delta' = 2 \arctan e^{-\delta}$$

where $\delta' = \Pi(\delta)$ is the *angle of parallelism* for the segment δ, that is, the angle between the lines **AB** and **AM**, perpendicular and parallel (respectively) to the line **BM**, as in Fig. 15.9B. This being the half-plane model, δ is represented by the part **AB** of the vertical radius **QB** of the semicircle that represents the line **BM**. The horizontal diameter **MM′** is the absolute line, whose points are 'at infinity'. The line **AM**, parallel to **BM**, is represented by another semicircle, with centre **O** and radii **OM** and **OA**. The angle of parallelism appears as the angle at **A** in the curvilinear triangle **BAM**, and again as $\angle$**QOA**. Since δ is the inversive distance between concentric circles with radii **QB** and **QA** (the latter not drawn), we have

$$e^{\delta} = \frac{\mathbf{QB}}{\mathbf{QA}} = \frac{\mathbf{QM}}{\mathbf{QA}} = \cot \tfrac{1}{2}\delta',$$

which is equivalent to 15.95.

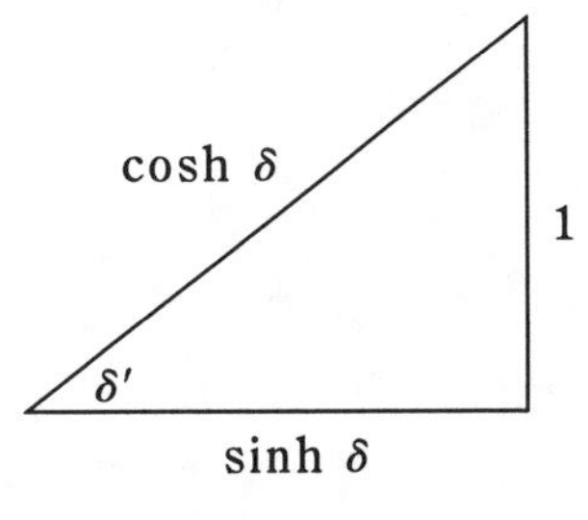

FIG. 15.9C

The right-angled triangle **AOQ** suggests a useful mnemonic (Fig 15.9C) for the six circular (i.e. trigonometrical) functions of δ' which are equal, in a peculiar order, to the six hyperbolic functions of δ (see pages 9, 208, 277): the sides of the triangle illustrate the theorem of Pythagoras in the form

$$1 + \sinh^2 \delta = \cosh^2 \delta,$$

and *the angle opposite to the side of length* 1 *is* δ'.

Comparing the projective and inversive aspects of hyperbolic geometry, we may declare that the hyperbolic plane is part of the real *projective* plane with a metric introduced by an 'absolute' conic, and it is also part of the real *inversive* plane with a different metric introduced by an 'absolute' circle (or line, since, in the vocabulary of inversive geometry, a line is simply another circle). Although we have found it convenient to describe the real inversive plane, with its points and circles, in terms of Euclidean distance, a purely axiomatic treatment is feasible (Ewald [1], p. 357; Mirsky [1], p. 40). Orthogonality can be defined by declaring two circles to be *orthogonal* if one of them belongs to a triad of mutually tangent circles while the other is determined by the three points of contact. Then inversion can be defined by saying that two points are *inverse* with respect to a circle σ if they are the intersection of two distinct circles orthogonal to σ.

Consider, in the Euclidean plane, a sequence of concentric circles whose radii are in geometric progression. Each circle, interchanging its two neighbours by inversion, is the mid-circle of those neighbours. From the inversive standpoint, the circles are evenly spaced with respect to inversive distance, and this property is invariant for inversion (Coxeter [1], p. 4). Thus repeated inversions enable one to multiply a given inversive distance by any whole number. The *mid-circle* (see page 265), which halves the inversive distance between any two disjoint circles, can be constructed as the locus of the point of contact of a variable pair of tangent circles which touch both the given circles. Repeated bisection, followed (if necessary) by a limiting process, will multiply the inversive distance by any real number, and the circle so obtained will be coaxal with the original pair of circles. Thus, to obtain a purely inversive definition for inversive distance, all that we still need is a special pair of circles whose inversive distance is precisely known. In Euclidean geometry, such a unit is provided by a pair of concentric circles with radii 1 and e; the simplest possible hyperbolic choice seems to be 15.93.

These ideas, including the above description of a loxodromic sequence (Fig. 15.9A and equation 15.94) belong to hyperbolic geometry as well as to Euclidean: they are 'absolute' in the sense of Bolyai (see page 17) and so are valid also on a sphere, like the sphere that led to 15.93.

The Euclidean plane, on which Fig. 15.9A is drawn, must be regarded as having been rendered inversive by the adjunction of a single point at infinity which lies on both the upper and lower parts of the curve (drawn as a broken line) which passes through the points of contact of σ_μ and $\sigma_{\mu+1}$ for all values of μ. This curve begins and ends at the two points of accumulation where the circles seem to become infinitesimal (although from the standpoint of inversive geometry they are all alike)! By applying two suitable inversions we can place this arrangement of circles on a geographical sphere in such a position that the two 'points of accumulation' are antipodal, say at the north and south poles. Now the curve becomes a loxodrome (or 'rhumb line') cutting all the meridians at the same angle ϕ, approximately $64°37'$ (Coxeter [11], pp. 115–116), and still passing through the points of contact of consecutive pairs of circles. Any one of the circles can be named σ_0, but it is natural to choose the circle whose centre lies on the equator. Then the remaining circles σ_ν have their centres in the northern or southern hemisphere according to the sign of ν.

APPENDIX

ANGLES AND ARCS IN THE HYPERBOLIC PLANE

H.S.M. Coxeter

Dedicated to H.G. Forder on his 90th birthday

1. Introduction. This essay may be regarded as a sequel to Chapter V of H.G. Forder's charming little book, *Geometry* [Forder **8***, pp. 80–95], which is summarized in Sections 2 and 3. Section 4 gives a fuller account of Lobachevsky's approach to hyperbolic trigonometry. Section 5 contains a simple proof that the natural unit of measurement is the length of a horocyclic arc such that the tangent at one end is parallel to the diameter at the other end. Section 6 frees Poincaré's half-plane model from its dependence on cross-ratio, by introducing a more natural concept: the *inversive distance* between two disjoint circles. Section 7 shows how an equidistant-curve of altitude a is represented by lines crossing the *absolute* line at angles $\pm\Pi(a)$. Finally, Section 8 provides an easy deduction of Poincaré's formula $|dz|/y$ for the hyperbolic line-element.

Math. Chronicle 9 (1980) 17–33.

*See page 316.

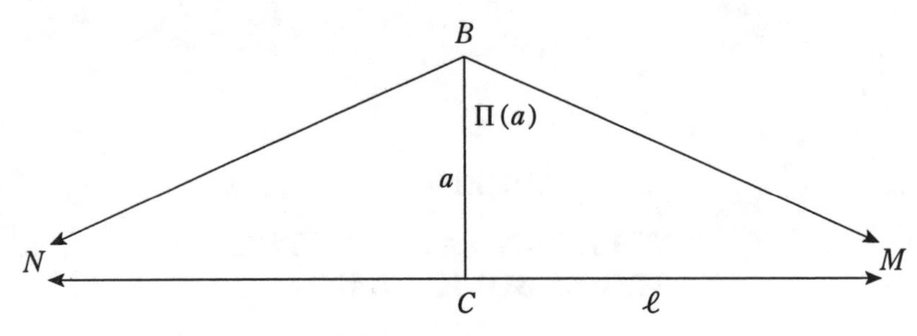

FIG. 1: Parallel rays

2. Asymptotic triangles. Felix Klein once said that "non-Euclidean geometry... forms one of the few parts of mathematics which... is talked about in wide circles, so that any teacher may be asked about it at any moment" [Klein **9**, p. 135]. The richest kind of non-Euclidean geometry is the one discovered independently about 1826 by J. Bolyai (1802–1860) and N.I. Lobachevsky (1793–1856). After 1858, when B. Riemann and L. Schläfli had discovered another kind, Klein named the old kind *hyperbolic*. This geometry shares with Euclid the first 28 propositions and many more. The first departure occurs when we consider a line ℓ and a point B not on it (see Figure 1). In the plane $B\ell$, the pencil of rays going out from B includes some that intersect the infinitely long line ℓ and others that do not. The two special rays that separate those intersecting ℓ from all the rest are said to be *parallel* to ℓ. In Euclid's geometry they are the two halves of a single line, but in hyperbolic geometry they form an angle. This angle is bisected by the perpendicular BC from B to ℓ. Lobachevsky called either half the *angle of parallelism* corresponding to the distance $a = BC$, and denoted it by $\Pi(a)$.

After proving that parallelism is an *equivalence relation* it becomes natural to describe the two rays parallel to ℓ as BM and BN where M and N are *ideal points*: the infinitely distant *ends* of the line $\ell = MN$. (It was D. Hilbert who first called them *ends*.) We speak of *asymptotic* triangles BCM, BCN, and call BMN a *doubly* asymptotic triangle. The limiting form of BMN when B recedes from C towards the end L of the ray CB is the *trebly* asymptotic triangle LMN (Figure 2): a *triangle* whose three angles are all zero while its three sides are all infinitely long!

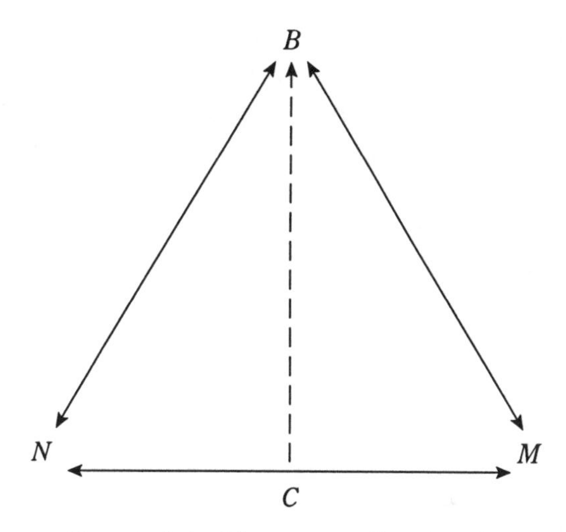

FIG. 2: A trebly asymptotic triangle

Gauss, in his famous letter of March 1832 to Bolyai's father, gave an extremely elegant proof that the area of any triangle ABC is proportional to its *angular defect* $\pi - A - B - C$ [see Coxeter **4**, pp. 297–299]. The second of the seven numbered steps in this proof is the statement that a trebly asymptotic triangle has a finite area. Admirers of Lewis Carroll [Dodgson **6**, p. 14] must face the sad fact that he rejected the possibility of hyperbolic geometry because he found it unthinkable that a triangle (or quadrangle) could retain a finite area when its sides were indefinitely lengthened.

This gap in Gauss's proof was finally closed in 1905 by Liebmann [**10**, p. 54; see also Coxeter **4**, p. 295]. For Liebmann saw how to dissect a finite quadrangle into an infinite sequence of small pieces that can be reassembled to form an asymptotic triangle.

3. The horocycle and the horosphere. Lobachevsky announced his newly discovered world in a lecture at Kazan on the 12th of February, 1826, which he published later as a little book, *Geometric investigations on the theory of parallel lines.* The Dover edition of Bonola's *Non-Euclidean Geometry* [Bonola **1**] includes, as an Appendix, G.B. Halsted's translation of that little book. There we can see how cleverly Lobachevsky derived hyperbolic trigonometry from spherical trigonometry (which is independent of considerations of parallel lines).

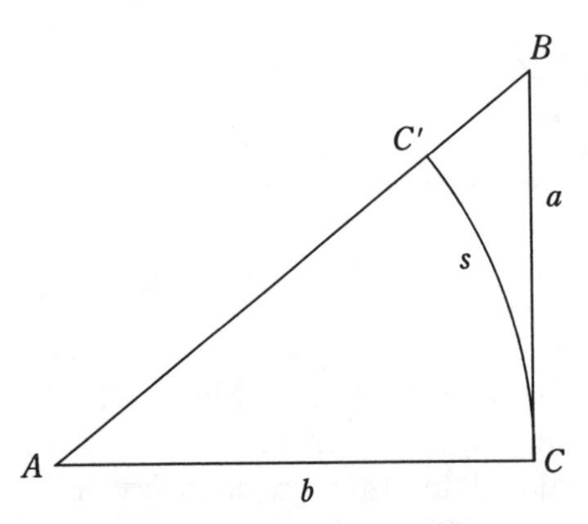

FIG. 3: An arc of a circle

Figure 3 shows an arc $s = CC'$ of a circle of radius $b = AC = AC'$. When C stays fixed while b tends to infinity, so that the centre A tends to the *end* N of the ray CA, the limiting form of the circle is an interesting curve called a *limiting*

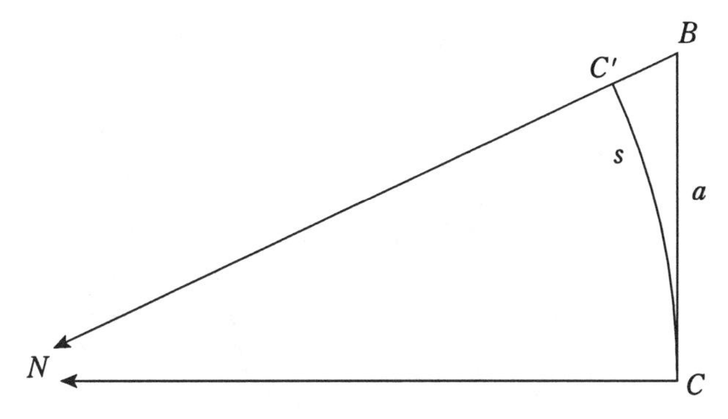

FIG. 4: An arc of a horocycle

curve or *boundary line* or *horocycle* (Figure 4), whose *diameters* form a pencil of parallel lines having N as their common *end* [Forder **8**, p. 85; Coxeter **4**, p. 300]. (The term *horocycle* has the same root as *horizon.*) In three-dimensional space, continuous rotation about the diameter NC yields a surface called a *horosphere,* which is the limiting form of a sphere whose centre recedes to infinity. All the lines through N, *diameters* of the horosphere, form a bundle of parallel lines, and each plane through any one of these lines cuts the horosphere along a horocycle.

Many of Lobachevsky's results are derived from his observation that the intrinsic geometry of the horosphere is Euclidean: the set of horocycles on the horosphere can be represented isometrically by the set of straight lines in the ordinary Euclidean plane. We shall find a modern proof in Section 6 [see also Liebmann **10**, p. 61; Coxeter **4**, p. 304].

4. Lobachevsky's asymptotic orthoscheme. In 1620, John Napier, Baron of Merchiston, obtained the $\binom{5}{3} = 10$ formulae which connect (in threes) the five *parts* of a right-angled spherical triangle ABC (right-angled at C): namely the three sides a, b, c and the acute angles A, B. For instance:

$$\cos c = \cos a \cos b = \cot A \cot B$$

[Donnay **7**, p. 40]. What Lobachevsky did was to base a special kind of asymptotic tetrahedron $BCAN$ on a hyperbolic triangle ABC by erecting a ray AN perpendicular to the plane ABC and completing the tetrahedron with parallel rays BN and CN so that the three edges BC, CA, AN are mutually perpendicular, BC is perpendicular to the plane ACN, and

$$\angle CBN = \Pi(a), \ \angle ACN = \Pi(b), \ \angle ABN = \Pi(c).$$

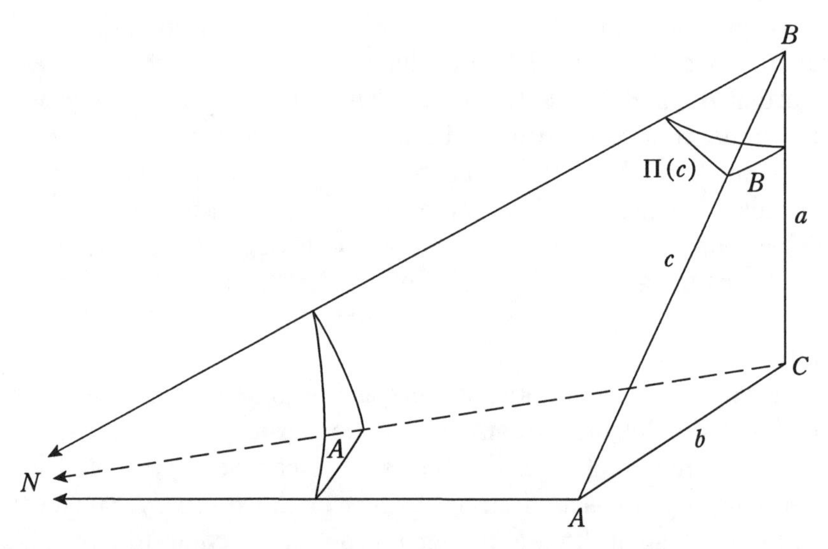

FIG. 5: An orthoscheme $BCAN$ with N at infinity

This tetrahedron $BCAN$ (Figure 5) is called an asymptotic *orthoscheme* [Coxeter **4**, p. 156].

Lobachevsky observed that the three planes, NBC, NCA, NAB, whose lines of intersection are parallel, cut out, from any horosphere with centre N, a ***horospherical triangle*** which is isometric to a Euclidean triangle. Since the first two of these three planes are perpendicular, this triangle is right-angled; therefore its two acute angles, which are the dihedral angles along the edges NA and NB of the orthoscheme, are complementary. But the dihedral angle along NA is equal to the angle A of the hyperbolic triangle ABC; therefore the dihedral angle along NB is $\frac{1}{2}\pi - A$.

It follows that the three planes BCA, BAN, BNC cut out, from a suitable ***sphere*** with centre B, a right-angled ***spherical*** triangle with sides B, $\Pi(c)$, $\Pi(a)$ and acute angles $\frac{1}{2}\pi - A$, $\Pi(b)$. In this way, Napier's ten spherical formulae (which remain valid in hyperbolic space) yield a corresponding set of ten hyperbolic formulae. For instance, Napier's

$$\cos c = \cot A \cot B$$

yields

$$\cos \Pi(a) = \tan A \cot \Pi(b). \tag{4.1}$$

At this stage, $\Pi(x)$ remains an unknown function of x, decreasing from $\Pi(0) = \frac{1}{2}\pi$ to $\Pi(\infty) = 0$, but after some rather formidable work Lobachevsky finds

$$\tan \tfrac{1}{2}\Pi(x) = e^{-x}. \tag{4.2}$$

At first he allows this e to be any positive number, but then he chooses a natural unit of measurement by identifying e with the base of Naperian logarithms.

Many simpler proofs of this famous formula have been discovered since Lobachevsky's time [for instance, Carslaw **2**, p. 109; Coxeter **4**, p. 453, Figure 16.7a; **5**, p. 395]. It obviously implies

$$\cos \Pi(x) = \tanh x, \ \cot \Pi(x) = \sinh x;$$

hence (4.1) becomes, in modern notation,

(4.3) $$\tanh a = \tan A \sinh b,$$

analogous to Napier's spherical formula $\tan a = \tan A \sin b$.

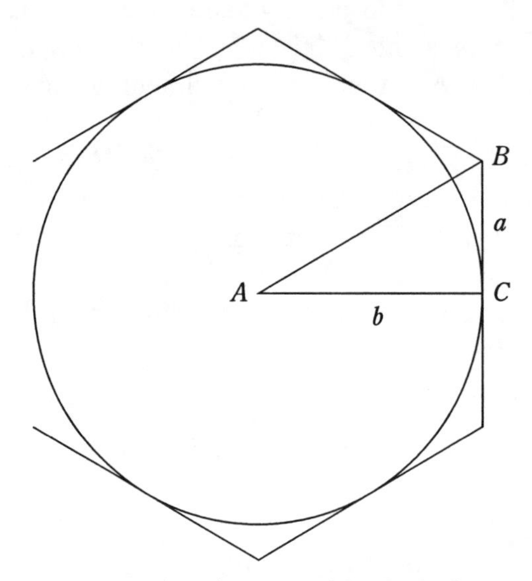

FIG. 6: A polygon and its incircle

5. The unit horocyclic arc. With this background, we can pass easily to some further results concerning circles and horocycles.

Consider (in the hyperbolic plane) a regular n-gon with centre A and an inscribed circle of radius $b = AC$, as in Figure 6. A typical vertex B forms, with A and C, a right-angled triangle ABC in which the side $a = BC$ is half a side of the polygon. By (4.3), in the form

$$\tanh a = \tan(\pi/n) \sinh b,$$

the perimeter of the n-gon is

$$2na = 2\pi \frac{a}{\tanh a} \frac{\tan(\pi/n)}{\pi/n} \sinh b.$$

We may now make a tend to zero and n to infinity (while fixing b) and deduce that the *circumference* of a circle of radius b is

$$2\pi \sinh b.$$

Consequently the *area* of a circle of radius r is

$$2\pi \int_0^r \sinh b \, db = 2\pi(\cosh r - 1) = 4\pi \sinh^2 \frac{r}{2}.$$

Restricting attention to what happens inside the angle A, as in Figure 3, we deduce that a circular *arc* of radius b subtending an angle A has length

$$s = A \sinh b \tag{5.1}$$

and that a *sector* of radius b and angle A has area

$$2A \sinh^2 \frac{b}{2} = s \tanh \frac{b}{2} \tag{5.2}$$

We see from (4.3) and (5.1) that

$$s = \frac{A}{\tan A} \tanh a.$$

Keeping BC fixed while pushing A away from C (to the *end* N, as in Figure 4) so that b tends to infinity and A to zero, we deduce that, for a horocyclic arc with *tangent* a,

$$s = \tanh a. \tag{5.3}$$

Also (5.2) shows that *the area of a horocyclic sector is equal to its arc* [Coxeter **5**, p. 250].

Keeping NC fixed while pushing B away from C (to the *end* L, as in Figure 7) so that a tends to infinity and $\tanh a$ to 1, we deduce that a horocyclic arc CC' has length 1 when *the tangent at* C *is parallel to the diameter through* C' [Carslaw **2**, p. 119; Coxeter

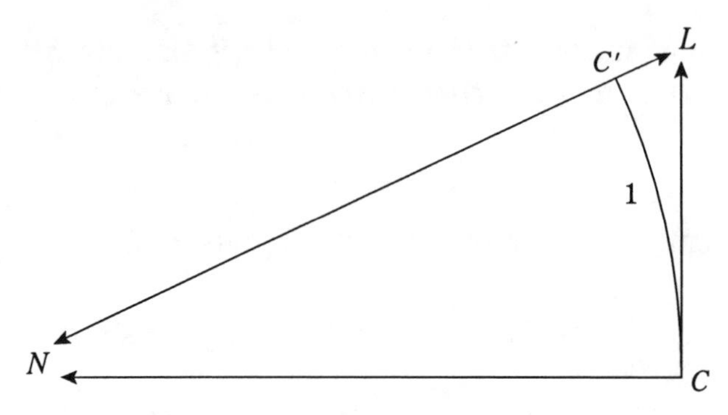

FIG. 7: A horocyclic arc of length 1

3, p. 250]. Also the area of this horocyclic sector NCC' is equal to 1. (An amusing consequence is that the curvilinear triangle formed by three mutually tangent horocyclic arcs of length 1 has area $\pi - 3$.)

6. Circles and cycles. Along with circles and horocycles, it is natural to consider also *hypercycles* or *equidistant-curves*. Such a curve is the locus of a point B at a constant distance a from a fixed line ℓ. We call a the *altitude* of the equidistant-curve, and ℓ the *axis*. Since B may be on either side of the axis, this curve (like a hyperbola) has two separate branches. However, since the isosceles birectangle $BB'C'C$ in Figure 8 has acute angles at B and B', both branches are concave as seen from any point between them. By varying a while keeping ℓ fixed, we obtain a pencil of *coaxial* equidistant-curves, which may be described as the orthogonal trajectories of the pencil of *ultraparallel* lines perpendicular to ℓ [Coxeter **4**, p. 300]. Analogously, *concentric horocycles* having a common centre N (at infinity) are the orthogonal trajectories of the pencil of *parallel* lines through N, and concentric *circles* are the orthogonal trajectories of a pencil of *intersecting* lines.

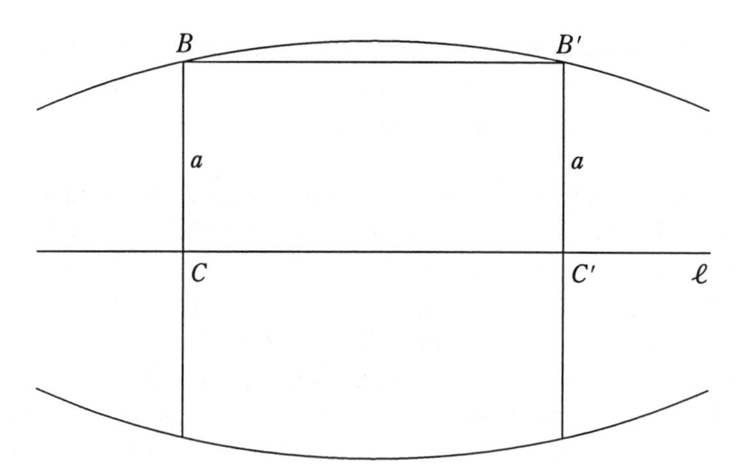

FIG. 8: An equidistant-curve of altitude a

The investigation of these *cycles* is facilitated by using one of Poincaré's Euclidean models for the hyperbolic plane. In this model the *ends* appear as the points of a horizontal line Ω, and the hyperbolic lines are represented by the circles and lines orthogonal to Ω, that is, by the circles with their centres on Ω, and the vertical lines. Since this model is conformal, the angles between two intersecting lines of the hyperbolic plane agree with the angles of intersection of the representative circles. The point of intersection of the two lines thus appears as a pair of points which are images of each other by reflection in Ω, or as a single point in the upper half-plane (if we choose to disregard the lower half-plane and represent lines by semicircles and vertical rays).

The length of a hyperbolic line-segment AB can be measured as the distance between lines through A and B perpendicular to AB, and this is the *inversive distance* between the representative circles: the natural logarithm of the ratio of the radii (greater to smaller) of any two *concentric* circles into which these representative circles can be inverted [Coxeter **4**, pp. 90, 303]. The inversive distance tends to zero when the circles approach tangency, and tends to infinity when one of the circles shrinks to a point, which is what happens when A (or B) becomes an *end*.

A coaxial pencil of disjoint circles having Ω for radical axis, being orthogonal to a pencil of intersecting circles with their centres *on* Ω, represents a pencil of concentric circles of the hyperbolic plane. As a limiting case, a coaxial pencil of *tangent* circles, touching Ω at a point N, being orthogonal to the pencil of circles touching the vertical line through N, represents a pencil of concentric *horocycles.* Inverting the Euclidean plane in a circle with centre N, we see that another pencil of concentric horocycles is represented by the pencil of lines *parallel* to Ω (in the Euclidean sense), that is, by all the horizontal lines except Ω itself. The vertical lines, being the diameters of these horocycles, are parallel in the hyperbolic sense as well as in the Euclidean sense. (The reader may wonder how their common *end* can lie on Ω; but we must remember that the plane of the representation, which we have called *Euclidean*, is more precisely an *inversive* plane, which is the Euclidean plane augmented by *just one* point at infinity, lying on *every* line, horizontal or vertical or oblique.) All the horocycles are symmetrical by reflection in each of these diameters; therefore any discrete set of vertical lines, evenly spaced, represents a set of diameters cutting out a set of points evenly spaced on any one of the horocycles. It follows that distances along the horocycle are proportional to the corresponding distances along the representative Euclidean line.

Figure 9 shows the Poincaré model for Figure 7. $OCC'L$ is a square of side 1 with O and L on Ω; the arc LC is a quadrant of a circle with centre O. As we have seen, the horizontal line CC' represents a horocycle. Since the arc LC, representing the *tangent* at C, touches the line $C'L$, which represents the *diameter* at C', this tangent and diameter are parallel. Hence, by

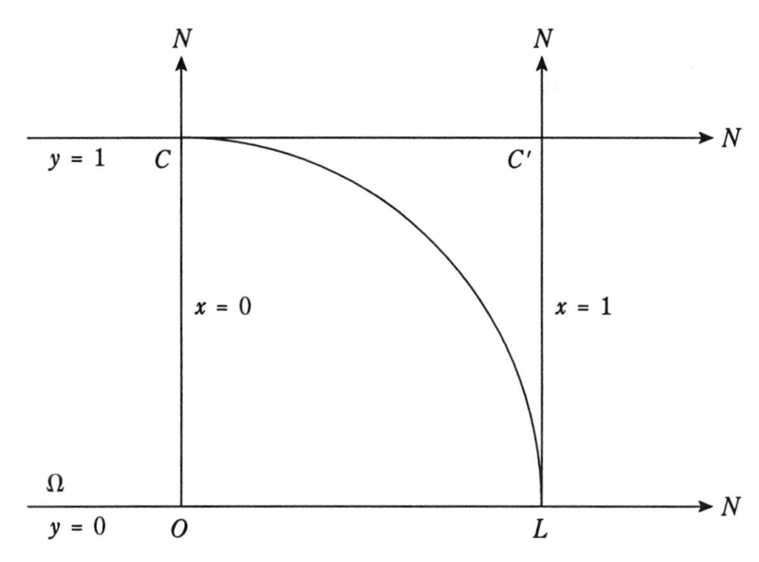

FIG. 9: Poincaré's version of Figure 7

Carslaw's theorem which was quoted near the end of Section 5, the horocyclic arc CC' is of length 1, so that, on this particular horocycle, hyperbolic distance is not merely proportional to Euclidean distance but equal to it. Rotating the whole figure continuously about the vertical line OC, we deduce the isometric representation of the horosphere on a Euclidean plane, which is Lobachevsky's observation mentioned at the end of Section 3.

7. The altitude of an equidistant-curve. A coaxial pencil of intersecting circles through two points M and N on Ω, being orthogonal to the pencil of disjoint circles having M and N for limiting points, represents a pencil of coaxial equidistant-curves whose axis is represented by the circle on MN as diameter. Inverting in a circle with centre N, we see that another pencil of coaxial equidistant-curves is represented by an ordinary pencil of lines, all passing through M', the inverse of

M. Thus any oblique line Γ, along with its image by reflection in Ω (see Figure 10) represents an equidistant-curve [Siegel **12**, p. 25].

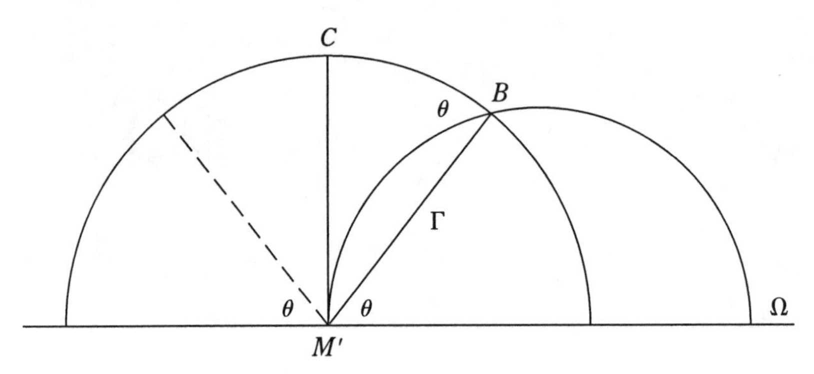

FIG. 10: Poincaré's version of Figure 8

It is natural to ask how the acute angle θ between Ω and Γ is related to the altitude a of the equidistant-curve. For this purpose, let any circle through M', with its centre on Ω, meet Γ again in B, so that $M'B$ is a chord, and let the circle through B with centre M' meet the vertical line through M' in C. The two arcs $M'B$ and BC obviously form an angle θ at their intersection B. Comparing Figure 10 with Figure 1, we see that

$$\theta = \Pi(a). \tag{7.1}$$

This shows that a is independent of the position of B on Γ, in agreement with our definition of the equidistant-curve.

8. The hyperbolic line-element. Taking Ω to be the x-axis of Cartesian coordinates, as in Figure 9 or Figure 11, we may represent any hyperbolic line ℓ by the y-axis, any point on this line by $(0, y)$, and a neighbouring point by $(0, y + dy)$. Circles

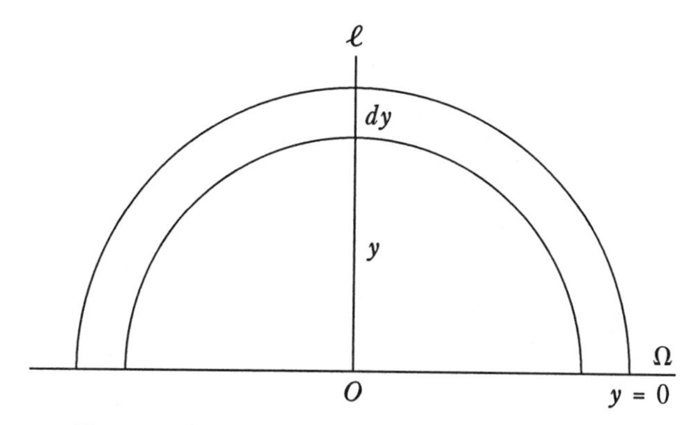

FIG. 11: An application of concentric circles

(or semicircles) with centre O represent lines perpendicular to ℓ, and the hyperbolic distance ds from $(0, y)$ to $(0, y + dy)$ can be measured as the inversive distance between these two circles. Since the circles are concentric, and dy is infinitesimal, this inversive distance is simply the logarithm of the ratio of their radii:

$$ds = \log \frac{y + dy}{y} = \log \left(1 + \frac{dy}{y}\right) = \frac{dy}{y}.$$

Since Poincaré's model is conformal, local measurements are the same in all directions. We have thus obtained his famous formula

(8.1) $$ds = \sqrt{dx^2 + dy^2}/y$$

[Poincaré **11**, p. 7; Liebmann **10**, p. 149]. In particular, along the horocycle $y = 1$,

$$ds = dx;$$

and along the equidistant-curve with polar equation (7.1), the arc from (r_1, θ) to (r_2, θ), corresponding to a distance b along the axis, is

$$\int_{r_1}^{r_2} \frac{dr}{r \sin \theta} = \log \frac{r_2}{r_1} \operatorname{cosec} \theta = b \cosh a,$$

where a is the altitude.

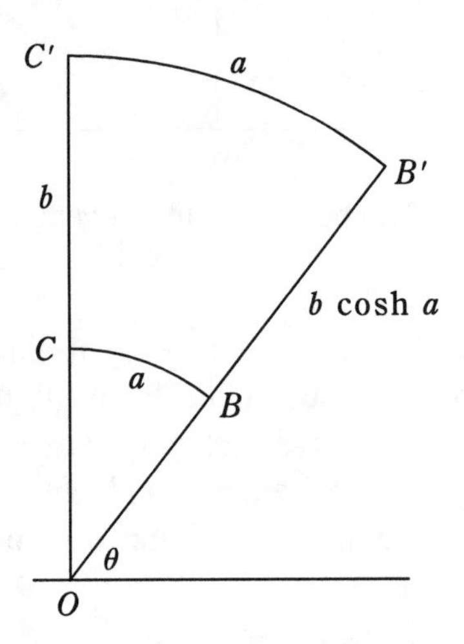

FIG. 12: A *sector* of an equidistant-curve

Jan van de Craats has remarked that the same kind of coordinates may by used to find the area of the *sector* $BB'C'C$ of an equidistant-curve, bounded by the arc BB', two diameters, and the *base* $b = CC'$ along the axis (see Figure 8). The element of length (8.1) yields the element of area

$$dx\, dy/y^2,$$

which is $dr\, d\phi / r \sin^2 \phi$ for polar coordinates (r, ϕ). Hence the area of $BB'C'C$ (Figure 12) is

$$\int_{r_1}^{r_2} \int_{\theta}^{\pi/2} \frac{dr\, d\phi}{r \sin^2 \phi} = \log \frac{r_2}{r_1} \cot \theta = b \sinh a.$$

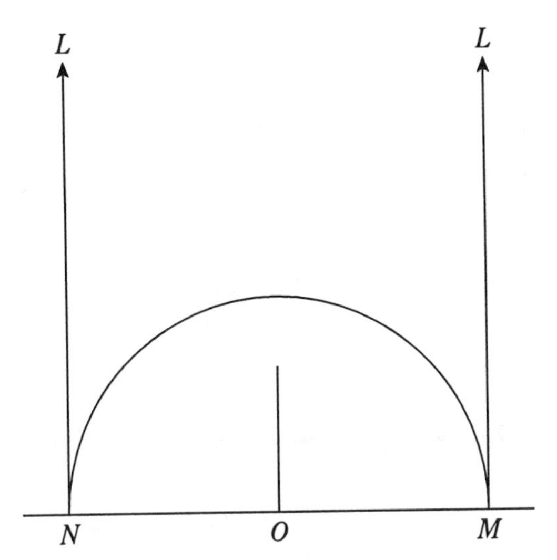

FIG. 13: A trebly asymptotic triangle

At the end of Section 2 we tried to deal with Gauss's assumption that a trebly asymptotic triangle has a finite area. Coordinates provide a quick verification if we represent such a triangle by the strip LMN above the circle $r = 1$, bounded by the ordinates $x = \pm 1$ (see Figure 13). Thus the area is

$$\int_{-1}^{1} \int_{\sqrt{1-x^2}}^{\infty} \frac{dx\,dy}{y^2} = \int_{-1}^{1} \frac{dx}{\sqrt{1-x^2}} = \pi.$$

References

1. Roberto Bonola, *Non-Euclidean Geometry*, Dover, New York, 1955.
2. H.S. Carslaw, *The Elements of Non-Euclidean Plane Geometry and Trigonometry*, Longmans-Green, London, 1916.
3. H.S.M. Coxeter, *Non-Euclidean Geometry*, 5th ed., University of Toronto Press, 1965.
4. H.S.M. Coxeter, *Introduction to Geometry*, 2nd ed., Wiley, New York, 1969.

5. H.S.M. Coxeter, *Parallel lines*, Canad. Math. Bulletin, 21.4 (1978), 385–397.
6. C.L. Dodgson, *Curiosa Mathematica*, 3rd ed., London, 1890.
7. J.D.H. Donnay, *Spherical Trigonometry*, Interscience, New York, 1945.
8. H.G. Forder, *Geometry*, Hutchinson, London, 1950.
9. Felix Klein, *Elementary Mathematics from an Advanced Standpoint*, vol. 2, Macmillan, New York, 1939.
10. Heinrich Liebmann, *Nichteuklidische Geometrie*, 2nd ed., Göschen, Berlin and Leipzig, 1912.
11. Henri Poincaré, *Théorie des groupes fuchsiens*, Acta Math. 1, (1883), 1–62.
12. C.L. Siegel, *Topics in complex function theory*, vol. 2, Wiley-Interscience, New York, 1971.

University of Toronto

BIBLIOGRAPHY (FOR THE WHOLE BOOK)

Baker [1]. *Principles of Geometry*: vol. I, Cambridge, England, 1929; vol II, 1930.

Baldus [1]. *Nichteuklidische Geometrie*, Leipzig, 1927.

Bellavitis [1]. "Saggio di geometria derivata," *Nuovi Saggi dell' Accademia di Padova*, 4 (1838).

Beltrami [1]. "Risoluzione del problema di riportare i punti di una superficie sopra un piano in modo che le linee geodetiche vengano rappresentate da linee rette," *Annali di Matematica* (1), 7 (1866), pp. 185–204.

——— [2]. "Saggio di interpretazione della geometria noneuclidea." *Giornale di Matematiche*, 6 (1868), pp. 284–312.

——— [3]. "Teoria fondamentale degli spazii di curvatura costante," *Annali di Matematica* (2), 2 (1868), pp. 232–255.

Bolyai [1]. *Tentamen juventutem studiosam in elementa matheseos purae, elementaris ac sublimioris methodo intuitiva evidentiaque huic propria, introducendi*, Maros-Vásárhelyini, 1832.

——— [2]. *Appendix scientiam spatii absolute veram exhibens: a veritate aut falsitate Axiomatis XI Euclidei (a priori haud unquam decidenda) independentem*, translated into English by G. B. Halsted, Austin, Texas, 1891.

Bonnesen [1]. "Geometriske konstruktioner på kuglefladen." *Nyt Tidsskrift for Matematik*, afd. B, 10 (1899), pp. 1–13, 25–35.

Bonola [1]. "Sulla introduzione degli enti improprii in geometria proiettiva," *Giornale di Matematiche*, 38 (1900), pp. 105–116.

——— [2]. *La geometria non-euclidea*, Bologna, 1906. References are made to the English translation by H. S. Carslaw, Chicago, 1912.

Boole [1]. "Notes on quaternions," *Philosophical Magazine* (3), 33 (1848), pp. 278–280.

Bose [1]. "On a new derivation of the fundamental formulae of hyperbolic geometry." *Tôhoku Mathematical Journal*, 34 (1931), pp. 291–294.

Busemann and Kelly [1]. *Projective Geometry and Projective Metrics*, New York, 1953.

Carslaw [1]. *The Elements of Non-Euclidean Plane Geometry and Trigonometry*, London, 1916.

Cartan [1]. *Leçons sur la Géométrie Projective Complexe*, Paris, 1931.

Cayley [1]. "On certain results relating to quaternions," *Philosophical Magazine* (3), 26 (1845), pp. 141–145.

——— [2]. "Recherches ultérieures sur les déterminants gauches," *Journal für die reine und angewandte Mathematik*, 50 (1855), pp. 299–313.

——— [3]. "A sixth memoir upon quantics," *Philosophical Transactions of the Royal Society of London*, 149 (1859), pp. 61–90.

——— [4]. Presidential Adress, *British Association for the Advancement of Science*, 53 (Southport, 1883), pp. 3–37.

Chasles [1]. *Traité de Géométrie Supérieure* (second edition), Paris, 1880.

——— [2]. *Traité des Sections Coniques*, Paris, 1865.

Clifford [1]. "Preliminary sketch of biquaternions," *Proceedings of the London Mathematical Society*, 4 (1873), pp. 381–395.

Coolidge [1]. *The Elements of Non-Euclidean Geometry*, Oxford, 1910.

Coxeter [1]. "The functions of Schläfli and Lobatschefsky," *Quarterly Journal of Mathematics*, 6 (1935), pp. 13–29.

——— [2]. "On Schläfli's generalization of Napier's Pentagramma Mirificum," *Bulletin of the Calcutta Mathematical Society*, 28 (1936), pp. 123–144.

——— [3]. "Excenter in hyperbolic geometry," *American Mathematical Monthly*, 51 (1944), pp. 600–601.

——— [4]. "Regular honeycombs in elliptic space," *Proceedings of the London Mathematical Society* (3), 4 (1954), pp. 471–501.

——— [5]. "Arrangement of equal spheres in non-Euclidean spaces," *Acta Mathematica Academiae Scientiarum Hungaricae*, 5 (1954), pp. 263–274.

——— [6]. *The Real Projective Plane* (third edition), New York, 1992.

——— [7]. "The affine plane," *Scripta Mathematica*, 21 (1955), pp. 5–14.

Coxeter [8]. "Hyperbolic triangles," *Scripta Mathematica*, 22 (1956), pp. 5–13.

——— [9]. *Introduction to Geometry*, (second edition) New York 1969.

——— [10]. "Inversive distance," *Annali di Matematica* (4), 71 (1966), pp. 73–83.

——— [11]. "Loxodromic sequences of tangent spheres," *Aequationes Mathematicae*, 1 (1968), pp. 104–121.

——— [12]. "Mid-circles and loxodromes," *Mathematical Gazette*, 52 (1968), pp. 1–8.

——— [13]. "Angles and arcs in the hyperbolic plane," *Mathematics Chronicle*, 9 (1980), pp. 17–33.

Cremona [1]. *Elements of Projective Geometry*, London, 1913.

Dedekind [1]. *Stetigkeit und irrationalen Zahlen*, Braunschweig, 1872. (There is an English translation by W. W. Beman, Chicago, 1901.)

Dehn [1]. "Die Legendre'schen Satze über die Winkelsumme im Dreieck," *Mathematische Annalen*, 53 (1900), pp. 404–439.

Donkin [1]. "On the geometrical interpretation of quaternions," *Philosophical Magazine* (3), 36 (1850), pp. 489–503.

——— [2]. "On the geometrical theory of rotation," *Philosophical Magazine* (4), 1 (1851), pp. 187–192.

Engel [1]. "Zur nichteuklidischen Geometrie," *Bericht über die Verhandlungen der Königlich sächsischen Gesellschaft der Wissenschaften zu Leipzig, Math.-Naturwiss. Classe*, 50 (1898), pp. 181–191.

Enriques [1]. *Lezioni di Geometria Proiettiva*, Bologna, 1904. References are made to the French translation by Paul Labérenne, Paris, 1930.

Ewald [1]. "Axiomatischer Aufbau der Kreisgeometrie," *Mathematische Annalen*, 131 (1956), pp. 354–371.

Fog [1]. "Om Konstruktion med Passeren alene," *Matematisk Tidsskrift* A (1935), pp. 16–24.

Forder [1]. *The Foundations of Euclidean Geometry*, Cambridge, England, 1927.

——— [2]. "On gauge constructions," *Mathematical Gazette*, 23 (1939), pp. 465–467.

Forder [3]. "On gauge constructions and a letter of Hjelmslev," *Mathematical Gazette*, 37 (1953), pp. 203–205.

Gauss [1]. *Werke*, vol. VIII, Göttingen, 1900.

Goursat [1]. "Sur les substitutions orthogonales et les divisions régulières de l'espace," *Annales Scientifiques de l'École Normale Supérieure* (3), 6 (1889), pp. 9–102.

Hamilton [1]. "A new species of imaginary quantities," *Proceedings of the Royal Irish Academy*, 2 (1843), pp. 424–434.

——— [2]. "On quaternions; or, on a new system of imaginaries in algebra," *Philosophical Magazine* (3), 25 (1844), pp. 490–495.

——— [3]. "Researches respecting quaternions," *Transactions of the Royal Irish Academy*, 21 (1848), pp. 199-296.

——— [4]. *Lectures on Quaternions*, Dublin, 1853.

Handest [1]. "Constructions in hyperbolic geometry," *Canadian Journal of Mathematics*, 8 (1956), pp. 389–394.

Heffter and Koehler [1]. *Lehrbuch der analytischen Geometrie*: vol. I, Leipzig, 1905; vol. II (by Heffter alone), 1923.

Hesse [1]. "De curvis et superficiebus secundi ordinis," *Journal für die reine und angewandte Mathematik*, 20 (1840), pp. 285–308.

Hessenberg [1]. "Ueber einen geometrischen Calcül," *Acta Mathematica*, 29 (1905), pp. 1–23.

Hilbert [1]. *Grundlagen der Geometrie*, Leipzig, 1913.

Hilbert and Bernays [1]. *Grundlagen der Mathematik*, vol. I, Berlin, 1934.

De la Hire [1]. *Sectiones conicae*, Paris. 1685.

Hjelmslev [1]. "Neue Begrundung der ebenen Geometrie," *Mathematische Annalen*, 64 (1907), pp. 449–474.

Hobson [1]. *A Treatise on Plane Trigonometry*, Cambridge, England, 1925.

Holgate [1]. *Projective Pure Geometry*, New York, 1934.

Jessen [1]. "Om Konstruktion med Passer og Lineal," *Matematisk Tidsskrift* A (1944), pp. 25–44.

Johnson [1]. "Directed angles in elementary geometry," *American Mathematical Monthly*, 24 (1917), pp. 101–105.

Johnson [2]. *Modern Geometry*, Cambridge, Mass., 1929.

Klein [1]. "Ueber die sogenannte Nicht-Euklidische Geometrie," *Mathematische Annalen*, 4 (1871), pp. 573–625.

—— [2]. "Ueber die sogenannte..." (Zweiter Aufsatz), *Mathematische Annalen*, 6 (1873), pp. 112–145.

—— [3]. *Vorlesungen über Nicht-Euklidische Geometrie*, Berlin, 1928.

Laguerre [1]. "Note sur la théorie des foyers," *Nouvelles Annales de Mathématiques*, 12 (1853), pp. 57–66.

Lieber [1]. *Non-Euclidean Geometry, or Three Moons in Mathesis*, Lancaster, Pa., 1940.

Liebmann [1]. *Nichteuklidische Geometrie*, Leipzig, 1923.

Lobatschewsky [1]. *Collection complète des œuvres géométriques de N.J. Lobatcheffsky*, Kazan, 1886.

———— [2]. *Zwei geometrische Abhandlungen*, Leipzig, 1898.

———— [3]. *Geometrical Researches on the Theory of Parallels*, translated into English by G. B. Halsted, Chicago, 1914.

Lütkemeyer [1]. *Ueber den analytischen Charakter der Integrale von partiellen Differentialgleichungen*, Göttingen, 1902.

McClelland and Preston [1]. *A Treatise on Spherical Trigonometry*: vol. I, London, 1890; vol. II, 1886.

McClintock [1]. "On the non-Euclidean geometry," *Bulletin of the New York Mathematical Society*, 2 (1893), pp. 21–33.

Maclaurin [1]. *A Treatise of Algebra*, London, 1761.

Milne [1]. *Homogeneous Coordinates*, London, 1924.

Minding [1]. "Wie sich entscheiden lässt, ob zwei gegebene krumme Flächen auf einander abwickelbar sind oder nicht; nebst Bemerkungen über die Flächen vn unveränderlichen Krümmungsmaasse," *Journal für die reine und angewandte Mathematik*, 19 (1839), pp. 370–387.

——— [2]. "Beitrage zur Theorie der kürzesten Linien auf krummen Flächen," *ibid.*, 20 (1840), pp. 323–327.

Minkowski [1]. "Space and Time," *Bulletin of the Calcutta Mathematical Society*, 1 (1909), pp. 135–141.

Möbius [1]. *Der barycentrische Calcul, ein neues Hülfsmittel zur analytischen Behandlung der Geometrie*, Leipzig, 1827.

Mohr [1]. *Euclides Danicus*, Copenhagen, 1928.

Mohrmann [1]. *Einführung in die Nicht-Euklidische Geometrie*, Leipzig, 1930.

Mokriščev [1]. "On the solvability of construction problems of second degree in the Lobatschewsky plane by means of hypercompass or compasses and horocompass," *Doklady Akademiya Nauk SSSR*, 91 (1953), pp. 453–456.

Moore [1]. "Sets of metrical hypotheses for geometry," *Transactions of the American Mathematical Society*, 9 (1908), pp. 487–512.

Mordoukhay-Boltovskoy [1]. *In memoriam Lobatschewsky* II, Kazan, 1927.

Mukhopadhyaya [1]. "Geometrical investigations on the correspondences between a right-angled triangle, a three-right-angled quadrilateral and a rectangular pentagon in hyperbolic geometry," *Bulletin of the Calcutta Mathematical Society*, 13 (1923), pp. 211–216.

Nestorowitsch [1]. "Sur l'équivalence par rapport à la construction du complexe MB et du complexe E," *Comptes Rendus (Doklady) de l'Académie des Sciences de l'URSS*, 22 (1939), pp. 224–227.

——————— [2]. "Sur la puissance constructive d'un complexe E sur le plan de Lobatchevski," *Comptes Rendus (Doklady) de l'Académie des Sciences de l'URSS*, 43 (1944), pp. 186–188.

——————— [3]. "Geometrical constructions with horocompass and rule in the Lobatschewsky plane," *Doklady Akademii Nauk SSSR*, 66 (1949), pp. 1047–1050.

——————— [4]. "On the equivalence of a hypercycle to an ordinary circle in constructions in the Lobatschewsky plane," *Doklady Akademii Nauk SSSR*, 69 (1949), pp. 731–735.

Neumann [1]. *Vorlesungen über Riemann's Theorie der Abel'schen Integrale*, Leipzig, 1884.

Owens [1]. "The introduction of ideal elements and a new definition of projective *n*-space," *Transactions of the American Mathematical Society*, 11 (1910), p. 141–171.

Pasch and Dehn [1]. *Vorlesungen über neuere Geometrie*, Berlin, 1926.

Picken [1]. "Euclidean geometry of angle," *Proceedings of the London Mathematical Society*, (2), 23 (1925), pp. 45–55.

Pieri [1]. "I principii della geometria di posizione, composti in sistema logico deduttivo," *Memorie dell'Accademia Reale di Torino* (2), 48 (1898), pp. 1–62.

Plücker [1]. *System der Geometrie des Raumes*, Düsseldorf, 1846.

——— [2]. *Neue Geometrie des Raumes*, Leipzig, 1868.

Poincaré [1]. "Théorie des groupes fuchsiens," *Acta Mathematica*, 1 (1882), pp. 1–62.

——— [2]. "Les géométries non-euclidiennes," *Revue Générale des Sciences*, 2 (1891), No. 23; English translation in *Nature*, 45 (1892), pp. 404–407.

Poncelet [1]. *Traité des Propriétés Projectives des Figures*, Paris, 1865.

Reye [1]. *Die Geometrie der Lage*, vol. II, Hannover, 1868.

Reyes y Prósper [1]. "Sur les propriétés graphiques des figures centriques," *Mathematische Annalen*, 32 (1888), pp. 157–158.

Richmond [1]. "The volume of a tetrahedron in elliptic space," *Quarterly Journal of Mathematics*, 34 (1903), pp. 175–177.

Riemann [1]. "Ueber die Hypothesen, welche der Geometrie zu Grunde liegen," *Abhandlungen der Königlichen Gesellschaft der Wissenschaften zu Göttingen*, 13 (1866), pp. 254–268. References are made to the English translation by W. K. Clifford, *Nature*, 8 (1873), pp. 14–17, 36–37.

Robb [1]. *Geometry of Time and Space*, Cambridge, England, 1936.

Robinson [1]. *The Foundations of Geometry* (Mathematical Expositions, No. 1), Toronto, 1940.

Robson [1]. *An Introduction to Analytical Geometry*, vol. I, Cambridge, England, 1940.

Russell [1]. *Principles of Mathematics*, London, 1937.

Schilling [1]. *Projektive und nichteuklidische Geometrie*, Leipzig, 1931.

——— [2]. "Die Extremaleigenschaften der ausserhalb des absoluten Kegelschnittes gelegenen Strecken in der projektiven Ebene mit hyperbolischer Geometrie," *Deutsche Mathematik*, 6 (1941), pp. 33–49.

Schläfli [1]. "On the mutiple integral $\int^n dx\,dy \ldots dz$, whose limits are $p_1 = a_1x + b_1y + \cdots + h_1z > 0$, $p_2 > 0, \ldots, p_n > 0$; and $x^2 + y^2 + \cdots + z^2 < 1$," *Quarterly Journal of Mathematics*, 2 (1858), pp. 269–301, and 3 (1860), pp. 54–68, 97–108.

Schoute [1]. *Mehrdimensionale Geometrie*, vol I, Leipzig, 1902.

Schur [1]. *Grundlagen der Geometrie*, Leipzig, 1909.

Seydewitz [1]. "Konstruktion und Klassifikation der Flächen zweiten Grades," *Archiv für Mathematik und Physik*, 9 (1848) pp. 102–158.

Smogorschevsky [1]. *Geometrical constructions in the Lobatschewsky plane*, Moscow and Leningrad, 1951.

Sommerville [1]. *Bibliography of Non-Euclidean Geometry*, London, 1911.

——— [2]. *The Elements of Non-Euclidean Geometry*, London, 1914.

——— [3]. *An Introduction to the Geometry of* n *Dimensions*, London, 1929.

——— [4]. "Metrical coordinates in non-Euclidean geometry," *Proceedings of the Edinburgh Mathematical Society* (2), 3 (1932), pp. 16–25.

Stäckel and Engel [1]. *Die Theorie der Parallellinien von Euklid bis auf Gauss*, Leipzig, 1895.

von Staudt [1]. *Geometrie der Lage*, Nürnberg, 1847.

——— [2]. *Beiträge zur Geometrie der Lage*, Nürnberg, 1856.

Steiner [1]. *Systematische Entwickelung der Abhängigkeit geometrischer Gestalten von einander*, Berling, 1832.

——— [2]. *Die geometrischen Konstruktionen ausgeführt mittels der geraden Linie und eines festen Kreises*, Leipzig, 1913.

Stephanos [1]. "Mémoire sur la représentation des homographies binaires par des points de l'espace avec application à l'étude des rotations sphériques," *Mathematische Annalen*, 22 (1883), pp. 299–367.

Struik [1]. *Lectures on Classical Differential Geometry*, Cambridge, Mass., 1950.

Study [1]. "Beiträge zur nichteuklidische Geometrie," *American Journal of Mathematics*, 29 (1907), pp. 101–167.

Sylvester [1]. "On the rotation of a rigid body about a fixed point," *Philosophical Magazine* (3), 37 (1850), pp. 440–444.

Vailati [1]. "Sulle relizioni di posizione tra punti d'una linea chiusa" and "Sulle proprietà caratteristiche delle varietà a una dimensione," *Rivista di Matematica*, 5 (1895), pp. 75–78, 183–185.

Veblen [1]. "A system of axioms for geometry," *Transactions of the American Mathematical Society*, 5 (1904), pp. 343–384.

Veblen and Young [1]. "A set of assumptions for projective geometry," *American Journal of Mathematics*, 30 (1908), pp. 347–380.

—————— [2]. *Projective Geometry*: vol. I, Boston, 1910; vol. II (by Veblen alone), 1918.

Whitehead [1]. *The Axioms of Projective Geometry*, Cambridge, England, 1906.

————— [2]. *The Axioms of Descriptive Geometry*, Cambridge, England, 1907.

Young [1]. "The geometry of chains on a complex line," *Annals of Mathematics* (2), 11 (1909), pp. 33–48.

Zeuthen [1]. *Die Lehre von den Kegelschnitten im Altertum*, Kopenhagen, 1886.

INDEX